Advances in

VIRUS RESEARCH

VOLUME 27

CONTRIBUTORS TO THIS VOLUME

Hans-W. Ackermann

A. G. Bukrinskaya

Anne M. Field

Raymond V. Gilden

John Hammond

C. H. Huang

Harvey Rabin

Darryl C. Reanney

Advances in

VIRUS   RESEARCH

Edited by

MAX A. LAUFFER

Andrew Mellon Professor of Biophysics
University of Pittsburgh
Pittsburgh, Pennsylvania

FREDERIK B. BANG

Department of Pathology
The Johns Hopkins University
Baltimore, Maryland

KARL MARAMOROSCH

Waksman Institute
 of Microbiology
Rutgers University
Piscataway, New Jersey

KENNETH M. SMITH

Cambridge, England

VOLUME 27

1982

ACADEMIC PRESS
A Subsidiary of Harcourt Brace Jovanovich, Publishers
New York   London
Paris   San Diego   San Francisco   São Paulo   Sydney   Tokyo   Toronto

ACADEMIC PRESS, INC.
111 Fifth Avenue, New York, New York 10003

United Kingdom Edition published by
ACADEMIC PRESS, INC. (LONDON) LTD.
24/28 Oval Road, London NW1 7DX

LIBRARY OF CONGRESS CATALOG CARD NUMBER: 53–11559

ISBN 0–12–039827–3

PRINTED IN THE UNITED STATES OF AMERICA

82 83 84 85   9 8 7 6 5 4 3 2 1

CONTENTS

Diagnostic Virology Using Electron Microscopic Techniques

ANNE M. FIELD

Studies of Japanese Encephalitis in China

C. H. HUANG

Plantago as a Host of Economically Important Viruses

JOHN HAMMOND

Penetration of Viral Genetic Material into Host Cell

A. G. BUKRINSKAYA

Comparative Biology and Evolution of Bacteriophages

DARRYL C. REANNEY AND HANS-W. ACKERMANN

Mechanisms of Viral Tumorigenesis

RAYMOND V. GILDEN AND HARVEY RABIN

CONTRIBUTORS TO VOLUME 27

Numbers in parentheses indicate the pages on which the authors' contributions begin.

HANS-W. ACKERMANN, *Department of Microbiology, Faculty of Medicine, Laval University, Quebec, P.Q., Canada G1K 7P4 (205)*

A. G. BUKRINSKAYA, *The D. I. Ivanovsky Institute of Virology, Academy of Medical Sciences, Moscow, USSR (141)*

ANNE M. FIELD, *Virus Reference Laboratory, Central Public Health Laboratory, Colindale Avenue, London NW9 5HT, England (1)*

RAYMOND V. GILDEN, *Biological Carcinogenesis Program, NCI-Frederick Cancer Research Facility, Frederick, Maryland 21701 (281)*

JOHN HAMMOND*, *Department of Botany and Plant Pathology, Purdue University, West Lafayette, Indiana 47907 (103)*

C. H. HUANG, *Institute of Virology, Chinese Academy of Medical Sciences, Beijing, People's Republic of China (71)*

HARVEY RABIN, *Biological Carcinogenesis Program, NCI-Frederick Cancer Research Facility, Frederick, Maryland 21701 (281)*

DARRYL C. REANNEY, *Department of Microbiology, La Trobe University, Bundoora, Victoria 3083, Australia (205)*

*Present address: USDA, Florist and Nursery Crops Laboratory, Beltsville, Maryland 20705.

KENNETH M. SMITH (1892–1981)

KENNETH M. SMITH, CBE, FRS

The Times, London, June 12, 1981, recorded in a detailed obituary the death on June 11 of Kenneth M. Smith, CBE, FRS, at the age of 88. Dr. Smith played a dominant role in the founding of *Advances in Virus Research* and served as an editor for nearly three decades until his death.

Prior to the middle of 1952, while in the United States as a Fellow at the National Institutes of Health, Dr. Smith and Mr. Kurt Jacoby, then vice-president and treasurer of Academic Press, engaged in extensive discussion concerning the establishment of a review journal in the field of virology. Dr. Smith and Mr. Jacoby then invited a biophysicist, Max A. Lauffer, to become with Dr. Smith a member of the original editorial team. The two editors planned Volume 1, which appeared in 1953 and set the tone for the publication.

Kenneth Manley Smith was born of English parentage in 1892 near Glasgow. He was educated at Dulwich College. He served in World War I until wounded and discharged and then became lecturer in Agricultural Entomology at Manchester University. He obtained the D.Sc. degree there in 1926. In 1927 Dr. Smith was appointed entomologist to the Potato Virus Research Station of the School of Agriculture at Cambridge University and was made a member of Downing College. He was awarded the Ph.D. degree by Cambridge in 1929. When the founding director, Dr. R. N. Salaman, of the Potato Virus Research Station retired, Dr. Smith succeeded him as director. The station later became the Virus Research Unit of the Agricultural Research Council, originally associated with the Molteno Institute of Cambridge University. When he was succeeded upon retirement as director by Dr. Roy Markham, Dr. Smith remained on the research staff for two years. Then in 1963, at the age of 70, he went to the University of Pittsburgh as Visiting Professor of Biophysics for two years; after that he served until 1969 as Visiting Professor of Botany at the University of Texas, Austin. Upon returning to his home in Cambridge in 1969, he continued to be active in science as an author and editor until his death.

Many honors came to Dr. Smith. He was a Fellow of the Royal Society, Leuwenhoek Lecturer of the Royal Society, and Master's Lecturer of the Royal Horticultural Society. He was an honorary member of the Association of Applied Biologists and an honorary life member of the Society for General Microbiology. In 1956 he was made a Commander of the Order of the British Empire. From 1945 to 1950 he was a Governor of Dulwich College, and in 1953 he was made an Honorary Fellow

of Downing College. Throughout his career he had extensive experience in the United States. In 1939 he was a Fellow at the Rockefeller
Institute for Medical Research at Princeton in the laboratory of Dr.
Louis O. Kunkel and in 1952 he was a Fellow at the National Institutes of Health in the laboratory of Dr. Ralph W. G. Wyckoff. As already mentioned, he spent seven years following retirement at the
University of Pittsburgh and at the University of Texas.

Kenneth Smith's scientific contributions were enormous. He was a
pioneer in the study of insect transmission of plant viruses and in the
resolution of the etiology of complex infections. Kenneth Smith was
the first to recognize several new plant viruses, among them tomato
bushy stunt virus, turnip yellow mosaic virus, and tobacco necrosis
virus. He contributed greatly to the study of the physical properties of
viruses—ultrafiltration, electron microscopy, and, in collaboration
with D.E. Lea, the action of ionizing radiations on viruses.

After World War II, Dr. Smith became interested in viral diseases of
insects. He was primarily responsible for discovering the cytoplasmic
polyhedrosis viruses. With Professor Robley Williams of the University of California at Berkeley he did important work on the morphology of the tipula iridescent virus. After his retirement, when in Pittsburgh, he became involved in ongoing work on the entropy-driven
polymerization of viral proteins, specifically that of cucumber virus 4.
At Pittsburgh, and later at Texas, he also carried out research on viruses of blue-green algae. His experimental work was published in a
vast number of scientific articles. In addition, he was the author of
many reviews published in prestigious journals, the author of chapters
in reference works, and the author of "A Textbook of Plant Virus Diseases," published in 1937 and revised in 1957 and 1972, "Insect Virology," first published in 1967, and "Plant Viruses," the fifth edition of
which was published in 1974. He wrote for the general reader as well.
"Viruses," published by Cambridge University Press in 1962, is an example.

Kenneth had many hobbies, particularly gardening and butterfly
collecting. In his youth he was a runner and tennis player and in his
later years an avid cyclist. After Kenneth returned to Cambridge following his postretirement stays in Pittsburgh and Austin, he developed serious problems in both hips. Sadly, he reported in correspondence with other editors that he had to trade his bicycle in for a
wheelchair. However, hip joint replacements were totally successful,
and in a footnote in one of his letters he reported joyfully that he had
bought a new bicycle. He was then in his eighties.

Kenneth was a charming, warmhearted, generous human being

with a wry sense of humor carefully concealed by his grave countenance from all but those who knew him well. While Kenneth was in Texas, a demented individual locked himself in the tower of a University building and shot with lethal accuracy at professors and students walking across campus. Kenneth was much alarmed by this incident because he walked that same path regularly. Shortly thereafter, Kenneth and the other editor met to discuss affairs of *Advances in Virus Research*. They agreed that, because of rapidly escalating developments in the field, it was time to invite a third editor active in the field of medical virology. Because *Advances in Virus Research* has always been international in its scope, serious consideration was given to inviting someone from Great Britain or continental Europe. However, the view was finally put forth that since the two original editors were in Austin and Pittsburgh, communication among the editors would be facilitated if one were chosen from North America. Kenneth concurred in the decision and agreed that an American be invited, but dryly added the condition that he be unarmed. It was thus that the late Dr. Frederik B. Bang became the third member of the editorial team.

Dr. Smith married Germaine Marie Noël in 1923. She and their son, Marcel, survive him.

The Editors and publishers cherish the memory of a great scientist and a delightful and true friend.

MAX A. LAUFFER
KARL MARAMOROSCH

FREDERIK B. BANG (1916–1981)

FREDERIK B. BANG

Dr. Frederik Barry Bang, an editor of *Advances in Virus Research* since 1969, died suddenly on October 3, 1981 at John F. Kennedy Airport en route to Europe. He was scheduled to present important papers in the Federal Republic of Germany and in Sweden on the contributions to human health of research in invertebrate pathology and marine biology.

Frederik Bang, the son of A. F. and Carol Klee Bang and the grandson of the Danish investigator who pioneered research on bovine brucellosis, a malady known to this day as Bang's disease, was born on November 5, 1916, in Philadelphia. He was educated at The Johns Hopkins University: A.B. (1935) and M.D. (1939). Following graduation from medical school, he interned for one year at the U. S. Marine Hospital in Baltimore. He then spent a year in the laboratory of Ernest Goodpasture at Vanderbilt University as a National Research Council Fellow in Medicine. From 1941 to 1946 he was assistant in the Department of Animal Pathology at the Rockefeller Institute for Medical Research, Princeton, New Jersey. His work at Princeton was interrupted from 1943 to 1946 by service in the southwest Pacific and in the Philippines as an officer in the U. S. Army Medical Corps, where he advanced to the rank of major. He returned to The Johns Hopkins University Medical School in 1946 as assistant professor of medicine and was promoted to associate professor in 1949. In 1953 he transferred to the School of Hygiene and Public Health at Hopkins as professor and chairman of the Department of Parasitology. There, from 1955 until his death, he was professor and chairman of the Department of Pathobiology. He served as director of The Johns Hopkins Center for Medical Research in Calcutta, India from 1961 to 1973, and, after the Center was transferred to Dacca, Bangladesh, he continued as director until 1976. Dr. Bang was an enthusiastic devotee of the Marine Biological Laboratory in Woods Hole, Massachusetts, where he served since 1978 as instructor in charge of the annual course on the comparative pathology of marine invertebrates. He and his family maintained a home in Woods Hole, where he was plannning to retire to continue research at the Marine Biological Laboratory.

Dr. Bang was honored as a Fulbright Scholar from 1955 to 1956, when he worked at the National Institute for Medical Research, London, in the laboratory of Christopher Andrewes. In the summers of 1961 and 1964 he was a Guggenheim Fellow in the Station Biologique in Roscoff, France.

Frederik was a relentless investigator, both in the laboratory and in the field. He published exhaustively—more than 250 articles and reviews. His scientific interests were extremely broad. They ranged from the feeding habits of mosquitoes to population biology, from chemotherapy to vitamin deficiency diseases, from cell culture to electron microscopy, from protozoan parasites to viruses, from famine to viral oncology, and from man to marine invertebrates. Dr. Bang published contributions on the malarial parasite, the gonococcus, the swine influenza bacterium, and *Schistosomiasis japonica*. His viral research involved lymphopathia venerea virus, eastern, western, and St. Louis encephalitis viruses, pseudorabies virus, many strains of influenza virus, poliomyelitis virus, New Castle disease virus, hepatitis virus, Rous sarcoma virus, the mammary tumor inciter, and a laryngotracheitis virus. His collaborators from many parts of the world included dozens of able and distinguished investigators, among whom was his wife. Despite the breadth of his concerns, his primary interest was the pathogenesis of virus diseases.

Dr. Bang was a member of the American Association for the Advancement of Science, the American Society of Tropical Medicine and Hygiene, the American Institute of Biological Sciences, the Interurban Clinical Club, the American Association of Immunologists, the American Society of Clinical Investigation, and the American Society of Experimental Pathology. He was also a member of the Society of Experimental Biology and Medicine, the Marine Biology Association of India, the International Epidemiological Association, the American Society for Microbiology, the American Association of Pathologists, the American Society of Parasitologists, the Society of Invertebrate Pathology, Phi Beta Kappa, and Sigma Xi. His editorial services included, in addition to *Advances in Virus Research, Cahiers de Biologie Marine*.

Dr. Bang married Betsy Garrett in 1940. Mrs. Bang, two daughters, Caroline and Molly, and one son, Axel Frederik, survive him.

Fred was a well-informed, demandingly logical, intellectually honest, fair-minded man. The Editors will always remember him as a delightful friend and a valued colleague.

MAX A. LAUFFER
KARL MARAMOROSCH

ADVANCES IN VIRUS RESEARCH, VOL. 27

DIAGNOSTIC VIROLOGY USING ELECTRON MICROSCOPIC TECHNIQUES

Anne M. Field

Virus Reference Laboratory
Central Public Health Laboratory
London, England

I. Introduction

Diagnosis of viral infections by observation of virus particles in thin sections of infected tissues has been a continuing but perhaps rather underused technique for the last 30 years. Observation of virus particles in suspension in metal-shadowed preparations had some diagnostic applications, but when the negative staining technique was introduced in 1959 and, as a result, virus particles were more readily recognizable, diagnostic use of the electron microscope became extremely practical. The presence in the world of smallpox infection and the consequent necessity for rapid differentiation between the smallpox virus and other viruses established the electron microscope as an essential tool in a few selected laboratories. Naturally these instruments were utilized for other virus diagnostic problems and gradually experience accumulated. Confirmation of electron microscopy as a good diagnostic technique for samples direct from patients came in the late 1960s with hepatitis B serum testing. In the 1970s the essentially noncultivable fecal viruses of hepatitis A and various diarrheal conditions were discovered and in these studies electron microscopy played an indispensable role.

The purpose of this article is to review the development of the use of electron microscopy in viral diagnosis in the last 20 years and to place it in context with other laboratory techniques. The field covered is confined to medical viral diagnosis, but parallel developments have taken place in both veterinary and botanical fields and techniques derived from both these sources are included where relevant. Viral diagnostic electron microscopy in the medical field has been reviewed by Doane (1974), by Doane and Anderson (1977), by Donelli *et al.* (1979), and by Hsiung *et al.* (1979).

II. Viral Morphology

Virus particles have characteristic morphologies (i.e., shape, substructure, and size) which, because they are fundamental properties, are important in viral classification. Viral structure is thus the basis for the use of the electron microscope for diagnosis and viruses having the same structure constitute a morphologic group, which may be a family of viruses or a genus. Within a group individual viruses cannot be differentiated on appearance and more sophisticated methods of antigenic analysis must be used, but it is often sufficient for diagnostic purposes merely to place a virus within such a group. Many diagnoses are made by recognition of the characteristic viral structure in the image displayed on the microscope screen and need not await confirmatory micrographs; for this an appreciation of the full range of viral morphology by the operator is essential.

The salient features of viral morphology as observed by negative staining and thin sectioning methods (Section III) are here briefly reviewed and illustrated diagrammatically in Fig. 1, and selected exam-

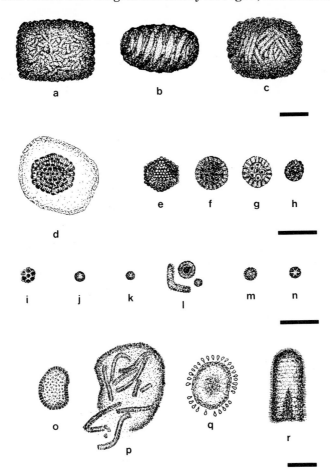

Fig. 1.   Systematic morphology: the negatively stained morphology of virus particles represented diagrammatically. a, Orthopoxvirus; b, parapoxvirus; c, molluscum contagiosum virus; d, herpesvirus; e, adenovirus; f, reovirus; g, rotavirus; h, papovavirus; i, calicivirus; j, picornavirus; k, parvovirus; l, hepatitis B antigen; m, Norwalk agent; n, astrovirus; o, orthomyxovirus; p, paramyxovirus; q, coronavirus; r, rhabdovirus. Bars = 100 nm.

ples are illustrated photographically (Figs. 2 to 12). Reviews of viral structure have recently been published (Rabin and Jenson, 1967; Madeley, 1972; Dalton and Haguenau, 1973; Cheville, 1975). This review is not intended to be comprehensive but stresses the features with diagnostic significance and is largely based on personal observations and the reviews quoted above. Terminology is according to recent viral classification (REPORT, 1979).

A. Systematic Morphology

1. Poxviruses

Poxviruses are the largest and most complex virus particles. The *Orthopoxvirus* genus includes vaccinia, variola, cowpox, and monkeypox viruses, all of which can infect man. Negatively stained particles are brick-shaped, 220 to 270 nm long, and 180 to 220 nm wide. When penetrated by stain the particles appear larger than when unpenetrated. The latter display a random arrangement of 9-nm-wide surface filaments. Members of the *Parapoxvirus* genus causing human infections are orf virus and milker's nodule virus. The negatively stained particles are ovoid, 220 to 300 nm long, and 150 to 180 nm wide. The single surface filament is narrower than on *Orthopoxvirus* particles and is arranged in a regular spiral which usually gives a criss-cross appearance because both sides of the particle are imaged. Stain-penetrated particles are identical with orthopoxviruses except in the ovoid shape and generally greater length and lesser width. A probable member of the poxvirus family is molluscum contagiosum virus which has brick-shaped particles resembling orthopoxviruses but with more rounded corners. The slightly narrower surface filaments are arranged rather more regularly than on orthopoxviruses. Stain-penetrated particles closely resemble the orthopoxviruses except in the rounded corners. In thin sections of infected cells it has been shown that all poxviruses mature within a cytoplasmic matrix. In mature particles an outer coat encloses two lateral bodies which lie on each side of a dense core. Particles are released by cell lysis.

2. Large Icosahedral Viruses

Because of their large size and complex structure the poxviruses cannot be confused with other viruses. Some of the larger icosahedral viruses, although of different families, can resemble each other under some circumstances, particularly if damaged. The families concerned are the Herpesviridae, Adenoviridae, Reoviridae (including reoviruses, rotaviruses, and orbiviruses), and Papovaviridae, arranged in descending size order.

Negatively stained herpesvirus particles have icosahedral 100- to 110-nm-diameter capsids bearing 162 capsomeres which are hexagonal when seen end-on and tubular in profile. Capsids are often surrounded by a membranous envelope bearing irregular short projections. This gives the virus a total diameter of 120 to 150 nm. Varying degrees of stain penetration of both capsid and envelope have been observed (Watson and Wildy, 1963; Tyrrell and Almeida, 1967). In thin sections of infected cells it has been shown that capsids are assembled in the nucleus where they acquire cores, which have variable morphology, and where paracrystalline arrays of capsids may be seen. Envelopes are formed around capsids at several sites: the nuclear membrane, cytoplasmic membranes, and cell membrane. Members of the group infecting humans are herpes simplex virus, cytomegalovirus, Epstein–Barr (EB) virus, and varicella zoster virus.

Adenoviruses negatively stained have icosahedral 70- to 80-nm-diameter capsids with a rigid angular appearance, becoming spherical when damaged. There are 252 capsomeres which are 7 to 9 nm in diameter. Twelve vertex capsomeres bear fibers (Valentine and Pereira, 1965), but these are rarely observed in the conditions used to prepare diagnostic specimens. In thin sections of infected cells it can be seen that capsids assemble and mature in the nucleus where they often form paracrystalline arrays. At least 33 antigenic types of adenovirus can infect humans.

The Reoviridae have particles with icosahedral symmetry and two concentric protein coats both bearing capsomeres. Members of the *Reovirus* genus, which includes three types capable of infecting humans, have particles which are 70 to 80 nm in diameter when negatively stained. The outer layer of capsomeres is rarely spontaneously lost, but when it is an inner, 50- to 55-nm-diameter particle, it remains (Fig. 2). Rotaviruses occur in feces and negatively stained particles are 65 to 75 nm in diameter when complete and 55 to 60 nm without the outer coat which strips off readily in natural conditions. Complete Rotavirus particles have a smooth circular outline but removal of the outer layer leaves a particle coated with capsomeres having the appearance of spokes of a wheel (Fig. 3). Colorado tick fever virus is an *Orbivirus* genus member capable of infecting humans. The virus has a substructure resembling the Reoviruses but with less clearly defined capsomeres on the surface of a 65- to 75-nm-diameter capsid (Palmer *et al.*, 1977). Thin sections show that all the Reoviridae replicate in the cytoplasm and usually mature from a matrix of viroplasm. Reoviruses

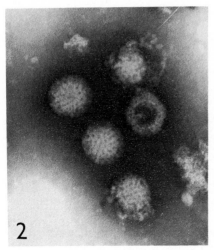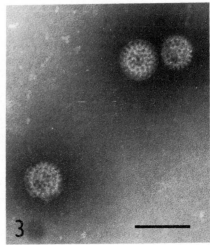

FIGS. 2 AND 3.   Reoviridae. Figure 2, reovirus particles from inoculated cell culture; Fig. 3, rotavirus particles from human feces. Negative stain. ×150,000. Bar = 100 nm.

frequently associate with the mitotic apparatus whereas rotaviruses replicate in close association with the endoplasmic reticulum. Viral release is by cell lysis except for orbiviruses which bud from cell membranes.

Papovaviruses when negatively stained show icosahedral symmetry, but with a skew arrangement, and the capsomeres are more prominent around one side of the particle than on the opposite side. Particles of the *Papillomavirus* genus are 50 to 55 nm in diameter. This genus includes human wart virus. Particles of the *Polyomavirus* genus have identical morphology but are smaller, 40 to 45 nm in diameter. This genus includes two human viruses, BK virus and JC virus. Filamentous forms and several isometric forms smaller than true virus particles can occur. In thin sections of infected cells capsids are seen to be assembled in the nucleus where they often form paracrystalline arrays. Particles may also be seen in cytoplasm, wrapped around by cellular membranes on entry to cells, while progeny virus particles are usually aligned on cytoplasmic membranes. Papillomaviruses tend to be more confined to the nucleus than polyomaviruses.

3. Small Isometric Viruses

Another collection of morphologically similar particles is the range of smaller sized (20- to 40-nm-diameter) isometric viruses, most of which do not have clearly recognizable surface subunits. This group includes

Caliciviridae, Picornaviridae, Parvoviridae, the hepatitis viruses, Norwalk agent, and astroviruses.

Members of the proposed family Caliciviridae negatively stained have roughly spherical particles, 29 to 40 nm in diameter, bearing 32 cup-shaped surface depressions arranged in icosahedral symmetry. In suitably oriented particles these give rise to a distinctive surface pattern of a hollow-centered star recently compared with the "Star of David" (Madeley, 1979). In thin sections the viruses are seen to mature in the cytoplasm in the cisternae of the endoplasmic reticulum and may form crystalline arrays (Love and Sabine, 1975). Human fecal calicivirus appears to be a member of this group (Fig. 4).

Negatively stained picornaviruses are spherical particles with a smooth surface and outline measuring 22 to 30 nm in diameter (Fig. 5). In infected cells they mature in the ground substance of the cytoplasm, sometimes forming paracrystalline arrays. Because they are only slightly larger than ribosomes individual particles may be difficult to recognize. The many antigenic types of poliovirus, Coxsackievirus, echovirus, and rhinovirus all infect humans. Rhinoviruses are more fragile than enteroviruses in negatively stained preparations and empty shells and partial shells are frequently observed. Enteroviruses have a peak buoyant density of 1.33 to 1.34 gm/ml in cesium chloride; the rhinovirus density is 1.38 to 1.41 gm/ml.

Members of the Parvoviridae when negatively stained closely resemble picornaviruses. They are spherical with a smooth surface and outline and the diameter ranges from 18 to 26 nm. In both families some particles may appear to have hexagonal outlines. Empty shells and partial shells are, as with rhinoviruses, commonly seen in parvovirus preparations (Fig. 6). Thin sections of infected cells reveal empty and complete parvovirus capsids in the nucleus and at later stages progeny virus particles are found in the cytoplasm embedded in matrix material or in membrane-bounded spaces. The peak buoyant density in cesium chloride is 1.39 to 1.42 gm/ml. Possible human members of this family have been observed in feces and sera.

Negatively stained hepatitis A virus particles are 27 nm in diameter, spherical, smooth surfaced, and smooth outlined (Fig. 7). In thin sections they are found in the cytoplasm. Possession of ribonucleic acid and other properties indicate that this virus is a member of the Picornaviridae.

Hepatitis B virus is pleomorphic. In negatively stained preparations of serum three particle types are seen: small round particles, 19 to 23 nm in diameter with a smooth surface and edge; long filamentous forms, 19 to 23 nm wide and varying lengths with a smooth surface or

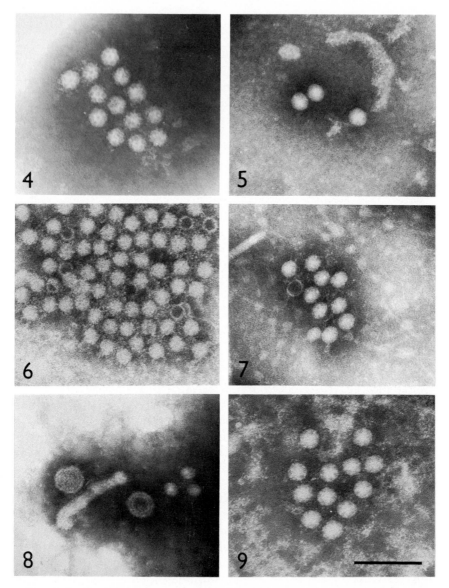

Figs. 4–9.   Small isometric viruses. Figure 4, calicivirus particles from human feces (courtesy of Mr. W. D. Cubitt); Fig. 5, picornavirus particles from inoculated cell culture; Fig. 6, parvovirus-like particles from human serum, aggregated by antibody; Fig. 7, hepatitis A virus particles from human feces, aggregated by antibody (courtesy of Dr. H. Appleton); Fig. 8, hepatitis B antigen particles from human serum; Fig. 9, astrovirus particles from human feces (courtesy of Mr. T. W. Lee). Negative stain. ×180,000. Bar = 100 nm.

with cross striations; Dane particles which are 42 nm in diameter and have a smooth outer layer enclosing a 27-nm spherical core (Fig. 8). The cores (HBcAg) can be extracted from the Dane particles and are antigenically different from the other pleomorphic particles (HBsAg) which are all antigenically similar (Almeida *et al.*, 1971). In thin sections of infected liver the HBcAg particles are seen in the nucleus and resemble parvoviruses. The three forms of HBsAg are seen in cisternae of endoplasmic reticulum in the cytoplasm (Huang and Neurath, 1979).

Norwalk agent and other morphologically similar viruses are at present unclassified. The particles bear some resemblance to caliciviruses but do not show the Star of David substructure. The major structural protein has properties similar to that of caliciviruses (Greenberg *et al.*, 1981). The outline of the particle is structured indicating the presence of some substructure. The particles are difficult to measure accurately because of the ragged outline but are approximately 27 to 34 nm in diameter. No thin section studies have been done. The agents are found in human feces (Kapikian *et al.*, 1972b).

Astroviruses are also unclassified at present. Negatively stained, they are 26- to 30-nm-diameter round particles with a smooth edge. A proportion of the particles bears a solid five- or six-pointed star on the surface (Madeley, 1979). No thin section studies of a full replicative cycle of the human virus are available but abortive cycles in cell culture show the virus located in cytoplasm (Kurtz *et al.*, 1979). Sheep astrovirus particles are cytoplasmic *in vivo* (E. Gray *et al.*, 1980). The particles are found in human feces (Fig. 9).

4. Viruses with a Fringe of Surface Projections

Viruses which are notable for bearing a fringed outer membrane include Orthomyxoviridae, Paramyxoviridae, Coronaviridae, Rhabdoviridae, and Marburg/Ebola viruses.

The orthomyxovirus particles are 80 to 120 nm in diameter and are pleomorphic in negatively stained preparations. Spherical and filamentous forms are the commonest. Spherical forms may show small blebs but these are probably preparation artifacts (Nermut, 1972). The surface of the limiting membrane is covered with 10-nm-long spike-like projections spaced 7 to 8 nm apart. This fringe has a regular appearance around the periphery of negatively stained particles (Fig. 10). End-on views of the projections give a regularly dotted appearance to the surface of particles. Some influenza C virus particles have a reticulate surface pattern (Apostolov and Flewett, 1969). The orthomyxovirus outer membrane is rarely penetrated by stain to reveal the internal component, which is a ribonucleoprotein helix, 9 nm in width, arranged in a tight coil (Murti *et al.*, 1980) (Fig. 11). Nuclei of sectioned cells show various granular and fibrillar structures in the early stages of infection but virus particles are recognized only in the cytoplasm where they mature by budding at the plasma membrane where it covers the already assembled nucleocapsid. This group includes the influenza viruses.

Paramyxoviruses are also pleomorphic, membrane-bound particles, with diameters 120 to 450 nm. Surface projections are approximately 8 nm long and are spaced at 8- to 10-nm intervals. In contrast with orthomyxoviruses these particles frequently fracture revealing the enclosed ribonucleoprotein which is a loosely arranged helix with a pitch of 5 nm and a width of 17 to 18 nm in the *Paramyxovirus* genus (parainfluenza viruses and mumps virus) and in the *Morbillivirus* genus [measles and subacute sclerosing panencephalitis (SSPE) viruses] (Fig. 12). In the *Pneumovirus* genus [respiratory syncytial virus (RSV)] the helix width is only 12 to 15 nm (Joncas *et al.*, 1969a; Norrby *et al.*, 1970; Bächi and Howe, 1973; Berthiaume *et al.*, 1974). In sections the tubular nucleocapsid is seen to accumulate in the cytoplasm and to be aligned beneath the cellular membrane which acquires projections in those regions. Finally mature virus particles bud through the membrane. The morbilliviruses also accumulate nucleocapsid in the nu-

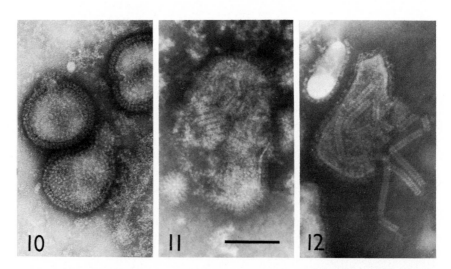

FIGS. 10–12. Orthomyxoviridae and paramyxoviridae. Figure 10, orthomyxovirus particles; Fig. 11, orthomyxovirus particle penetrated by stain and showing the helical nucleocapsid; Fig. 12, paramyxovirus particle and helical nucleocapsid. Negative stain. ×150,000. Bar = 100 nm.

cleus. SSPE virus has lost the budding maturation stage in the natural host.

Coronaviruses have pleomorphic membrane-bound particles with diameters ranging from 75 to 160 nm. They are surrounded by club-shaped surface projections approximately 20 nm long which however are easily lost. In thin sections of infected cells particles are seen to assemble in the cytoplasm and mature by budding into vesicles, accumulating there before being released by cell lysis. The particles in thin sections consist of a translucent core surrounded by a 20-nm-thick membrane bearing 15- to 20-nm-long projections. The human coronavirus affects the respiratory tract. Another possible member of the family is the human enteric coronavirus which, in common with some other animal enteric coronaviruses, has narrow surface spikes rather than club-shaped projections. These spikes are 20 nm long and frequently have knobs and extra T pieces at the distal ends (Caul *et al.,* 1977).

Rhabdoviruses are bullet-shaped particles 130 to 250 nm long and 70 to 80 nm wide. The group includes rabies and vesicular stomatitis virus. Rabies virus is very fragile and in negatively stained preparations most particles are pleomorphic and have variable lengths. Particles penetrated by the stain exhibit internal cross-striations at 4.5- to 5-nm intervals which is the coiled ribonucleoprotein helical component. Some particles exhibit a reticular surface structure but all have surface projections which are 8 to 10 nm long. In thin sections viral matrix material can be seen in the cytoplasm. The site of virus maturation depends on the virus and the cells used; rabies particles mostly bud from intracytoplasmic membranes but can bud in the viral matrix from *de novo* membranes.

Marburg and Ebola viruses, so far unclassified, have structural similarities to rhabdoviruses. In effect they are very long bullet-shaped particles which are frequently curved into hooked and circular forms and are 80 nm wide and 130 to 2500 nm long or even longer. Surface projections are 10 nm long and a helical inner structure with 40 to 50 nm diameter has 5 to 6 nm perodicity. In sections nucleocapsids are formed in the cytoplasm and particles mature by budding at the plasma membrane (Murphy *et al.,* 1978).

5. *Viruses without Distinctive Morphology by Negative Staining*

A collection of viruses which are not often encountered in diagnostic electron microscopy, largely because the negatively stained morphology is not distinctive, includes the Togaviridae, Arenaviridae, Bunyaviridae, and Retroviridae.

Togaviruses infecting man include Sindbis and Chikungunya in the *Alphavirus* genus, dengue, yellow fever, and Japanese B encephalitis in the *Flavivirus* genus, and rubella in the *Rubivirus* genus. With negative staining the viruses are difficult to recognize but they consist of a 40- to 70-nm-diameter envelope closely applied to a 25- to 35-nm nucleocapsid which probably has icosahedral symmetry. Most particles bear surface projections with little regularity. Viruses are more clearly seen in thin sections of infected cells where they multiply in the cytoplasm and mature by budding at the plasma membrane (alphaviruses), the endoplasmic reticulum and Golgi zone membranes (flaviviruses), and into vacuoles and at the plasma membrane (rubiviruses). Particles in thin sections exhibit an electron-dense core surrounded by a translucent zone bordered by the viral envelope (Holmes *et al.,* 1969; Matsumara *et al.,* 1971).

Arenaviruses are spherical or pleomorphic, 110- to 130-nm-diameter particles with club-shaped 10-nm-long surface projections. The most distinctive morphology is shown in thin sections where it can be seen that the limiting membrane encloses a varying number of 20- to 25-nm-diameter, ribosome-like particles. In the infected cells a mass of ribosomes accumulates in a dense cytoplasmic matrix and the virus particles mature by budding through the cell membrane (Murphy and Whitfield, 1975). Viruses of the Tacaribe complex, Lassa fever virus and lymphocytic choriomeningitis virus are human pathogens.

Bunyaviruses are spherical or oval, 90- to 100-nm-diameter particles with an envelope bearing surface projections. The envelope encloses helical ribonucleoprotein 2 to 2.5 nm wide. The particles assemble in the cytoplasm and mature by budding through smooth surfaced vesicles in the Golgi region. Human agents are Bunyamwera and California encephalitis virus.

The *Oncovirus* subgroup of the Retroviridae were originally described morphologically as A, B, C, and D type virus particles by Bernhard (1960). When negatively stained the particles are spherical, enveloped, 80 to 100 nm in diameter, and have surface projections which are seen best in the type B *Oncovirus* genus. Often particles have surface blebs or tails but these are probably preparation artifacts. Beneath the envelope the core is probably icosahedral and contains the ribonucleoprotein which may be helical. In thin sections the particle outer envelope encloses a nucleoid which resembles an A type particle and is eccentrically placed in mature type B oncoviruses and is central in type C oncoviruses. Mature type B particles are 90 to 200 nm in diameter and type C particles are 85 to 110 nm diameter (Dalton, 1972; de Harven, 1974). Although viruses of this family are found in mammals, the morphological evidence for human oncoviruses is very slight. The other subfamily, the Spumaviridae, includes the foamy viruses and a possible human member. Negatively stained particles are enveloped, about 90 nm in diameter, with 12- to 15-nm-long surface projections. In thin sections of infected cells virus particles are observed in the cytoplasm budding into cytoplasmic vacuoles and through the plasma membrane. The moderately electron-dense core is 40 to 60 nm in diameter and is separated from the 70-nm envelope by empty space bridged by striations; spikes on the surface increase the total diameter to 90 nm (Clarke *et al.*, 1969).

B. Measuring Virus Particles

Although size is an important aid to identification it is unwise to rely on the accuracy of size estimates; the sizes quoted for virus particles in the preceding section are only approximate as measurement is subject to many variables. In practice it is useful to have a size marker in the instrument so that a rough estimate of virus particle size can be made using the image on the fluorescent screen. The microscope itself must be calibrated to ensure the accuracy of any given magnification as manufacturing tolerance is generally only within 5%. At high magnifications, such as those used for virus work, calibration based on cross-grating replicas is inaccurate and calibration using measurements of the lattice spacing in negatively stained catalase crystals is considered more reliable (Wrigley, 1968). Even small height changes of the specimen in the microscope affect magnification (Agar *et al.*, 1974) and this creates problems as it is almost impossible to ensure that specimen grids will stay perfectly flat and will be positioned at exactly the height of the calibration grid. Another cause of possible error is that lens currents vary over a period and this too can affect magnification.

In the photographic darkroom careless enlarger setting affects the magnification of prints and photographic paper dimensions can change. Thus measurements are best made on the negatives except when particle images are so small that this in itself induces errors. The thickness of stain surrounding particles is variable in negatively stained preparations, consequently particles may be flattened and so appear to be larger (Nermut, 1977); in addition excessive photographic contrast of the image can affect apparent particle size by obscuring the edge of the particle. In thin sections virus particles are usually smaller than in negatively stained preparations as a result of the processing used (Glauert, 1975).

A comparison of sizes obtained for bacteriophage particles by negative staining, thin sectioning, and freeze drying accompanied by metal shadowing, showed that negative staining gave the best correlation with the size derived from X-ray analysis (Earnshaw *et al.*, 1978).

C. *Morphologic Variation*

The preceding descriptions of negatively stained virus particles refer to viruses untreated as far as possible. The shape of the more delicate viruses can be altered by high-speed centrifugation (Polson and Stannard, 1970), and on the grid by stretching of the support film (Ronald *et al.*, 1977). Myxovirus pleomorphism is probably a preparation and storage artifact (Nermut, 1972). Chemicals used in purification and concentration of viruses may affect morphology (Almeida *et al.*, 1979). Even the negative stain can alter viral morphology: phosphotungstic acid is probably the least damaging in this respect but even this can disrupt the helical paramyxovirus ribonucleoprotein into short lengths (Hosaka, 1968) while uranyl acetate shrinks bacteriophage heads and causes the tails to swell (Ackermann *et al.*, 1974b) and potassium borotungstate splits myxovirus surface projections (Flewett and Apostolov, 1967).

Fixation of foot and mouth disease virus with glutaraldehyde before negative staining produced empty particles and increased their diameter by 25% (Sangar *et al.*, 1973). The internal details of enveloped herpesviruses were also obscured by either glutaraldehyde or osmium tetroxide fixation, probably because negative stain no longer penetrated the fixed envelope (Vernon *et al.*, 1976; Field, unpublished observations). Murphy *et al.* (1970) found that glutaraldehyde or osmium tetroxide fixation obscured the envelope detail of arenaviruses; glutaraldehyde-fixed coronaviruses lost the clarity of the surface projections in negatively stained preparations (Caul *et al.*, 1977). On the other hand rotavirus structure was stabilized by formaldehyde fixation (Woode *et al.*, 1976) and retroviruses were less pleomorphic if glutaraldehyde fixed and critical point dried than if unfixed and air dried (Gonda *et al.*, 1978). Influenza virus particles fixed in osmium tetroxide were less pleomorphic than when unfixed but the detailed surface structure was not so clear (Reuss *et al.*, 1967).

The intentional disruption of virus particles on the grid with detergent has been used to study their internal structure (Almeida and Brand, 1975).

III. Techniques for Electron Microscopy of Viruses

A. Negative Stain

The basis of negative staining is that the electron-dense stain surrounds the virus particles and flows into surface crevices, giving a clear image of the outside surface of the particles. Sometimes stain penetrates to the interior of the capsids and an image of the outer shell in profile results (Horne, 1975). The technique was first described in 1959 (Brenner and Horne, 1959) and was soon applied successfully to viral diagnostic specimens. The usual method for viral diagnosis is to mix virus suspension and negative stain in equal volumes, place a drop of mixture onto a Formvar–carbon-coated grid, remove excess fluid by touching the edge of the grid to filter paper, and allow the grid to air-dry. Alternatively virus suspension can be applied to the grid and dried before stain is added. The stain used most commonly is phosphotungstic acid.

Samples must be rich in virus to overcome the limitations of the technique. Some samples, even diagnostic specimens from patients, contain sufficient virus to be examined without concentration; however, others need concentration and a simple method is to pellet virus by ultracentrifugation of a clarified sample (Almeida *et al.*, 1967b). A further refinement is purification by density gradient centrifugation. Virus may be concentrated from samples by adding Lyphogel, which is a polyacrylamide hydrogel capable of absorbing water, salts, and small molecules to leave virus particles in a greatly reduced volume (Ashcavai and Peters, 1971; Whitby and Rodgers, 1980). Virus can be concentrated from very large volumes by membrane filtration (Torrella and Morita, 1979) or can be adsorbed to a polyelectrolyte and eluted to a smaller volume (Chaudhary and Westwood, 1972). Virus particles can be released from infected cells of solid tissues, cell cultures, and organ cultures by lysing the concentrated cells in a small volume of distilled water (Almeida *et al.*, 1967c; Almeida and Tyrrell, 1967) and, if necessary, further concentration of virus in such cell lysates can be effected by one of the techniques mentioned above.

Various methods have been used to remove contaminating salts from samples. An agar diffusion technique devised by Kellenberger and Arber (1957) was developed by Kelen *et al.* (1971) and by Anderson and Doane (1972a). It utilizes an agar substrate and either the grid with a microdrop (approximately 0.01 ml) of virus suspension on it is placed on the agar or the microdrop is applied to the agar surface and the grid is floated on the microdrop. When the drop has dried the fluid, salts, and low-molecular-weight substances have diffused into the agar leaving virus particles and larger debris on the grid ready for negative staining. Salts may also be removed by careful washing of grids after air-drying the viral sample onto them (Cartwright *et al.*, 1969; Bond and Hall, 1972). Sucrose can be dialyzed from viral samples already applied to grids by using a wick of filter paper to run a buffer solution continuously across the grid (Webb, 1973).

The pseudoreplica method concentrates virus from samples and removes salts simultaneously (Smith and Melnick, 1962). The viral sample is applied to the surface of a small block of agar and allowed to dry. The agar surface is covered with Formvar solution, and the resultant virus-coated film is floated onto negative stain and mounted on a grid. The technique has been used successfully with diagnostic samples (Burtonboy *et al.*, 1978; Lee *et al.*, 1978) but is perhaps unnecessarily complicated.

Although excellent viral morphology results from spraying virus particles onto grids, it is not to be recommended for diagnostic samples since the technique is relatively insensitive, needing very high concentrations of virus, and the resultant aerosol could be dangerous (Horne, 1967; England and Reed, 1980).

It is generally accepted that the threshold concentration of virus necessary for detection in negatively stained preparations is 10^5 to 10^6 particles per milliliter (Galasso, 1967; Monroe and Brandt, 1970; Ball and Harris, 1972; Chaudhary and Westwood, 1972).

If viral samples are too well purified it may be difficult to obtain adherence to grids, but adding a wetting agent such as bovine serum albumin or bacitracin may solve this problem (Horne, 1967; Gregory and Pirie, 1972).

Of the many different negative stains available phosphotungstic acid at pH 6.4 has been found to be the most reliable for general diagnostic work. Silicotungstate gives a less granular background and preserves paramyxovirus structure better than phosphotungstate (Bloth and Norrby, 1967). Phosphotungstate often gave better definition of the viruses encountered in veterinary diagnosis when used at pH 6 as opposed to pH 7 (Spadbrow and Francis, 1969). Caul *et al.* (1977) found phosphotungstate superior to ammonium molybdate or uranyl acetate for examination of fecal coronaviruses. Flewett and Apostolov (1967) found potassium borotungstate damaged myxovirus surface projections.

B. Negative Stain Immune Electron Microscopy

The morphology of negatively stained virus particles is sufficient for grouping purposes but it is necessary to use immune electron microscopy (IEM) to differentiate morphologically identical but antigenically distinct viruses. When specific antiserum is added to a suspension of virus particles molecules of antibody attach to the particles and can be seen after negative staining. When optimal proportions of virus and antiserum are used the virus particles are agglutinated by the antibody into immune complexes. Methods and applications have been reviewed by Almeida and Waterson (1969a), Doane (1974), and Doane and Anderson (1977).

The simplest method for negative stain IEM is to mix small volumes (i.e., 0.1 ml) of viral suspension and antiserum, incubate at 37°C or room temperature for 1 hour, dilute to a reasonable volume, and ultracentrifuge to pellet the immune complexes for negative staining (Almeida and Waterson, 1969a). When reagents are scarce smaller volumes (0.01 to 0.02 ml) can be mixed and incubated, then microdrops are removed to an agar surface and grids are floated on the drops. When the fluid has dried the grids are treated with negative stain (Kelen *et al.*, 1971). Alternatively, microdrops of virus–serum mixture may be applied directly to the grids, but before negative staining repeated washing is necessary to remove salts and low-molecular-weight substances which would otherwise obscure the immune reaction (Milne and Luisoni, 1977). Unfortunately it is necessary to use salt-containing fluids for immune electron microscopy to ensure optimal combination of virus with antibody (Ball and Brakke, 1968). For routine virus identification specific antibodies can be incorporated into agar in microtiter plate wells and the plates can be stored with a grid on top of the agar in each well. When required, microdrop samples of the virus to be identified are added on top of the grids in the wells and allowed to dry into the agar. Grids are removed and treated with negative stain (Doane, 1974). Gel immunodiffusion test precipitin lines can be cut out, homogenized, and negatively stained to show virus–antibody complexes (Beale and Mason, 1968). A method of specific attraction of virus particles to antiserum-treated grids was developed for plant viruses (Derrick, 1973) and further refined by Shukla and Gough (1979) who used staphylococcal Protein A to enhance antibody coating of the grid. The technique has recently been used successfully in rotavirus diagnosis (Nicolaieff *et al.*, 1980) but it was noted that coronaviruses and small round viruses also present in the samples were not seen on the rotavirus–antiserum-treated grids. Virus particles specifically adsorbed onto antiserum-coated grids may be treated further with antisera to investigate their antigenic nature (Milne and Luisoni, 1977).

Complement causes immune lysis of some viruses and may change the appearance of immune complexes so sera are best heat inactivated before use to eliminate these effects (Almeida and Waterson, 1969a). Addition of a second antiserum specific for the immunoglobulin in the original immune complex increases the total antibody layer around the virus particles making detection of complexes easier (Saif *et al.*, 1977). Antibody may be conjugated to ferritin molecules which makes identification of antibody easier (Brzosko *et al.*, 1970; Patterson, 1975), although antibody molecules can be seen readily without a marker when negatively stained. Ferritin labeling has been found useful when a second layer of antibody is used to identify the species of immunoglobulin involved in immune complexes (Locarnini *et al.*, 1977).

The basic necessities for satisfactory negative stain IEM are virus particles in large numbers, free from cell debris and free of antibody, and an antibody preparation which is also free of immune complexes. If the viral antigen is a sample from a patient some antibody may be present. Although it is possible to use such antigens by careful grading of the amount of antibody seen in immune complexes interpretation is more difficult. False clumping of virus particles occurs particularly during abrupt pH changes (Floyd, 1979). Rheumatoid factor can induce mixed clumping of nonidentical viruses to give misleading results in typing tests by IEM (Stannard *et al.*, 1980). It is best to use viruses unfixed because some lose antigenicity when fixed (Narayan *et al.*, 1973), although others retain activity (Woode *et al.*, 1976; Chaudhary *et al.*, 1979).

The application of negative stain IEM has been particularly useful for the study of the antigenic nature of some of the newly discovered noncultivable viruses. For example human parvovirus-like particles have been compared antigenically by this technique (Paver *et al.*, 1975) as have rotaviruses from humans and from veterinary samples (Woode *et al.*, 1976). The system can be reversed and, using known viral antigens, specific viral antibodies can be detected in human sera. Again the technique is most useful when the antigen is a noncultivable virus such as hepatitis A virus (Dienstag *et al.*, 1976a) or Norwalk agent (Parrino *et al.*, 1977).

Negative stain IEM can be used to detect viruses in clinical samples or after culture *in vitro* since with certain viruses it increases the sensitivity of negative stain visualization some 100 times (Doane,

1974). Such enhanced sensitivity depends upon the titer of the antiserum used (Lamontagne *et al.*, 1980) and Zissis *et al.* (1978) found negative stain IEM did not in fact increase sensitivity of rotavirus detection. When single virus particles are readily recognizable, such as rotaviruses, it is probably more sensitive to have many scattered single particles in an ordinary negative stain rather than a much smaller number of virus aggregates in immune preparations. Perhaps the main advantage of IEM in virus detection is the specific aggregation of virus particles of unremarkable morphology so that their viral nature can be appreciated. This was the method used to identify rubella virus (Best *et al.*, 1967) and the small round virus particles such as hepatitis B antigen (Almeida *et al.*, 1969b), rhinoviruses (Kapikian *et al.*, 1972a), and hepatitis A virus (Feinstone *et al.*, 1973).

Incorporation of atypical forms along with typical virus particles in the same specific immune complex demonstrates their common viral antigenicity. This was shown for the tubular forms of human polyomaviruses (Albert and Zu Rhein, 1974) and for filamentous forms of rotaviruses (Holmes *et al.*, 1975). Conversely negative stain IEM demonstrated that HBcAg particles differed antigenically from the three forms of HBsAg because they were not associated in the same immune complexes (Almeida *et al.*, 1970). Amorphous material possessing viral antigenicity can also be identified using negative stain IEM methods (Almeida *et al.*, 1981).

C. Thin Sections

While tissues can be homogenized to extract virus particles for negative staining it is often preferable to search for viruses *in situ* in thin sections. When particles are scanty the thin section technique may be more sensitive than negative staining. Some viruses have a more distinctive morphology in thin sections than when negatively stained. Thin sectioning allows observation of the pathogenesis of the infection as well as identification of the viral cause.

There are many standard schedules for tissue fixation and embedding and methods will not be reviewed here. Examination of thin sections for viruses entails the use of relatively high magnifications in the electron microscope and methods should be chosen which will satisfy this condition. For samples where results are required urgently there are rapid embedding methods (Doane and Anderson, 1977). Examination of thick sections of resin-embedded material stained with toluidine blue for light microscopy may reduce sampling error and so reduce time spent examining thin sections. Even with such selection the thin

sectioning technique has very limited potential in rapid viral diagnosis.

Confirmation of viral infection by electron microscopy on tissues
originally processed for light microscopy is frequently useful. For example, virus particles have been seen in thin sections of tissues which
had been stored in formalin over a long period (Zu Rhein and Chou,
1965; Hashida and Yunis, 1970). Paraffin sections can be marked to
indicate cells with inclusions and the appropriate areas of the section
reembedded for electron microscopic examination of the same cells
(Blank et al., 1970; Rossi et al., 1970; Pinkerton and Carroll, 1971;
Bhatnagar et al., 1977). Cytologic preparations have been reembedded
in a similar fashion and virus particles seen in cells (Takeda, 1969;
Coleman et al., 1977b). Cells positive for viral antigens by immunofluorescence have also been reembedded and the same cells have
been found to contain the expected virus particles (Epstein and
Achong, 1968). Tissue and cell structure is generally adversely affected
in reprocessed samples, because the original processing for light microscopy is not suitable for electron microscopy, but viral structures
rarely disintegrate so far as to be unrecognizable. However, initial use
of Bouin's fixative did destroy herpesvirus structure (Cockson and
Holmes, 1977).

D. Thin Section Immune Electron Microscopy

Viral antigens can be detected in thin sections of infected cells by
IEM with suitably labeled specific antibodies. Certain factors limit the
technique for diagnostic virology: prolonged fixation and the use of
standard concentrations of fixatives reduce antigenicity and also limit
the penetration of reagents into fixed cells, thus making it difficult to
investigate intracellular antigens (Smit et al., 1974; Brown and Thormar, 1976). The average diagnostic schedule is not oriented toward
light fixation of specimens immediately followed by intricate processing. Various methods are available to process already heavily fixed
tissues but the preservation of fine structure is generally poor
(Miyamoto et al., 1971; Hadler and Dourmashkin, 1975; Mohanty,
1975; Wendelschafer-Crabb et al., 1976; Bohn, 1980; Sisson and Vernier, 1980). Reactions can be attempted on already thin-sectioned material but nonspecific staining is a problem (Thomson et al., 1967;
Hansen et al., 1979; Takamiya et al., 1979). Trypsin digestion has restored viral antigenicity to formalin-fixed material for immunofluorescence and might be useful for thin section IEM (Huang et al., 1976;
Swoveland and Johnson, 1979; Johnson et al., 1980).

The antibody markers used in thin section IEM are usually ferritin or peroxidase and the methods have been reviewed by Howe *et al.* (1974) and by Kurstak and Kurstak (1974). Cytochrome c has also been used to label antibodies (Singer, 1974) and peroxidase–antiperoxidase methods based on those used in light microscopy have recently been developed (Hsu and Ree, 1980).

Reprocessing immune light microscopy preparations for thin sections or thin section IEM has confirmed reaction specificities (Epstein and Achong, 1968; Chapman, 1970; Kumanishi and Hirano, 1978).

IV. APPLICATION OF THE TECHNIQUES TO SAMPLES FROM PATIENTS

Virus particles are sometimes present in such large numbers in clinical specimens that they can be detected directly by electron microscopy and negative staining methods in particular can be used to provide a rapid diagnosis. There are of course limitations in sensitivity and in the fact that the technique gives only a morphological grouping in the first instance. However, IEM can be used in certain circumstances to give further identification.

A. *Lesions of Skin and Mucous Membranes*

Viral skin lesions may contain a large number of virus particles and samples are thus highly suitable for electron microscopy. Before the advent of negative staining techniques viruses extracted from such samples were examined, with variable success, after metal shadowing (Nagler and Rake, 1948; Van Rooyen and Scott, 1948; Melnick *et al.*, 1952) and tissue biopsies have been thin sectioned to demonstrate virus particles (cf. Sutton and Burnett, 1969; Kimura *et al.*, 1972). However with negative staining a rapid diagnosis can be made. Vesicular fluid is a suitable starting material for this technique; equally good are dried smears of lesion scrapings, rehydrated in a minimal quantity of distilled water. Crude extracts of solid tissue in distilled water may also be utilized (Macrae *et al.*, 1969). Because of the high viral content of the lesions it is usually unnecessary to concentrate virus before negative staining. Samples are best unfixed as this facilitates extraction of virus from the cells and fixation before negative staining may hinder virus recognition, especially of herpesviruses. Because of variation of viral content in different samples it is advisable to prepare specimens from more than one lesion (Cruickshank *et al.*, 1966; Harkness *et al.*, 1977).

Application of negative staining techniques to diagnosis of viral skin

lesions, particularly to the differential diagnosis of smallpox (a poxvirus as opposed to the herpesvirus of varicella zoster), was a major
factor in establishing electron microscopy in diagnostic virology (Peters *et al.*, 1962; Williams *et al.*, 1962). Although smallpox has been
eradicated other diagnostic problems remain for solution by electron
microscopy. Human monkeypox infections were diagnosed by electron
microscopy and in some cases other laboratory tests were negative
(Breman *et al.*, 1980). Orf virus grows only with difficulty and electron
microscopy is the only practical diagnostic method; similarly, laboratory diagnosis of molluscum contagiosum is possible only by electron
microscopy. Compared with virus isolation and gel precipitin tests
electron microscopy was conspicuously effective in diagnosis of varicella zoster virus infections (Cruickshank *et al.*, 1966; Macrae *et al.*,
1969). Although poxviruses and herpesviruses are morphologically dis-
tinguishable, individual orthopoxviruses (vaccinia, variola, and cowpox) cannot be differentiated morphologically nor can the herpesviruses
(herpes simplex and varicella zoster). Negative stain IEM is not useful
in this diagnostic situation because of the small sample size and because other laboratory tests can be more simply used to give precise
identification (Macrae *et al.*, 1969).

The viral etiology of skin warts has been repeatedly demonstrated by
electron microscopy. In thin sections paracrystalline arrays of papillomavirus particles have been observed (Strauss *et al.*, 1949; Bunting,
1953) and these thin section appearances have been correlated with
intranuclear inclusions seen by light microscopy (Almeida *et al.*, 1962).
The virus particles reacted with wart virus antiserum in thin section
IEM preparations (Viac *et al.*, 1978). Negative staining was successfully applied to homogenates of skin warts (Williams *et al.*, 1962) and
negative stain IEM was used to study antigenicity of extracted virus
(Almeida *et al.*, 1969a). Some skin warts contain only a small number
of virus particles in thin sections (Maciejewski *et al.*, 1973) and genital
warts usually have a low viral content. Oriel and Almeida (1970) suggested that the best method to extract virus from biopsies of genital
warts was by light grinding to disrupt only the surface layers rather
than complete homogenization of the tissue. A recent study of the wart
virus lesions of epidermodysplasia verruciformis showed that negative
staining was slightly more sensitive than thin sections to detect virus
in the early malignant lesions where viral content is very low (Yabe
and Sadakane, 1975). Thin sections have shown typical papillomavirus
particles in lesions of focal epithelial hyperplasia, a rare condition of
the oral mucosa (Praetorius-Clausen and Willis, 1971; Hanks *et al.*,

1972; Van Wyk *et al.*, 1977), but as yet there has only been one report of a successful negative stain diagnosis (Goodfellow and Calvert, 1979). Laryngeal wart thin sections have shown scanty papillomavirus-like particles some of which were not entirely convincing in the published micrographs (Dmochowski *et al.*, 1964; Boyle *et al.*, 1971; Spoendlin and Kistler, 1978), but these findings were confirmed when papillomavirus antigen was found in laryngeal papillomas by light microscopy peroxidase–antiperoxidase techniques (Lack *et al.*, 1980). Papillomavirus particles have also been observed in thin sections of atypical genital warts resembling early dysplasia in the cervix and vagina (Laverty *et al.*, 1978; Morin and Meisels, 1980).

Particles resembling paramyxovirus nucleocapsid have been reported in thin sections of cells in measles skin rash biopsies (Kimura *et al.*, 1975), in skin lesions of discoid lupus (Hashimoto and Thompson, 1970), in Behçet skin lesion biopsies (Tawara *et al.*, 1976), and in warts, Bowen tumours, and basal cell carcinomas (Maciejewski *et al.*, 1973). These findings must be viewed with caution as artifacts which resemble nucleocapsid are not uncommon. This will be discussed in Section VI,B.

B. Nasopharyngeal Secretions

Nasopharyngeal secretions have been examined for virus particles after dilution in distilled water and negative staining. In samples revealing paramyxovirus particles studied by Doane *et al.* (1967) all yielded parainfluenza virus type 1 in cell culture. Joncas *et al.* (1969b) examined lysates of cells in nasopharyngeal secretions by negative staining and also used Doane's method and found that several samples contained paramyxoviruses. These were differentiated into respiratory syncytial virus (RSV) and other paramyxoviruses according to the width of the nucleocapsid. Isolation studies confirmed all the RSV identifications. However adenoviruses and picornaviruses isolated from these samples were never detected by electron microscopy. In a large survey of routine specimens examined by electron microscopy and virus isolation it was shown that electron microscopy was comparatively insensitive for the detection of myxoviruses and paramyxoviruses (Pavilanis *et al.*, 1971) although Valters *et al.* (1975) found that negative stain IEM increased the sensitivity of electron microscopy for detection of viruses in throat swabs. The difficulty of differentiating the paramyxoviruses seen has made it necessary to continue either routine virus isolation or immunofluorescence, which has proved the most useful rapid and specific diagnostic method providing suitable samples are available (Gardner and McQuillin, 1980; Minnich and Ray, 1980).

Partially enveloped herpesvirus particles observed in negatively stained preparations of concentrated throat washings from a patient excreting virus were identified as EB virus by culture methods (Lipman *et al.*, 1975) and Lee *et al.* (1978) detected herpesvirus particles by the pseudoreplica method in throat swabs of five infants which were identified as cytomegalovirus in cultures from all samples.

Many patients excreting rotavirus in their feces have respiratory as well as gastrointestinal symptoms, but a study of nine such patients using negative staining methods failed to detect rotaviruses in throat swabs or nasopharyngeal secretions (Lewis *et al.*, 1979).

Saliva has often been subjected to negative staining and negative stain IEM studies for hepatitis B particles. Occasionally the search has been successful (Kistler *et al.*, 1973; Bancroft *et al.*, 1977) though the photographic evidence is not always completely convincing, probably because so few particles are present. A recent report correlated the presence of Dane particles in negatively stained density gradient fractions of saliva with the presence of DNA polymerase. Small round particles resembling those of HBsAg were also present but negative stain IEM was not done to prove their identity (Macaya *et al.*, 1979).

Norwalk agent particles have been found in a sample of vomit concentrated 100-fold by ultracentrifugation before negative staining. However, three more vomit samples positive for Norwalk agent by radioimmunoassay were negative by electron microscopy (Greenberg *et al.*, 1979).

C. Serum

The transmission of hepatitis B in blood was known for many years before the first electron micrographs of negatively stained small round and long particles of HBsAg in serum were published (Bayer *et al.*, 1968). Shortly after this negative stain IEM was used to aggregate the particles so rendering them more recognizable (Almeida *et al.*, 1969b; Hirschman *et al.*, 1969). This was immediately useful in diagnosis and sera began to be widely examined by negative stain IEM after concentration. Circulating immune complexes were observed in some sera which were examined without adding antibody (Almeida and Waterson, 1969b). Sera with such complexes frequently gave false-negative results in the other relatively crude detection tests then available (Krohn *et al.*, 1970; Cossart *et al.*, 1971). It has been suggested that electron microscopy is more sensitive than the modern radioimmune and hemagglutination assays for detection of immune complexes (Trepo *et al.*, 1974). Dane described the larger particles bearing his name in 1970 (Dane *et al.*, 1970) and Almeida *et al.* (1971) showed that the cores of the Dane particles could be extracted and, by negative stain IEM, were antigenically distinct from the other particles. Positive sera always contain the small round particles; long particles are next in frequency and Dane particles are the least common (Dane *et al.*, 1970; Stannard *et al.*, 1973). The extra sensitivity of modern tests has rendered electron microscopy less useful in routine detection of hepatitis B antigens but it is a relatively easy way to identify sera containing Dane particles and their presence can be correlated with high risk of hepatitis transmission. Gel precipitin techniques are still used in routine testing and confirmation of the specificity of these tests by negative staining of extracted gel lines is useful.

Hepatitis A infection is not blood transmitted but there is one report of virus-like particles of variable size in human sera in the acute stages (Zanen-Lim, 1976). The samples examined were gel precipitin lines formed between patients' sera and animal sera raised against them, a system susceptible to nonspecific reactions. Some particles illustrated resembled common artifacts found in human sera which will be discussed in Section VI,A.

It is uncertain if non-A, non-B hepatitis has one etiologic agent or many. Studies are underway to examine sera by negative staining for possible particles but the results which have been obtained are inconsistent with each other. In one study seemingly specific gel precipitin lines between sera were examined by an unusual method: the gel containing the line was fixed in osmium and embedded in araldite blocks before negative staining. No significant particles were seen (Tabor *et al.*, 1979). In another study particles 60 nm in diameter with 40-nm-diameter cores were seen in some sera (Coursaget *et al.*, 1979). Hantz *et al.* (1980) reported the presence of particles closely resembling those of HBsAg in size and shape; however these were antigenically distinct from HBsAg, were present in very low concentration, and were not reactive in negative stain IEM with antisera which gave gel precipitin lines with the same sera. The gel lines were not examined. Yoshizawa *et al.* (1980) reported the presence in very low concentrations of virus-like particles approximately 27 nm in size detectable only by negative stain IEM using large volumes of sera from patients with non-A, non-B hepatitis. Similar particles were observed in chimpanzee sera after experimental infection. Mori *et al.* (1980), in a negative staining study of density gradient fractions of non-A, non-B sera, showed hexagonal

32-nm enveloped particles with 22-nm hexagonal cores as well as free 22- to 24-nm particles, but no IEM was done.

Theoretically, any virus causing massive viremia could be detected in serum by electron microscopy. Ebola virus was seen in human serum in the course of one infection (Bowen *et al.*, 1978).

When testing sera by negative stain IEM for HBsAg we occasionally observed parvovirus-like particles with a diameter of approximately 23 nm. The human HBsAg detector serum routinely used had antibodies to the parvovirus-like particles and so formed them into immune complexes closely resembling those of HBsAg small round particles (Cossart *et al.*, 1975). These parvovirus-like agents are antigenically distinct from those found in feces (Paver *et al.*, 1975) and no diseases, except possibly a short febrile illness (Shneerson *et al.*, 1980) and onset of hypoplastic crisis in sickle cell anemia (Pattison *et al.*, 1981), have yet been associated with them. Particles resembling coronaviruses have also been seen in sera during routine HBsAg testing by negative staining techniques (Zuckerman *et al.*, 1970; Stannard *et al.*, 1973), but these particles are probably forms of serum lipoproteins (Ackermann *et al.*, 1974a).

D. Urine

Negative staining examination of urine after suitable concentration has revealed herpesvirus particles in patients excreting cytomegalovirus (Paradis *et al.*, 1969). Montplaisir *et al.* (1972) found electron microscopy was more successful if large volumes, 30 to 50 ml, were used. They also noted that cytomegalovirus isolation in suitable cell cultures, although slower than electron microscopy, was much more sensitive. This was confirmed, comparing isolation with pseudoreplica methods, by Lee *et al.* (1978), who observed however that electron microscopy was most sensitive in tests on children younger than 6 months; presumably in these congenital infections large numbers of cytomegalovirus particles are excreted. In contrast Henry *et al.* (1978) found that negative staining of ultracentrifugation pellets from 5-ml samples of urine was more sensitive than virus isolation, however, they were using a detection level of only four or five virus particles to a grid indicating prolonged examination in the electron microscope. The same authors examined thin sections of cells excreted in urine but saw no herpesvirus particles. Immune electron microscopy has not been used to differentiate between herpesviruses seen in urine. We have seen herpesvirus particles in urine from which herpes simplex virus has been grown; without the culture results the particles might have been assumed to be cytomegalovirus (Field and Gardner, unpublished observations). Our experience with urine of immunosuppressed adults has been that cytomegalovirus is frequently cultured but rarely seen (Coleman *et al.*, 1973).

We first observed papovaviruses, of the polyomavirus subgroup, in concentrated urine samples from an immunosuppressed renal transplant recipient in 1971 (Gardner *et al.*, 1971). Since then we and other groups of workers have seen similar particles in urine of various categories of patients: renal transplant recipients (Coleman *et al.*, 1973; Lecatsas *et al.*, 1973; Dougherty and Di Stefano, 1974; Hogan *et al.*, 1980a), patients under treatment for malignancy (Reese *et al.*, 1975; Gardner, 1977), and pregnant women (Coleman *et al.*, 1977a, 1980; Lecatsas *et al.*, 1978). Sometimes the virus particles are coated with an antibody-like substance and specific antiviral antibodies can be detected in urine by negative stain IEM and other tests (Gardner *et al.*, 1971; Reese *et al.*, 1975). When sufficient virus is present in urine it is possible to identify the precise type of virus by negative stain IEM provided the particles are not already coated with antibody (Gardner, 1977).

All our attempts to detect polyomavirus particles in urine by negative stain electron microscopy have been accompanied by virus culture and cytologic studies. To confirm cytologic detection of virus the urine cells with viral inclusions were embedded for thin sectioning. Although ultrastructural preservation of the cells was poor, polyomavirus particles were clearly identified (Coleman *et al.*, 1977b). Cytology is often more sensitive than either virus isolation or negative stain electron microscopy and reprocessing in this way has increased the diagnostic potential of the electron microscope (Coleman *et al.*, 1980).

As well as typical polyomavirus particles Lecatsas and Prozesky (1975) observed filamentous and minispherical forms in one urine. Recently papillomavirus particles were seen in urine from pregnant women (Lecatsas and Boes, 1979).

Urine samples can be contaminated by feces and this may be the origin of rotavirus particles which were seen in one sample of urine of nine examined from babies excreting rotaviruses in feces (Chrystie *et cl.*, 1975).

Urine of two patients in the acute phase of non-A, non-B hepatitis contained virus-like particles 60 nm in diameter with 40-nm cores (Coursaget *et al.*, 1979).

E. Feces

In 1972 Anderson and Doane described an agar filtration technique to rid samples of undesirable salts and to concentrate virus particles onto the grids. A sample chosen to illustrate the technique was a fecal extract and the micrograph showed particles resembling rotaviruses though the identification made was reovirus. This may be the first picture published of human rotavirus (Anderson and Doane, 1972a) though it is not clearly stated in the text that the feces were human. In the same year the detection by negative stain IEM of Norwalk virus particles in fecal extracts from volunteers in gastroenteritis experiments was reported (Kapikian *et al.*, 1972b). The observation in 1973 of rotaviruses in thin sections of duodenal biopsies of infants with diarrhea (Bishop *et al.*, 1973) was closely followed by the observation of rotaviruses in fecal extracts by negative staining methods (Flewett *et al.*, 1973; Bishop *et al.*, 1974; Middleton *et al.*, 1974). All these illustrated the suitability of electron microscopy for the diagnosis of viral agents in diarrhea and opened up a new field of fecal virology which continues to expand (Flewett, 1979; Holmes, 1979).

Fecal extracts may be examined by negative staining without concentration but this is successful only if virus content is very high. It is more usual to employ some method, usually ultracentrifugation, to concentrate virus (Flewett *et al.*, 1973). Portnoy *et al.* (1977) compared three methods: direct examination of uncentrifuged fecal extracts; a pseudoreplica method using clarified fecal extracts; and virus concentration by ultracentrifugation. They found that the last two techniques were more sensitive than direct examination and that the small viruses were more readily detected by ultracentrifugation than by the pseudoreplica method. Concentration of enteroviruses from fecal extracts by adsorption onto a polyelectrolyte followed by elution into a smaller volume for negative staining was successful for detection of 10^6 poliovirus particles per milliliter (Chaudhary and Westwood, 1972). Concentration of virus in fecal extracts by selective removal of water, salts, and low-molecular-weight substances into Lyphogel had a sensitivity equal to or greater than ultracentrifugation and the viral morphology was not affected (Whitby and Rodgers, 1980). Ammonium sulfate precipitation of viruses from fecal extracts has been found useful and the morphology of coronaviruses was particularly well preserved (Caul *et al.*, 1978). Narang and Codd (1980a) found that low-speed centrifugation of lightly clarified fecal extracts onto grids placed at the base of specially shaped tubes was sufficient for most fecal diagnostic work although smaller virus particles (20 to 35 nm) did not appear on the grids unless present in large aggregates (Narang and Codd, 1980b). Comparing Narang and Codd's method with other techniques, Roberts *et al.* (1980) found that leaving the prepared centrifuge tubes on the bench gave the same results as the low-speed centrifugation and they suggested the aggregates of virus seen were released from infected cells which had settled on the grid to be lysed in the subsequent staining. Narang and Codd's experiments show that the comparatively high centrifugal forces used in most laboratories to clarify fecal extracts might rid the preparation of much of the desired virus particularly if it is aggregated. Juneau (1979) suggested incorporating normal human immunoglobulin into agar used in the agar-filtration techniques thus making virus particles in fecal extracts more easily recognized by coating them with antibody. However, if there is too much antibody on the particles viral structure may be obscured and identification becomes difficult. It has also been shown that human antibodies are capable of indiscriminate coating of viruses, virus-like particles, and bacteriophages (Almeida et al., 1974; Locarnini et al., 1974).

The diagnosis of rotavirus infection has considerable clinical value. Tests other than electron microscopy to detect rotaviruses are available (WHO Scientific Working Group, 1980) but are all limited by a requirement for high quality viral antisera. Rotaviruses have a type-specific antigen on the outer layer of capsomeres and a group-specific antigen on the inner layer. Loss of the outer layer is common and immune detection methods are imprecise as a result. Rotaviruses may already be coated with antibody when excreted (Watanabe and Holmes, 1977) which also renders detection by immune methods more difficult. This has ensured a continued role for electron microscopy in rotavirus diagnosis. In addition electron microscopy is a nonselective method and in the search for rotaviruses other agents may be revealed.

Aberrant rotavirus capsids in tubular form are sometimes observed. Negative stain IEM studies showed these were antigenically identical to the spherical capsids (Holmes et al., 1975). Human rotavirus subtypes have been identified by various techniques including negative stain IEM (Zissis and Lambert, 1978).

Adenovirus particles have been observed by negative staining in feces of patients with gastroenteritis and frequently are noncultivable (Bruce White and Stancliffe, 1975; Bryden et al., 1975). It has recently been demonstrated that they belong to a new adenovirus serotype (Johansson et al., 1980).

Coronavirus-like particles have been seen in human feces but their presence often cannot be correlated with illness (Caul et al., 1975; Mathan et al., 1975; Schnagl et al., 1978; Clarke et al., 1979). Morphologically they differ from the classical coronavirus by having greater pleomorphism and narrower surface projections (Caul and Egglestone, 1977; Caul et al., 1977). The viral nature of these particles has not yet been established conclusively. Although coronavirus-like particles were seen in the abortive replication cycle in cell culture the thin sections failed to show typical coronavirus maturation stages (Caul and Egglestone, 1977). The viral nature of the fecal particles has been questioned by Dourmashkin *et al.* (1980) who sectioned a fecal extract pellet which by negative staining contained typical pleomorphic coronavirus-like particles. No typical coronaviruses were seen in the thin sections although large numbers of fringed membrane-bound objects were present. However, it is possible that the pleomorphism of the negatively stained particles reflects some abnormality, such as lack of core material, which would give rise to this appearance in thin sections.

Norwalk agent was first described in feces from volunteers with experimental gastroenteritis (Kapikian *et al.*, 1972b). Unlike the smooth-surfaced parvoviruses and picornaviruses, negatively stained Norwalk particles appear to have a structured surface and edge. Particles are 27 to 34 nm in diameter with a density of 1.36 to 1.41 gm/ml. Particles similar in appearance and antigenically related were observed in feces in a gastroenteritis outbreak in Montgomery County and other morphologically similar but antigenically unrelated particles were reported from Hawaii (Thornhill *et al.*, 1977). Particles resembling Norwalk morphologically and antigenically, 27 to 30 nm in diameter and with density 1.38 gm/ml, were excreted by patients with gastroenteritis following the consumption of oysters in Australia (Cross *et al.*, 1979; Murphy *et al.*, 1979). Two further agents morphologically identical to Norwalk, with densities 1.35 to 1.37 and 1.37 to 1.4 mg/ml, but both antigenically distinct from Norwalk were found in feces of gastroenteritis patients in Japan (Taniguchi *et al.*, 1979; Kogasaka *et al.*, 1980). Morphologically similar particles have also been seen in the United Kingdom and, based on the examination of these particles and of Norwalk agent, it has been suggested that there are morphological similarities with caliciviruses (Caul *et al.*, 1979).

Viruses morphologically indistinguishable from classical caliciviruses were observed in human feces (Madeley and Cosgrove, 1976) but were not associated with illness until more recently when they were implicated in winter vomiting disease and gastroenteritis (McSwiggan *et al.*, 1978; Chiba *et al.*, 1979, 1980; Cubitt *et al.*, 1979, 1980; Suzuki *et al.*, 1979). Negative stain IEM has been used in these studies to detect the appearance of antibodies to the agents seen and to investigate the antigenic nature of the caliciviruses.

Particles measuring 26 to 30 nm with a solid star-shaped pattern on their surface have been termed astroviruses. The surface pattern differs from the hollow Star of David pattern on the 29- to 33-nm-diameter calicivirus. Madeley (1979) has described the differential morphology in detail. Astroviruses were first observed in negative staining studies of feces of babies in maternity ward outbreaks of gastroenteritis and in other infants with gastroenteritis (Madeley and Cosgrove, 1975).

Parvovirus-like particles with a diameter of 22 nm were first observed in human feces by Paver *et al.* (1973). For their detection it was found necessary to use negative stain IEM with postinfection human sera to agglutinate the particles. Identification of parvoviruses depends upon the smooth-surfaced morphology; the size, which is slightly smaller than the morphologically similar enteroviruses; and the density, which is higher than enterovirus density. Because of overlap in both size and density ranges between these two groups precise identification is often impossible. Fecal parvovirus-like agents are noncultivable whereas enteroviruses can usually be grown in cell cultures. Appleton *et al.* (1977) described 26-nm parvovirus-like particles with density 1.38 to 1.40 gm/ml in feces from a school outbreak of winter vomiting disease (the Ditchling agent) and other 25- to 26-nm parvovirus-like particles with density 1.40 gm/ml (Appleton and Pereira, 1977) in feces of patients with gastroenteritis after eating cockles (cockle agent). Negative stain IEM showed that these two agents differed antigenically, both were unrelated to Norwalk but Ditchling was related to the W agent, an earlier reported parvovirus-like particle (Paver *et al.,* 1973). Another school outbreak of gastroenteritis was associated with a 23- to 26-nm parvovirus-like particle but no density estimations or cross-reactions with other parvovirus-like agents were described for the Paramatta agent (Christopher *et al.,* 1978). Although Norwalk was assumed to be the cause of the Australian oyster-associated gastroenteritis, parvovirus-like particles, 22 to 25 nm in diameter, were also seen in many of the fecal samples and similar particles were seen in one oyster sample (Murphy *et al.,* 1979).

Negative staining examination of feces of hepatitis patients, originally in a search for HBsAg and latterly for hepatitis A virus, has revealed interesting particles. Cross *et al.* (1971) found 15- to 25-nm-diameter small round particles and 35- to 45-nm-diameter particles resembling Dane particles. By gel diffusion and negative stain IEM there was no antigenic similarity between the smaller particles and the small round particles of serum HBsAg but slight cross-reactivity between the larger particles and the Dane particles of HBsAg was observed. Moodie *et al.* (1974) observed that gut digestive enzymes would degrade all hepatitis B antigen with the exception of the Dane particle cores and in fact there have been no convincing electron microscopy reports of hepatitis B particles in feces.

The causative agent of hepatitis A was revealed as a 27-nm-diameter small, round, smooth-surfaced virus particle by Feinstone *et al.* (1973) who used volunteers' fecal samples and negative stain IEM with convalescent sera to detect the virus. Excretion of the particles was time related to symptoms. These findings have been confirmed by others (Locarnini *et al.*, 1974; Gravelle *et al.*, 1975) and particles seen in different outbreaks have been found to be antigenically identical. Coulepis *et al.* (1980) showed that maximal virus excretion occurred just before the onset of symptoms, fell slightly in the 5 days after onset while patients had dark urine, and then reduced steadily until by 2 weeks after onset virus was only just detectable by electron microscopy.

F. Breast Milk

Breast milk has been tested for HBsAg by negative stain IEM after concentration of the samples by ultracentrifugation and small round particles have been seen (Boxall *et al.*, 1974), but no strict tests were done to exclude the presence of occult blood.

Following the analogy of the mouse mammary tumor virus human breast milk has been surveyed for retroviruses. Particles resembling type B oncoviruses were observed by Moore *et al.* (1969) using thin sections and negative staining in parallel with biochemical studies on fractionated milk. Some samples of human breast milk degraded the structure of true type B oncovirus particles added to them (Sarkar and Moore, 1972) rendering the particles unrecognizable. Chopra *et al.* (1973) found type D oncovirus-like particles in breast milk using negative staining but could not correlate electron microscopy and biochemical findings.

G. Cerebrospinal Fluid

It is possible to demonstrate viral antigens by immunofluorescence in cells in cerebrospinal fluid (CSF) in viral encephalitis or meningitis (Dayan and Stokes, 1973; Taber *et al.*, 1973). A paramyxovirus has been seen in CSF by negative staining and mumps virus was isolated from the sample (Doane *et al.*, 1967). Thin sections of CSF cells in presumed mumps virus meningitis showed cytoplasmic collections of tubules resembling paramyxovirus nucleocapsid (Herndon *et al.*, 1974).

H. Tissues

Most human tissues have been examined by electron microscopic methods for virus particles at some time. However, it is comparatively rare for such investigations to be included in routine viral diagnosis. Sample selection is a considerable problem for both thin-sectioned and negatively stained preparations of tissues. Immune electron microscopy on sectioned material presents considerable technical difficulties and viral content of tissue homogenates may be too low for negative stain IEM so a virus seen in the tissue cannot always be sufficiently well identified for diagnostic purposes. Immunofluorescence and other light microscopy immune methods have greater diagnostic potential because they are simpler, larger samples are used, and precise virus identification is possible. Nevertheless, electron microscopy has been important in revealing viral etiology, often for the first time, and it has frequently been the impetus for development of the more convenient light microscopy techniques. Of recent years examination of tissues for viruses by electron microscopy has concentrated upon three major areas: the brain, the liver, and tumors. These applications illustrate well the techniques, problems, and achievements in viral diagnosis by electron microscopy of tissues.

1. Brain

Ultrastructural studies of virus infections of the human brain have been reviewed recently (Mirra and Takei, 1976).

The light microscopy neuropathology of brain affected by the rare condition progressive multifocal leukoencephalopathy (PML) was first described by Åström et al. (1958). Lesions were later shown by thin section electron microscopy to contain papovavirus particles by Zu Rhein and Chou (1965). Extracts of formalin-fixed brain were negatively stained and typical papovavirus particles were observed which were 41 nm in diameter and were clearly members of the polyomavirus subgroup (Howatson et al., 1965). Virus particles were also demonstrated in unfixed PML brain homogenates by negative staining (Schwerdt et al., 1966). It was not until 1971 that a new polyomavirus, JC virus, was cultivated from PML brain (Padgett et al., 1971). The following year two isolates of a polyomavirus antigenically similar to simian virus 40 (SV40) were reported from PML brains (Weiner et al., 1972) but all subsequent strains isolated from such material have been identified as JC virus (Padgett et al., 1976). In thin sections of brain typical spherical polyomavirus particles are found in the nuclei and cytoplasm of oligodendrocytes and filamentous forms of the virus are also frequently present (Fig. 13). Brain homogenates may be so rich in virus that concentration by ultracentrifugation before negative staining is unnecessary. Filamentous particles have not been seen in negatively stained brain extracts even when they were plentiful in the thin sections (Field, unpublished observations). When JC virus had been grown *in vitro* and specific antisera were prepared, virus particles extracted from infected brains were identified by negative stain IEM and parallel immunofluorescence studies were performed on brain sections (Narayan *et al.*, 1973). For successful negative stain IEM virus content of the brain sample must be high and virus particles must be free of cell debris. Because JC virus is relatively difficult to cultivate this technique has great potential. Although diagnosis of the polyomavirus infection by recognition of the typical particles in thin sections and negative stains of formalin-fixed brain is straightforward, it has proved impossible to identify the virus antigenically by negative stain IEM after formalin fixation (Narayan *et al.*, 1973; Padgett *et al.*, 1976; Field and Gardner, unpublished observations).

Laboratory diagnosis of infection with the measles-like virus of subacute sclerosing panencephalitis is generally based upon detection of measles antibodies in serum and in cerebrospinal fluid. In rare

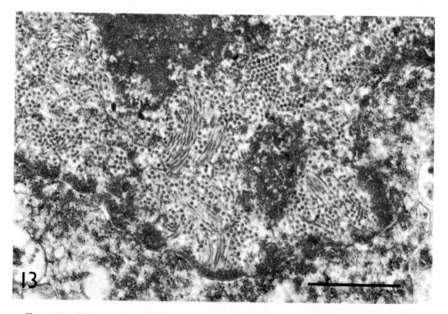

Fig. 13. Thin section of PML brain showing both spherical and filamentous polyomavirus particles within the nucleus. ×25,000. Bar = 1 μm.

cases when brain biopsy is performed the usual laboratory diagnosis is by immunofluorescence with specific measles virus antiserum or, less desirably, measles convalescent human serum. Confirmation by electron microscopy may be sought because virus culture, although possible (Chen *et al.*, 1969; Horta-Barbosa *et al.*, 1969), is technically difficult. Thin sections of SSPE brain have been found to contain collections of 17- to 19-nm-diameter tubular paramyxovirus nucleocapsids in nuclei and cytoplasm of oligodendrocytes and neurons (Herndon and Rubinstein, 1968) (Fig. 14). The nuclear particles are clearly tubular and have no surface coating but cytoplasmic nucleocapsid is usually coated with a granular substance. There have been only two reports of clearly recognizable paramyxovirus nucleocapsid in negatively stained SSPE brain homogenates (Dayan and Cumings, 1969; Dayan and Almeida, 1975). We have examined concentrated homogenates from nine SSPE brains, of which four contained typical SSPE viral tubules in thin sections and two contained measles antigen by immunofluorescence, but in none have we seen any paramyxovirus nucleocapsid by negative staining (Richmond and Field, unpublished observations). Antigenic identification of the virus seen in brain is thus dependent

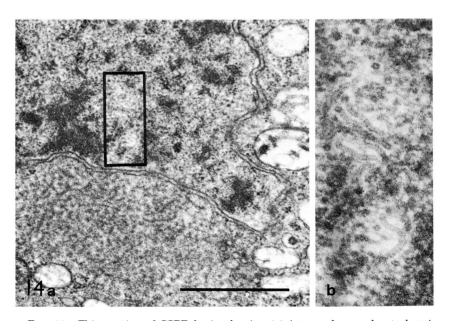

FIG. 14.   Thin section of SSPE brain showing (a) intranuclear and cytoplasmic paramyxovirus nucleocapsids, ×30,000, bar = 1 μm; (b) enlarged boxed area from (a) of intranuclear nucleocapsids, ×100,000. (Courtesy of Mrs. J. E. Richmond.)

upon thin section IEM with all its attendant technical difficulties (Jenis *et al.*, 1973). SSPE brain cells cultured *in vitro* have been shown to contain paramyxovirus nucleocapsids in nuclei and cytoplasm (Chen *et al.*, 1969; Katz *et al.*, 1969) and typical negatively stained paramyxovirus helix has been seen in extracts of these cultured cells (Iwasaki and Koprowski, 1974).

In cases of suspected herpes encephalitis laboratory confirmation of herpes simplex is sometimes urgently required and the most sensitive and rapid method for this is immunofluorescence on brain biopsy material. Herpesvirus particles are found in thin sections (Fig. 15) in nuclei and cytoplasm of neurons and glial cells (Harland *et al.*, 1967; Roy and Wolman, 1969; Baringer and Swoveland, 1972; Viloria and Garcia, 1976). The thin section technique is too slow for rapid diagnosis and as virus-infected cells are distributed unevenly sample selection is a major problem. Virus can be seen in negatively stained brain homogenates but this procedure is comparatively insensitive (Flewett, 1973; Ross, 1973; Joncas *et al.*, 1975). In our experience virus particles are extremely sparse even after ultracentrifugation of brain homogenates (Field, Porter, and Richmond, unpublished observations). Other herpesviruses besides herpes simplex virus can infect brain: McCormick

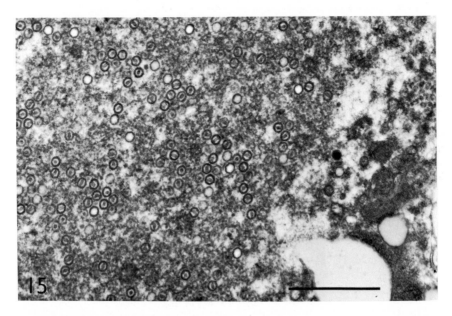

FIG. 15. Thin section of herpes encephalitis brain showing herpesvirus particles. ×25,000. Bar = 1 μm.

et al. (1969) saw herpesvirus particles in glial cells in thin sections of the brain from a patient with encephalomyelitis and cultured varicella zoster virus. In the absence of virus culture complete identification of an observed herpesvirus is difficult. Thin section IEM is technically difficult although possible (Kumanishi and Hirano, 1978); negative stain IEM is unsatisfactory because there is insufficient virus and even by immunofluorescence there are problems with antigenic cross-reactions with other herpesviruses (Emmons and Riggs, 1977).

Electron microscopy has not on the whole proved useful for the laboratory diagnosis of rabies infection of human brain. The methods more generally used are histology, immunofluorescence, and virus isolation. The Negri bodies in histological sections are sometimes but not always identical with the cytoplasmic viral factory areas of thin sections (Morecki and Zimmerman, 1969; Lemercier *et al.*, 1970; Vallat *et al.*, 1977) and rabies virus particles may bud from factory sites or from diverse cytoplasmic membranes of neurons. Factory sites and virus particles are easily recognized in thin sections. Cell culture-derived rabies virus is difficult to identify in negatively stained preparations because of its fragility, indicating that attempts at rapid diagnosis by this technique on brain samples would be unsatisfactory.

Adenovirus encephalitis was investigated by Chou *et al.* (1973) who found inclusion bodies in histological sections and typical, mostly intranuclear, adenovirus particles in neurons and glial cells in thin sections. Adenovirus particles were also seen in negatively stained brain homogenate, from which adenovirus type 32 was later isolated.

Investigating eastern equine encephalomyelitis, Bastian *et al.* (1975) isolated the virus, found typical histology, and for the first time demonstrated typical togavirus particles in thin sections of human brain.

Particles resembling paramyxovirus nucleocapsids have been observed in thin sections of multiple sclerosis brain lesions, mostly sited in nuclei with probable leakage to the cytoplasm (Prineas, 1972; Lhermitte *et al.*, 1973; Watanabe and Okazaki, 1973) and sometimes exclusively in the cytoplasm (Narang and Field, 1973; Pathak and Webb, 1976). However, similar particles have been seen in unrelated conditions of the brain, in normal brain, and in other organs; immunofluorescence and IEM studies have failed to identify the particles as viral. Suggestions have been made that abnormal condensation of nuclear chromatin and multiple invaginations of cytoplasmic membranes might be the cause of these virus-like structures (Baringer and Swoveland, 1972; Dubois-Dalcq *et al.*, 1973; Tanaka *et al.*, 1974, 1975b; Hayano *et al.*, 1976; Kirk and Hutchinson, 1978; Lehrich and Arnason, 1978). Although a paramyxovirus (parainfluenza virus type 1) was retrieved from multiple sclerosis brains by cell fusion techniques, no virus particles were seen in thin sections of the original brains, although some cytoplasmic nucleocapsid was seen at the eleventh pass of brain cells in culture (Ter Meulen *et al.*, 1972). Not surprisingly, in view of the experience with SSPE brain, no paramyxovirus particles have been reported in negatively stained extracts of multiple sclerosis brain.

Particles identified by Bastian (1971) as papovaviruses were seen in cytoplasm and extracellularly in a human choroid papilloma, but the particles were not morphologically characteristic and the lack of intranuclear particles was unusual. Similar particles in other human choroid papillomas were shown to be glycogen (Carter *et al.*, 1972). Early reports claimed ultrastructural evidence for association of a papovavirus with the SSPE paramyxovirus (Koprowski *et al.*, 1970; Oyanagi *et al.*, 1970) but none of the published micrographs has convincingly demonstrated papovavirus and the particles were never intranuclear and were seen only in cultured brain cells and not in the original tissue. A large number of 45-nm-diameter granules with irregular edges were observed in astrocyte cell processes in a Creutzfeldt–Jakob diseased brain and were identified as papovaviruses although the morphology was not convincing (De Reuck *et al.*, 1976). Kirk and Hutchinson (1978) believe that these cytoplasmic papovavirus-like structures are probably reticulosomes and related structures normally present in cells.

Creutzfeldt–Jakob disease has some characteristics of a virus infection and brain tissue has been examined by electron microscopy for a possible etiologic agent. The histology, characteristic of a spongioform encephalopathy, is the usual means of laboratory diagnosis. In thin sections tubular particles and variably sized round and hexagonal particles with and without cores have been reported (Vernon *et al.*, 1970; Bots *et al.*, 1971; Narang, 1975), but none has been demonstrated to be viral. Recently objects resembling spiroplasmas have been observed in thin sections of Creutzfeldt–Jakob brains (Bastian, 1979; A. Gray *et al.*, 1980).

2. Liver

Infection of the liver with hepatitis B virus is accompanied by ultrastructural changes. Parvovirus-like particles were first described in hepatocyte nuclei in thin sections by Nowosławski *et al.* (1970) and these observations were soon confirmed (Nelson *et al.*, 1970; Scotto *et al.*, 1970) (Fig. 16). For some time it was assumed that these particles were identical to the small round particles of HBsAg in negatively stained serum and immunofluorescence and thin section IEM using

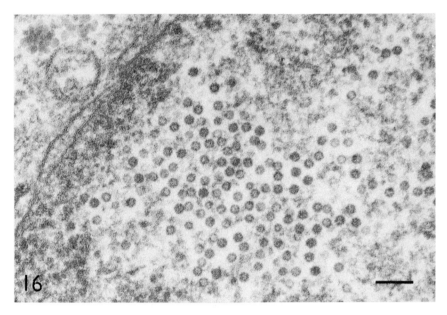

FIG. 16. Thin section of human liver showing intranuclear HBcAg particles.
×100,000. Bar = 100 nm. (Courtesy of Mrs. J. E. Richmond.)

human antisera tended to confirm this impression (Gerber et al., 1972; Huang et al., 1972). Meanwhile cytoplasmic particles resembling the pleomorphic forms of serum HBsAg were seen in liver cells in thin sections (Huang, 1971; Stein et al., 1971). By immunofluorescence, sera specific for HBsAg revealed antigen in cytoplasm and sera specific for HBcAg traced a nuclear antigen and so it was suggested that the nuclear, parvovirus-like particles must be HBcAg and the pleomorphic cytoplasmic particles HBsAg (Gerber et al., 1974; Gyorkey et al., 1974). Thin section IEM has confirmed this and has also shown HBcAg in cytoplasmic maturing Dane particles (Huang and Neurath, 1979). Examination of liver homogenates with negative staining had already shown the three morphologic forms of HBsAg and when 25- to 27-nm round particles were also seen it was suggested that these might be the intranuclear, parvovirus-like particles since they were certainly antigenically different from the particles of HBsAg (Almeida et al., 1970; Huang and Groh, 1973a). Extracts of heavily infected human liver are a good source of HBcAg for detection of antibodies by negative stain IEM (Cohen and Cossart, 1977; Cohen, 1978).

Recent reports have described intranuclear, roughly spherical, 27-nm-diameter particles in human liver thin sections in non-A, non-B hepatitis (Gmelin *et al.*, 1980; Grimaud *et al.*, 1980). These particles are however somewhat unconvincing as virus because of their size range and irregular outlines. Particles with similar morphology to the HBcAg have been extracted from non-A, non-B liver but they are antigenically distinct (Hantz *et al.*, 1980).

Herpesvirus particles have been seen in the nuclei and cytoplasm of liver cells of a patient who died with infectious mononucleosis and extensive hepatic necrosis (Chang and Campbell, 1975). Arenavirus particles were described in liver thin sections of a patient with Lassa fever (Winn *et al.*, 1975) and of another patient with Argentine hemorrhagic fever (Junin virus) (Maiztegui *et al.*, 1975). Ebola virus particles have been observed in large numbers in hepatic cells and bile canaliculi in human infection (Ellis *et al.*, 1978).

3. Tumors

The benign human tumors molluscum contagiosum and warts have been shown by electron microscopy to have poxvirus and papovavirus etiology, respectively, and have already been described in Section IV,A. Papovavirus-like particles measuring 45–50 nm were seen in a cell line derived from a nephroblastoma (Wilms' tumor) at the fortieth and subsequent pass levels. The particles were recognized in nuclei in thin sections but were less convincing in negatively stained preparations. Attempts to confirm the findings on other Wilms' tumor cell lines were fruitless and no virus particles were seen in the original tumor (Smith *et al.*, 1969).

Although herpesvirus particles of EB virus have frequently been observed in cultured Burkitt lymphoma cells by thin section electron microscopy since they were first described by Epstein *et al.* (1964), there has been no report of such particles in the original tumor. EB virus particles have also been seen in lymphoblastoid cell lines derived from nasopharyngeal carcinoma (Vuillaume and de Thé, 1973), in leukemic buffy coat cell cultures (Zeve *et al.*, 1966), and in lymphoid cell lines derived from lymph nodes from patients with various cancers (Jensen *et al.*, 1967), and in one of these cases herpesvirus particles were seen in a few cells of the original lymph node biopsy. Epstein–Barr virus particles have also been seen in lymphoid cell lines from patients with infectious mononucleosis (Moses *et al.*, 1968; Steel and Edmond, 1971), with hepatitis (Douglas *et al.*, 1969), and from an apparently normal patient (Moore *et al.*, 1967). Generally thin section methods have been used to detect the virus. Negative staining has been used (Hummeler *et al.*, 1966) but appears to be less sensitive (Moses *et al.*, 1968; Hillman *et al.*, 1977). The sophisticated immunofluorescence

tests for the various EB virus antigens have supplanted electron microscopy in diagnosis particularly as even uncultured Burkitt tumor
cells carry one of these antigens. Katayama *et al.* (1974) suggested the
morphology of Burkitt tumor cells in thin sections was so characteristic, even though no virus was seen, that electron microscopic diagnosis
was superior to histology. Full identification of particles seen in cultured lymphoid cell lines is not always attempted, but herpesvirus
particles seen in cell lines from patients with Kaposi's sarcomas were
associated with cytomegalovirus in some instances (Giraldo *et al.*,
1972).

The search for retroviruses in human tumors, cell cultures of tumors,
and in placentas by electron microscopy was reviewed by De Harven
(1974) who reluctantly concluded that the morphologic evidence for
human types A, B, and C oncoviruses at that time was extremely
slight. Since 1974 the situation has not changed markedly. Tests for
biochemical markers for the presence of a retrovirus are now frequently
performed in parallel with electron microscopy. Such parallel studies
have utilized thin sectioning (Birkmayer *et al.*, 1974; Warnaar *et al.*,
1976) and negative staining (Mak *et al.*, 1974) of fractionated cell
extracts.

Retrovirus-like particles were seen in cultured myeloid cells from a
patient with acute myelogenous leukemia but not in the original tumor,
although there was biochemical evidence for retrovirus in the uncultured cells (Gallagher and Gallo, 1975). This virus was later shown to
be closely related to simian sarcoma-associated virus isolated from a
woolly monkey fibrosarcoma.

The placenta has provided somewhat better morphological evidence
for human retroviruses (Dalton *et al.*, 1974; Imamura *et al.*, 1976) but
the viral morphology was not absolutely comparable with other mammalian C-type particles (Dalton *et al.*, 1974; Dirksen and Levy, 1977).
Similar particles were detected in cultured testicular tumor cells but
parallel biochemical studies were negative (Bronson *et al.*, 1979).
Human embryonic cells in culture have been studied and particles
resembling type C oncoviruses have been observed (Chandra *et al.*,
1970; Panem *et al.*, 1975).

Types A, B, and C oncoviruses have somewhat variable structure in
both thin sections and negatively stained preparations (Sarkar and
Moore, 1972; Sarkar *et al.*, 1975) which makes morphological diagnosis
difficult. The budding stage of maturation, which is perhaps the most
convincing evidence of the viral nature of these particles, has rarely
been observed in human tissues. The negatively stained morphology,
particularly of type C oncoviruses, is so variable that this method is

unsuitable for diagnosis although Lo and Howatson (1978) have suggested detergent treatment before negative staining to standardize
particle shape.

The other retrovirus subgroup, the Spumaviridae, have more distinctive morphology. Although not seen in the original tumor a
spumavirus was identified in thin sections of cultured nasopharyngeal
carcinoma cells (Achong *et al.*, 1971).

V. Application of the Techniques to Laboratory Samples

Electron microscopy alone rarely gives a sufficiently specific diagnosis of viral infection on samples taken directly from the patient and
other techniques are generally employed to reach a more conclusive
result. In the course of laboratory tests on such samples electron microscopy can be used to detect the presence of virus in inoculated cell
cultures, in the precipitin lines of gel immunodiffusion tests, and following passage of an agent in laboratory animals. Morphology can be
very helpful in the preliminary identification of the virus, and, provided suitable controls are examined, electron microscopy can also detect contamination with endogenous viruses from cell cultures. Density
gradient studies to characterize a virus are often monitored by electron
microscopy, particularly if the virus is noncultivable. Simple negative
staining techniques are most useful and negative stain IEM can be
used for more specific identification of viruses seen. Thin sections are
useful to detect viruses which do not have clear negatively stained
morphology, such as rubella virus, and thin sections are also useful in
examination of tissues from inoculated laboratory animals.

A. *Cell Cultures*

Viruses inoculated into cell cultures usually take some days to grow
and produce typical cytopathic effects. Negative staining can give early
confirmation of the presence of a virus and preliminary identification
by its morphology. This is a useful diagnostic aid in most circumstances and particularly so when the diagnosis is urgent as in dangerous infections such as Ebola (Bowen *et al.*, 1977; Johnson *et al.*, 1977).
But early electron microscopic examination of cultures, before
cytopathic effects are well advanced, is often fruitless (Field, unpublished observations). However, the use of negative stain IEM in-

creased the sensitivity of virus detection so that cultures examined only 24 hours after inoculation showed virus particles (Edwards et al., 1975). Rhinoviruses are particularly difficult to detect in cell cultures by negative staining (Tyrrell and Almeida, 1967) and Kapikian et al. (1972a) found negative stain IEM was better able to detect these viruses in crude cell culture extracts. In practice diagnostic use of IEM is seldom feasible because of the large number of possible serotypes.

We have found that examination of inoculated cell cultures by negative staining is useful for rapid differentiation of viruses when initial biological observations cannot group them. Examples are differentiation of myxoviruses from paramyxoviruses, adenoviruses from herpesviruses, and vaccinia virus (an orthopoxvirus) from enteroviruses.

Electron microscopy of viruses which grew in cell cultures without clearly discernible cytopathic effects was used in initial work with human coronaviruses (Almeida and Tyrrell, 1967; McIntosh et al., 1967; Tyrrell and Almeida, 1967) and human polyomaviruses (Gardner et al., 1971). This procedure has remained useful in our laboratory to monitor growth of the human polyomaviruses both on isolation and further passage in cell cultures.

Viruses which are endogenous in cell cultures are a particular hazard in diagnostic laboratories and electron microscopy by negative staining and thin sectioning is useful for the detection of these agents (Anderson and Doane, 1972b). Frequently endogenous viruses cause no cytopathic effects to arouse suspicion of their presence. For those who work with monkey kidney cell cultures the simian paramyxovirus SV5 and the simian polyomavirus SV40 have been particular problems, but these agents can be easily detected in negatively stained preparations. On the other hand simian foamy virus, a retrovirus, can be easily detected by its cytopathic effect but by electron microscopy is more difficult, thin sections being more sensitive than negative staining (Anderson and Doane, 1972b). Polyomaviruses have been detected by electron microscopy in Vero cell lines (Waldeck and Sauer, 1977; Gardner and Field, unpublished observations) and in pig kidney cell lines (Newman and Smith, 1972; Tischer et al., 1974). BHK-21 hamster cell lines carry the hamster R virus detectable in thin sections (Shipman et al., 1969). Parvoviruses were detected in many continuous cell lines by techniques which included electron microscopy (Hallauer et al., 1971). It was assumed that many of these originated from trypsin used when subculturing, but recently parvoviruses detected in cultured cells had as their source the calf serum used in culture media (Nettleton and Rweyemamu, 1980).

B. Gel Diffusions

Virus particles may be purified by gel electrophoresis. The particles are concentrated into certain regions of the gel which can then be extracted for electron microscopy. Weintraub *et al.* (1962) purified plant viruses from plant sap by this technique and examined the gel fractions by metal shadowing to prove the purity of the harvests. Ahmad-Zadeh *et al.* (1968) used negative staining techniques on gel electrophoresis eluates to prove separation of adenovirus soluble antigens into hexons, fibers, and pentons.

Gel immunoprecipitin lines formed between viral antigens and homologous antibodies can be treated in the same way and the specificity of the lines becomes evident when the expected virus particle–antibody complexes are seen. Beale and Mason (1968) investigated the antigenic nature of full and empty poliovirus particles this way. Huang and Groh (1973b) applied the technique to HBsAg–antibody and HBcAg–antibody reactions. However, Almeida *et al.* (1974) found that immunoprecipitin lines between fecal extracts and human sera were frequently devoid of recognizable virus particles. When gel precipitin lines were examined for herpesviruses in thin sections by Konn *et al.* (1969), of the three lines formed between EB virus and a rabbit antiserum, herpes particles, which were surrounded by antibody, were seen in only one line.

C. Density Gradients

Fractions from density gradient centrifugation can be examined in the electron microscope for virus particles to determine their buoyant density. The cesium or sucrose must be removed before examination by application of the methods for eliminating salts described in Section III,A. Use of the technique depends upon the presence in the fractions of enough virus for particles to be detectable either by negative staining or by negative stain IEM. A typical study was that of Torikai *et al.* (1970) on a parvovirus using negative stain on the gradient fractions: particles banding at a density of 1.43 gm/ml were complete and were not penetrated by stain; particles in the 1.34 gm/ml fraction were partially penetrated by stain but were intact; at 1.30 gm/ml only shells completely penetrated by stain were seen. Parallel studies showed the particles with 1.43 gm/ml density contained normal nucleic acid while those at 1.34 gm/ml had little if any nucleic acid.

Cultivable virus can be located in density gradient harvests by its infectivity but for noncultivable viruses electron microscopy is a practical detection method which has been widely applied in density determination of agents such as the fecal parvovirus-like particles (Paver *et al.*, 1974; Appleton *et al.*, 1977; Appleton and Pereira, 1977), Norwalk agent (Kapikian *et al.*, 1973), other fecal agents resembling Norwalk (Kogasaka *et al.*, 1980), and HBsAg (Bond and Hall, 1972). The interpretation of results in the absence of infectivity experiments can be difficult as shown by the discrepancy in original estimates of the density of hepatitis A virus. Some estimates gave a peak density around 1.4 gm/ml, consistent with parvoviruses (Feinstone *et al.*, 1974), while other estimates were 1.34 gm/ml, consistent with enterovirus density (Provost *et al.*, 1975b; Maynard *et al.*, 1975). Eventually, general agreement was reached that the latter estimates were correct (Moritsugu *et al.*, 1976; Schulman *et al.*, 1976). The confusion arose because both enteroviruses and parvoviruses display multiple peaks in density gradients and the ranges overlap significantly.

D. Laboratory Animals

Laboratory animals are sometimes inoculated with known viruses in order to develop model systems for the study of human disease. Examples are picornavirus-induced hepatitis in mice (Burch *et al.*, 1973), adenovirus infection of mouse adrenal glands as a model for Allison's disease (Hoenig *et al.*, 1974), and parainfluenza type 1 infection of mouse brain as a model for multiple sclerosis (Tanaka *et al.*, 1975a). Electron microscopy of affected tissues in these experimental infections can ultimately be useful in providing examples of what might be expected in the human diagnostic situation.

Laboratory animals infected with viruses which do not grow in cell cultures may generate sufficient virus for use as reagents in diagnostic tests. Production of HBcAg from chimpanzee livers for use in detecting antibodies by complement fixation (Hoofnagle *et al.*, 1973) was suggested by thin sectioning electron microscopy studies which demonstrated intranuclear HBcAg particles in liver cells, the antigenicity being confirmed by immunofluorescence (Barker *et al.*, 1973). Similarly, thin section studies on livers of marmosets infected with hepatitis A virus showed cytoplasmic picornavirus-like particles which could be extracted for use as antigen to detect antibodies by negative stain IEM (Provost *et al.*, 1975b) and by complement fixation (Provost *et al.*, 1975a).

Specimens from animals infected with virus-containing material may be examined by electron microscopy, using all the same techniques as for human samples. The GB hepatitis agent passaged in marmosets has been examined using serum, liver, and feces in negative staining and negative stain IEM studies (Almeida *et al.*, 1976; Dienstag *et al.*, 1976b; Appleton, 1977), but no firm conclusions resulted. Current descriptions of non-A, non-B hepatitis agents in chimpanzees are similarly conflicting (Bradley *et al.*, 1979; Shimizu *et al.*, 1979; Tsiquaye *et al.*, 1980; Yoshizawa *et al.*, 1980).

VI. Artifacts

A. *Negative Stains*

Since any sample examined by negative staining in diagnostic virology is likely to contain some cellular debris it is important to appreciate that this can give rise to artifacts which can be confused with virus particles. Cellular membranes may bear projections comparable in size with the surface projections of orthomyxoviruses, paramyxoviruses, and rhabdoviruses (Cunningham and Crane, 1966; Berg *et al.*, 1969). Most cellular membrane projections tend to be globular rather than spike-like, particularly those on mitochondrial internal membranes, and with practice it is easy to differentiate such artifacts from virus particles. Particles with 5- to 10-nm projections are occasionally seen in sera and have been tentatively identified as coronaviruses, but Ackermann *et al.* (1974a), by enzyme digestion experiments, proved they were composed of lipoprotein. It is worth noting that negatively stained cell debris is indistinguishable from mycoplasmas (Wolanski and Maramorosch, 1970), thus the negative staining technique should not be used for the detection of mycoplasma contamination of cell cultures and thin sectioning and even scanning electron microscopy are the methods of choice (Boatman *et al.*, 1976).

Small lipoprotein particles in some sera have diameters about 20 to 23 nm and closely resemble the small round particles of HBsAg (Solaas, 1978). While individual particles are difficult to distinguish from HBsAg they tend to cluster together into a palisade which differs from HBsAg immune complexes because the edges of adjacent particles tend to flatten against one another (Fig. 17).

Virus-like artifacts in feces are common and are a serious problem since confirmation of the viral nature of the particles seen by culture is rarely possible. Bacterial cell walls often display substructure similar to arrays of small virus particles (Dalen, 1978). The smaller isometric bacteriophages and viruses from edible plants which might be expected to make an occasional appearance in feces are often comparable in size,

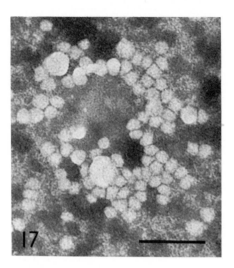

FIG. 17. Artifact in human serum resembling hepatitis B antigen. Negative stain. ×180,000. Bar = 100 nm.

appearance, and buoyant density with the small fecal viruses (Tikhonenko, 1970; Brown and Hull, 1973). Even the development of antibodies to an agent is not proof that it is a human virus as human convalescent sera do contain antibodies to fecal bacteriophage particles (Almeida *et al.*, 1974; Locarnini *et al.*, 1974).

B. *Thin Sections*

Brief consideration has already been given to papovavirus-like and paramyxovirus-like particles in thin sections resulting from artifacts (Section IV,H,1). Kirk and Hutchinson (1978) have explained cytoplasmic particles in both these categories as normal cellular components. Cytoplasmic paramyxovirus-like tubules in dilated cisternae of endoplasmic reticulum have often been described. In high-quality micrographs Schürch and Fukuda (1974) demonstrated continuity between the tubules and the cisternal membranes, proving that the tubules arose as invaginations of the membrane and were cellular rather than viral. Eady and Odland (1975) confirmed this in wounded tissue and suggested the phenomenon occurred in regenerating cells. Intranuclear filaments which sometimes appear to be tubular have also been identified as paramyxovirus nucleocapsid, but the interpretation generally favored is that this is a postmortem chromatin change (Blin-

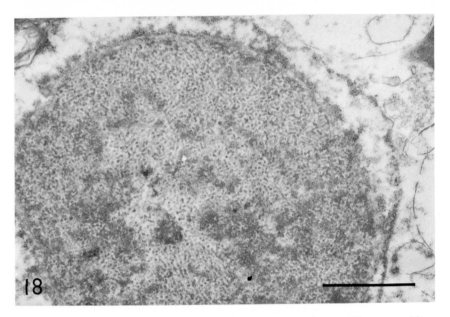

FIG. 18. Thin section of human brain showing intranuclear artifacts resembling paramyxovirus nucleocapsids. ×25,000. Bar = 1 μm.

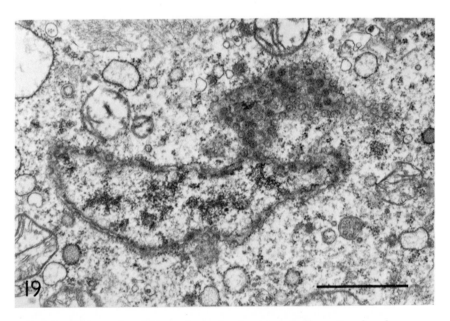

FIG. 19. Thin section of human brain showing tangentially sectioned nuclear pores which resemble herpesvirus particles. ×25,000. Bar = 1 μm. (Courtesy of Mrs. J. E. Richmond.)

zinger *et al.,* 1974; Hayano *et al.,* 1976) (Fig. 18). In addition, tangentially sectioned nuclear pores can resemble herpesviruses (Fig. 19).

Densely stained spherical cytoplasmic particles approximately 20 nm in diameter have been observed in muscle cells and variously interpreted as picornaviruses (Gyorkey *et al.,* 1978) or, with histochemical proof, as glycogen (Collins and Gilbert, 1977; Green *et al.,* 1979) or, when histochemistry disproved glycogen, as arrays of ribosomes (Oshiro *et al.,* 1976).

Dalton (1975) described 30- to 60-nm-diameter structures which resembled viruses and could be observed in thin sections of cultured cells, both intra- and extracellularly. He demonstrated that these originated from the fetal bovine serum used in the media and showed that none of the particles was viral.

VII. Electron Microscopy and Other Diagnostic Methods Compared

A. *Virus Detection*

Electron microscopy is a relatively good test in virus diagnostic work because positive results are convincing, there is photographic evidence, and results obtained from samples direct from the patient are not complicated by possibilities of cross-contamination. But it does suffer from a low sensitivity and from a lack of automation for large-scale studies compared with some of the other available tests. It is not the method of choice in the diagnosis of herpes simplex encephalitis because, although rapid, electron microscopy is insensitive compared with immunofluorescence on brain tissue (Flewett, 1973; Ross, 1973). It is, however, the method of choice for varicella zoster diagnosis because of the difficulty of culturing virus from the generally unsuitable specimens provided and the lack of sensitivity of other available tests (Macrae *et al.,* 1969). Electron microscopy is useful in cytomegalovirus detection only in very young children and the need for parallel virus isolation procedures is apparent, for even cytology is less sensitive than virus culture (Montplaisir *et al.,* 1972; Henry *et al.,* 1978; Lee *et al.,* 1978).

When virus particles are antibody coated and unable to multiply in cell culture, electron microscopy has the advantage as a detection method. This can be seen in studying polyomaviruses in urine where virus isolation and negative stain electron microscopy otherwise have roughly equivalent sensitivities (Gardner *et al.,* 1971; Coleman *et al.,*

1973). Cytology is more sensitive in polyomavirus detection than either of the other two techniques and reprocessed cytologically positive cells can be thin-sectioned to confirm the presence of polyomavirus (Coleman *et al.*, 1980). A new immunofluorescence test promises greater sensitivity for the detection of urine polyomavirus (Hogan *et al.*, 1980b) and a peroxidase–antiperoxidase light microscopy technique for polyomaviruses in tissues should also prove useful (Gerber *et al.*, 1980).

Electron microscopy was initially sufficiently sensitive to confirm positive results by complement fixation and gel immunodiffusion tests for HBsAg (Cossart *et al.*, 1971). When the more sensitive radioimmunoassay (RIA) and hemagglutination techniques were developed as routine tests for hepatitis B, the use of electron microscopy was largely discontinued except for specialized applications such as the assessment of Dane particle content of positive sera. Immunofluorescence and immunoperoxidase light microscopic detection of hepatitis B antigens in liver are more sensitive than thin section electron microscopy (Roos *et al.*, 1976), but false-positive reactions can be a problem (Omata *et al.*, 1980).

For the detection of hepatitis A virus RIA and enzyme-linked immunosorbent assay (ELISA) seem to be equally as good as negative stain IEM (Hollinger *et al.*, 1975; Purcell *et al.*, 1976; Locarnini *et al.*, 1978; Mathiesen *et al.*, 1978).

Electron microscopy has been of major diagnostic importance in the detection of noncultivable fecal viruses, but for large-scale investigations of rotavirus-associated nonbacterial gastroenteritis more suitable tests have been developed. An early report indicated that gel immunoelectrophoresis was not very sensitive for detecting rotaviruses but complement fixation was almost as sensitive as electron microscopy (Spence *et al.*, 1975). The immunofluorescence test on cells after centrifugation with rotavirus-containing fecal extracts was not as sensitive as electron microscopy, perhaps because particles lacking the outer capsomere layer were not taken into the cells (Banatvala *et al.*, 1975; Bryden *et al.*, 1977). Direct immunofluorescence on the fecal extract, however, was as specific as electron microscopy and slightly more sensitive (Yolken *et al.*, 1977). RIA and ELISA tests have greater sensitivity than electron microscopy (Middleton *et al.*, 1977; Birch *et al.*, 1979; Sarkkinen *et al.*, 1979; Seigneurin *et al.*, 1979) and usually specificity is good, though ELISA false positives have been observed (Yolken and Stopa, 1979).

Similarly, the RIA for Norwalk is specific, at least as sensitive as negative stain IEM, and more sensitive than immune adherence assay (IAHA) (Greenberg *et al.*, 1978, 1979). But in a recent series of tests negative stain IEM proved more sensitive than RIA in detecting Australian strains of Norwalk agent (Grohmann *et al.*, 1980).

An ELISA to detect noncultivable adenoviruses has recently been described and is almost as sensitive as electron microscopy but might not detect antibody-coated particles (Johansson *et al.*, 1980). Adenovirus in tonsils was detected by the presence of viral nucleic acid in the tissue with greater sensitivity than by virus isolation (Lord *et al.*, 1980), and similar studies with wart virus have been reported (Krzyzek *et al.*, 1980).

The great advantage of electron microscopy not possessed by RIA, ELISA, and other such immune tests is that detection of a range of agents is possible in a single test. This advantage is inevitably lost if specific IEM methods are used to concentrate virus from specimens.

B. Seroidentification of Viruses

Negative stain IEM can be used to serotype viruses but it is scarcely justifiable to use this difficult technique for viruses which can be cultivated and thus serotyped more easily by conventional means. Despite this, IEM typing has been described for adenoviruses (Luton, 1973), picornaviruses (Chaudhary *et al.*, 1971; Hughes *et al.*, 1977), and ortho- and paramyxoviruses (Kelen and McLeod, 1974). Papovaviruses were also serotyped by negative stain IEM because, although some grew well in cell culture and were easy to type by conventional means, others were not as amenable (Almeida *et al.*, 1969a; Field *et al.*, 1974).

Negative stain IEM has been a standard method of comparing strains of the noncultivable fecal viruses such as hepatitis A viruses (Locarnini *et al.*, 1974; Gravelle *et al.*, 1975), rotaviruses (Woode *et al.*, 1976; Zissis and Lambert, 1978), parvovirus-like particles (Paver *et al.*, 1975; Appleton and Pereira, 1977), and Norwalk group agents (Thornhill *et al.*, 1977; Kogasaka *et al.*, 1980). The development of alternative tests for the detection of these agents has also facilitated serotyping and, for example, Norwalk-like agents found in an Australian outbreak of oyster-associated gastroenteritis were confirmed as Norwalk by RIA (Murphy *et al.*, 1979).

C. Antibody Detection

Negative stain IEM utilizing known viral particle antigens has been used in recent years to assess the antibody response. Certain precautions are necessary and ideally the antigen should consist of virus particles which are well separated and clear of all attached cell debris.

If particles are already antibody coated interpretation of results becomes difficult. These tests have been used to detect antibody response to EB virus (Henle et al., 1966), rabies virus (Chaudhary et al., 1979), papovaviruses (Ogilvie, 1970; Gardner et al., 1971), hepatitis A virus (Feinstone et al., 1973; Gravelle et al., 1975; Locarnini et al., 1977; Coulepis et al., 1980), hepatitis B virus (Almeida et al., 1971; Cohen, 1978), rotavirus (Kapikian et al., 1974), Norwalk agent (Kapikian et al., 1972b; Parrino et al., 1977; Thornhill et al., 1977; Murphy et al., 1979), fecal calicivirus (Chiba et al., 1979; Cubitt et al., 1979; Suzuki et al., 1979), and astrovirus (Kurtz et al., 1977). The interpretation of the Norwalk negative stain IEM for antibody detection has been most difficult because sera taken in both the acute and the convalescent phases of infection contain antibody and careful grading of the amount of antibody coating the particles has been necessary to demonstrate rising titers.

Other tests originally developed to detect these agents, such as RIA and ELISA, can be reversed if suitable amounts of antigen are available to detect antibodies as, for example, in hepatitis B (Cohen, 1978), in hepatitis A (Purcell et al., 1976; Mathiesen et al., 1978), in Norwalk (Greenberg et al., 1978) and, by immunofluorescence, in astrovirus infections (Kurtz and Lee, 1978). Generally these tests are at least as sensitive for antibody detection as negative stain IEM and have advantages for mass screening. Because of the limited excretion period of hepatitis A virus the detection of antibody, particularly immunoglobulin M, is more likely to be used to establish the diagnosis. Specificity of some of these tests may be a problem and negative stain IEM can be used to monitor this aspect.

Immunoglobulin M (IgM) molecules have a distinctive shape and can be differentiated from immunoglobulin G (IgG) in negatively stained preparations of immune complexes between virus particles and serum immunoglobulin fractions (Almeida et al., 1967a; Svehag and Bloth, 1967; Green, 1969). It is necessary to use fractionated serum; otherwise, the IgM structure would be obscured by any IgG present. Thus negative stain IEM can be used to detect specific antiviral IgM response, a useful criterion for recent infection. This test has been used successfully with papillomavirus (Goffe et al., 1966), polyomavirus (Flower et al., 1977), and hepatitis A virus (Locarnini et al., 1977; Coulepis et al., 1980). Used in conjunction with other tests for IgM directed toward HBcAg, Cohen (1978) noted that IgM was undetectable in IAHA and complement fixation tests. Gibson et al. (1981) used a variety of tests for polyomavirus IgM and found that negative stain IEM, although less sensitive, was the only completely reliable test (Figs. 20 and 21).

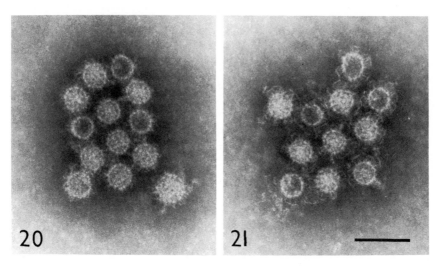

FIGS. 20 and 21.   Figure 20, polyomavirus particles coated with immunoglobulin G. Figure 21, polyomavirus particles coated with immunoglobulin M. Negative stain. ×150,000. Bar = 100 nm.

D. *Electron Microscopy in Surveillance of Other Diagnostic Methods*

The emergence of diagnostic tests which may be more sensitive than electron microscopy and more suitable for large-scale screening of specimens has still demanded the continuing use of electron microscopy to monitor the performance of these tests by confirmation of results on selected specimens. Electron microscopy also plays a role in quality control of the viral reagents utilized in these tests.

VIII. SAFETY

There is now a general awareness of the need for greater safety in the laboratory, not only for the laboratory worker but also for engineers servicing equipment, and all potentially infectious material must be inactivated before examination by electron microscopy. The electron beam may kill viruses, but only in the area of the specimen grid irradiated, and the high vacuum conditions will not inactivate virus. Chemical fixation prior to embedding and thin sectioning does kill viruses but some agents are not inactivated. Negative staining procedures do not inactivate viruses (Horne and Wildy, 1963). The whole aim of electron microscopy in viral diagnosis is to examine recognizable virus particles and some of the most effective ways of sterilizing the starting material have an adverse effect on negatively stained viral morphology, as discussed in Section II,C.

Exposure to ultraviolet irradiation inactivates most viruses (Cameron et al., 1979; Hughes et al., 1979), but papovaviruses are comparatively resistant (Shah et al., 1976; Cameron et al., 1979). Negatively stained viral morphology is usually unaffected by ultraviolet irradiation although poliovirus particles were penetrated by stain after prolonged exposure (Katagiri et al., 1967). Our standard virus inactivation procedure is to expose negatively stained grids to a high intensity, short wavelength, ultraviolet lamp at a distance of 6.5 cm from the source where the emission registers 700 to 800 μW/cm^2. The grid is irradiated for 5 minutes on each side. The lamp has been tested for its capacity to kill vaccinia virus and human polyomaviruses under these conditions (unpublished observations).

Inactivating viruses in tissues for embedding and thin sectioning must be done with a fixative which also preserves viral and cellular structure. Glutaraldehyde and formalin are most commonly used and are quite effective virucidal agents at the usual fixative concentrations, although the markedly lower concentrations needed for thin section IEM studies may not be active (Borick, 1968; Graham and Jaeger, 1968; Bowen et al., 1969; Sabel et al., 1969; Saitanu and Lund, 1975). Papovaviruses tend to be more resistant (Tevethia and Tevethia, 1976) and viruses in tissues may resist fixative action longer than when tested in suspension (Cunliffe et al., 1979).

The agents of slow virus CNS diseases such as Creutzfeldt–Jakob disease are highly resistant to formalin (Gajdusek and Gibbs, 1976) and special care must be exercised when handling any material from such cases.

Care must also be taken when handling the various toxic chemicals which are routinely used in electron microscopy (Drury, 1980).

IX. SUMMARY AND CONCLUSIONS

The supreme advantage of electron microscopy in virus diagnosis is that any virus, if present in the sample in sufficient quantity, will be recognized. The technique is flexible because it is nonselective and this applies to both negative staining and thin sectioning. The disadvantages are the high cost of the electron microscope, the need for highly trained operators, the comparative lack of sensitivity for virus detection, and the relatively small number of specimens which can be examined in a given time. As a means of opening up new fields of diagnostic virology electron microscopy has been preeminent, but after this initial stage it tends to be replaced by techniques based on newer more biochemical concepts, especially for large-scale diagnostic work.

The slowly developing techniques of electron microscopy itself seem at present to have little to offer diagnostic virology. Even high-resolution scanning electron microscopy is incapable of revealing virus particle–cell interactions in a way which can be utilized in diagnostic work. The scanning transmission mode of operation, which can induce image contrast changes electronically, may enhance studies with unstained sections and perhaps facilitate thin section IEM. It might even alter ways of examining virus particles in suspension, but early results are not particularly encouraging.

The immense contribution of electron microscopy to diagnostic virology in the last 10 years should not be underestimated. The whole concept of a virus diagnostic laboratory has changed from one in which most diagnoses are serological but with an occasional virus being isolated from a large number of samples tested, to a laboratory in which some virus isolation work continues together with a great deal of useful diagnostic work on infections with viruses which are never cultured in the laboratory. This change of outlook has been largely brought about by discoveries made with the electron microscope.

For the future, electron microscopes already heavily used in virus diagnostic work will continue to be used in this field and new discoveries will be made from time to time as in the past. The newly established laboratory, especially in financially less well endowed parts of the world, will probably utilize diagnostic tests in kit form rather than electron microscopy to search for clinically important noncultivable viruses as a first priority. However, the electron microscope has now become an established and essential part of any large virus diagnostic laboratory.

ACKNOWLEDGMENTS

I should like to thank Dr. M. S. Pereira for critically reviewing the manuscript, Dr. H. Appleton, Dr. S. D. Gardner, and Mrs. J. E. Richmond for helpful discussions, Mrs. E. C. Paddon for preparing the electron micrographs, Mr. A. A. Porter for Fig. 1, and Mrs. M. Joy for secretarial assistance.

REFERENCES

Achong, B. G., Mansell, P. W. A., Epstein, M. A., and Clifford, P. (1971). *J. Natl. Cancer Inst.* **46**, 299–307.

Ackermann, H.-W., Cherchel, G., Valet, J.-P., Matte, J., Moorjani, S., and Higgins, R. (1974a). *Can. J. Microbiol.* **20**, 193–203.

Ackermann, H.-W., Jolicoeur, P., and Berthiaume, L. (1974b). *Can. J. Microbiol.* **20,** 1093–1099.

Agar, A. W., Alderson, R. H., and Chescoe, D. (1974). "Practical Methods in Electron Microscopy" (A. M. Glauert, ed.), Vol. 2. Elsevier, Amsterdam.

Ahmad-Zadeh, C., Herzberg, M., Paccaud, M. F., and Regamey, R. H. (1968). *Pathol. Microbiol.* **32,** 254–266.

Albert, A. E., and Zu Rhein, G. M. (1974). *Int. Arch. Allergy Appl. Immunol.* **46,** 405–416.

Almeida, J. D., and Brand, C. M. (1975). *J. Gen. Virol.* **27,** 313–318.

Almeida, J. D., and Tyrrell, D. A. J. (1967). *J. Gen. Virol.* **1,** 175–178.

Almeida, J. D., and Waterson, A. P. (1969a). *Adv. Virus Res.* **15,** 307–338.

Almeida, J. D., and Waterson, A. P. (1969b). *Lancet* **2,** 983–986.

Almeida, J. D., Howatson, A. F., and Williams, M. G. (1962). *J. Invest. Dermatol.* **38,** 337–345.

Almeida, J. D., Brown, F., and Waterson, A. P. (1967a). *J. Immunol.* **98,** 186–193.

Almeida, J. D., Waterson, A. P., and Drewe, J. A. (1967b). *J. Hyg.* **65,** 467–474.

Almeida, J. D., Waterson, A. P., and Plowright, W. (1967c). *Arch. Gesamte Virusforsch.* **20,** 392–396.

Almeida, J. D., Oriel, J. D., and Stannard, L. M. (1969a). *Microbios* **1,** 225–232.

Almeida, J. D., Zuckerman, A. J., Taylor, P. E., and Waterson, A. P. (1969b). *Microbios* **1,** 117–123.

Almeida, J. D., Waterson, A. P., Trowell, J. M., and Neale, G. (1970). *Microbios* **2,** 145–153.

Almeida, J. D., Rubinstein, D., and Stott, E. J. (1971). *Lancet* **2,** 1225–1227.

Almeida, J. D., Gay, F. W., and Wreghitt, T. G. (1974). *Lancet* **2,** 748–750.

Almeida, J. D., Deinhardt, F., Holmes, A. W., Peterson, D. A., Wolfe, L., and Zuckerman, A. J. (1976). *Nature (London)* **261,** 608–609.

Almeida, J. D., Bradburne, A. F., and Wreghitt, T. G. (1979). *J. Med. Virol.* **4,** 269–277.

Almeida, J. D., Skelly, J., Howard, C. R., and Zuckerman, A. J. (1981). *J. Virol. Methods* **2,** 169–174.

Anderson, N., and Doane, F. W. (1972a). *Appl. Microbiol.* **24,** 495–496.

Anderson, N., and Doane, F. W. (1972b). *Can. J. Microbiol.* **18,** 299–304.

Apostolov, K., and Flewett, T. H. (1969). *J. Gen. Virol.* **4,** 365–370.

Appleton, H. (1977). *Nature (London)* **267,** 729–730.

Appleton, H., and Pereira, M. S. (1977). *Lancet* **1,** 780–781.

Appleton, H., Buckley, M., Thom, B. T., Cotton, J. L., and Henderson, S. (1977). *Lancet* **1,** 409–411.

Ashcavai, M., and Peters, R. L. (1971). *Am. J. Clin. Pathol.* **55,** 262–268.

Åström, K.-E., Mancall, E. L., and Richardson, E. P. (1958). *Brain* **81,** 93–111.

Bächi, T., and Howe, C. (1973). *J. Virol.* **12,** 1173–1180.

Ball, E. M., and Brakke, M. K. (1968). *Virology* **36,** 152–155.

Ball, F. L., and Harris, W. W. (1972). *Proc. Soc. Exp. Biol. Med.* **139,** 728–733.

Banatvala, J. E., Totterdell, B., Chrystie, I. L., and Woode, G. N. (1975). *Lancet* **2,** 821.

Bancroft, W. H., Snitbhan, R., Scott, R. McN., Tingpalapong, M., Watson, W. T., Tanticharoenyos, P., Karwacki, J. J., and Srimarut, S. (1977). *J. Infect. Dis.* **135,** 79–85.

Baringer, J. R., and Swoveland, P. (1972). *J. Ultrastruct. Res.* **41,** 270–276.

Barker, L. F., Chisari, F. V., McGarth, P. P., Dalgard, D. W., Kirschstein, R. L., Almeida, J. D., Edgington, T. S., Sharp, D. G., and Peterson, M. R. (1973). *J. Infect. Dis.* **127,** 648–662.

Bastian, F. O. (1971). *Lab. Invest.* **25,** 169–175.

Bastian, F. O. (1979). *Arch. Pathol. Lab. Med.* **103**, 665–669.
Bastian, F. O., Wende, R. D., Singer, D. B., and Zeller, R. S. (1975). *Am. J. Clin. Pathol.* **64**, 10–13.
Bayer, M. E., Blumberg, B. S., and Werner, B. (1968). *Nature (London)* **218**, 1057–1059.
Beale, A. J., and Mason, P. J. (1968). *J. Gen. Virol.* **2**, 203–204.
Berg, P. A., Muscatello, U., Horne, R. W., Roitt, I. M., and Doniach, D. (1969). *Br. J. Exp. Pathol.* **50**, 200–208.
Bernhard, W. (1960). *Cancer Res.* **20**, 712–727.
Berthiaume, L., Joncas, J., and Pavilanis, V. (1974). *Arch. Gesamte Virusforsch.* **45**, 39–51.
Best, J. M., Banatvala, J. E., Almeida, J. D., and Waterson, A. P. (1967). *Lancet* **2**, 237–239.
Bhatnagar, R., Johnson, G. R., and Christian, R. G. (1977). *Can. J. Comp. Med.* **41**, 416–419.
Birch, C. J., Lehman, N. I., Hawker, A. J., Marshall, J. A., and Gust, I. D. (1979). *J. Clin. Pathol.* **32**, 700–705.
Birkmayer, G. D., Miller, F., and Marguth, F. (1974). *J. Neural. Transm.* **35**, 241–254.
Bishop, R. F., Davidson, G. P., Holmes, I. H., and Ruck, B. J. (1973). *Lancet* **2**, 1281–1283.
Bishop, R. F., Davidson, G. P., Holmes, I. H., and Ruck, B. J. (1974). *Lancet* **1**, 149–151.
Blank, H., Davis, C., and Collins, C. (1970). *Br. J. Dermatol.* **83**, 69–80.
Blinzinger, K., Anzil, A. P., and Jellinger, K. (1974). *Acta Neuropathol.* **28**, 69–73.
Bloth, B., and Norrby, E. (1967). *Arch. Gesamte Virusforsch.* **21**, 71–77.
Boatman, E., Cartwright, F., and Kenny, G. (1976). *Cell Tissue Res.* **170**, 1–16.
Bohn, W. (1980). *J. Gen. Virol.* **46**, 439–447.
Bond, H. E., and Hall, W. T. (1972). *J. Infect. Dis.* **125**, 263–268.
Borick, P. M. (1968). *Adv. Appl. Microbiol.* **10**, 291–312.
Bots, G. T. A. M., Man, J. C. H., de, and Verjaal, A. (1971). *Acta Neuropathol.* **18**, 267–270.
Bowen, E. T. W., Simpson, D. I. H., Bright, W. F., Zlotnick, I., and Howard, D. M. R. (1969). *Br. J. Exp. Pathol.* **50**, 400–407.
Bowen, E. T. W., Platt, G. S., Lloyd, G., Baskerville, A., Harris, W. J., and Vella, E. E. (1977). *Lancet* **1**, 571–573.
Bowen, E. T. W., Lloyd, G., Platt, G., McArdell, L. B., Webb, P. A., and Simpson, D. I. H. (1978). *Proc. Int. Colloq. Ebola Virus Infect. Other Haemorrh. Fevers, Antwerp,* Dec. 6–8, 1977, pp. 95–202.
Boxall, E. H., Flewett, T. H., Dane, D. S., Cameron, C. H., MacCallum, F. O., and Lee, T. W. (1974). *Lancet* **2**, 1007–1008.
Boyle, W. F., McCoy, E. G., and Fogarty, W. A. (1971). *Ann. Otol. Rhinol. Laryngol.* **80**, 693–698.
Bradley, D. W., Cook. E. H., Maynard, J. E., McCaustland, K. A., Ebert, J. W., Dolana, G. H., Petzel, R. A., Kantor, R. J., Heilbrunn, A., Fields, H. A., and Murphy, B. L. (1979). *J. Med. Virol.* **3**, 253–269.
Breman, J. G., Ruti, K., Steniowski, M. V., Zanotto, E., Gromyko, A. I., and Arita, I. (1980). *Bull. W.H.O.* **58**, 165–182.
Brenner, S., and Horne, R. W. (1959). *Biochim. Biophys. Acta* **34**, 103–110.
Bronson, D. L., Fraley, E. E., Fogh, J., and Kalter, S. S. (1979). *J. Natl. Cancer Inst.* **63**, 337–339.
Brown, F., and Hull, R. (1973). *J. Gen. Virol.* **20**, 43–60.
Brown, H. R., and Thormar, H. (1976). *Acta Neuropathol.* **36**, 259–267.
Bruce White, G. B., and Stancliffe, D. (1975). *Lancet* **2**, 703.

Bryden, A. S., Davies, H. A., Hadley, R. E., Flewett, T. H., Morris, C. A., and Oliver, P. (1975). *Lancet* **2**, 241–243.

Bryden, A. S., Davies, H. A., Thouless, M. E., and Flewett, T. H. (1977). *J. Med. Microbiol.* **10**, 121–125.

Brzosko, W. J., Madalinski, K., Krawczynski, K., and Nowosławski, A. (1970). *Lancet* **1**, 1058.

Bunting, H. (1953). *Proc. Soc. Exp. Biol. Med.* **84**, 327–332.

Burch, G. E., Tsui, C. Y., and Harb, J. M. (1973). *Br. J. Exp. Pathol.* **54**, 249–254.

Burtonboy, G., Lachapelle, J.-M., Tennstedt, D., and Lamy, M.-E. (1978). *Ann. Dermatol. Venereol.* **105**, 707–712.

Cameron, K. R., Tomkins, L. M., Eglin, R. P., Ross, L. J. N., Wildy, P., and Russell, W. C. (1979). *Arch. Virol.* **62**, 31–40.

Carter, L. P., Beggs, J., and Waggener, J. D. (1972). *Cancer (Philadelphia)* **30**, 1130–1136.

Cartwright, B., Smale, C. J., and Brown, F. (1969). *J. Gen. Virol.* **5**, 1–10.

Caul, E. O., and Egglestone, S. I. (1977). *Arch. Virol.* **54**, 107–117.

Caul, E. O., Paver, W. K., and Clarke, S. K. R. (1975). *Lancet* **1**, 1192.

Caul, E. O., Ashley, C. R., and Egglestone, S. I. (1977). *Med. Lab. Sci.* **34**, 259–263.

Caul, E. O., Ashley, C. R., and Egglestone, S. I. (1978). *FEMS Microbiol. Lett.* **4**, 1–4.

Caul, E. O., Ashley, C., and Pether, J. V. S. (1979). *Lancet* **2**, 1292.

Chandra, S., Liszczak, T., Korol, W., and Jensen, E. M. (1970). *Int. J. Cancer* **6**, 40–45.

Chang, M. Y., and Campbell, W. G. (1975). *Arch. Pathol.* **99**, 185–191.

Chapman, J. A. (1970). *Lab. Pract.* **19**, 477–481.

Chaudhary, R. K., and Westwood, J. C. N. (1972). *Appl. Microbiol.* **24**, 270–274.

Chaudhary, R. K., Kennedy, D. A., and Westwood, J. C. N. (1971). *Can J. Microbiol.* **17**, 477–480.

Chaudhary, R. K., Cho, H. C., and Monette, M. T. (1979). *Can. J. Microbiol.* **25**, 1209–1211.

Chen, T. T., Watanabe, I., Zeman, W., and Mealey, J. (1969). *Science* **163**, 1193–1194.

Cheville, N. F. (1975). "Cytopathology in Viral Diseases" (J. L. Melnick, ed.), Vol. 10, Monographs in Virology Series. Karger, Basel.

Chiba, S., Sakuma, Y., Kogasaka, R., Akihara, M., Horino, K., Nakao, T., and Fukui, S. (1979). *J. Med. Virol.* **4**, 249–254.

Chiba, S., Sakuma, Y., Kogasaka, R., Akihara, M., Terashima, H., Horino, K., and Nakao, T. (1980). *J. Infect. Dis.* **142**, 247–249.

Chopra, H., Ebert, P., Woodside, N., Kvedar, J., Albert, S., and Brennan, M. (1973). *Nature (London) New Biol.* **243**, 159–160.

Chou, S. M., Roos, R., Burrell, R., Gutmann, L., and Harley, J. B. (1973). *J. Neuropathol. Exp. Neurol.* **32**, 34–50.

Christopher, P. J., Grohmann, G. S., Millsom, R. H., and Murphy, A. M. (1978). *Med. J. Aust.* **1**, 121–124.

Chrystie, I. L., Totterdell, B., Baker, M. J., Scopes, J. W., and Banatvala, J. E. (1975). *Lancet* **2**, 78–79.

Clarke, J. K., Gay, F. W., and Attridge, J. J. (1969). *J. Virol.* **3**, 358–362.

Clarke, S. K. R., Caul, E. O., and Egglestone, S. I. (1979). *Postgrad. Med. J.* **55**, 135–142.

Cockson, A., and Holmes, M. (1977). *Lab. Pract.* **26**, 175–177.

Cohen, B. J. (1978). *J. Med. Virol.* **3**, 141–149.

Cohen, B. J., and Cossart, Y. E. (1977). *J. Clin. Pathol.* **30**, 709–713.

Coleman, D. V., Gardner, S. D., and Field, A. M. (1973). *Br. Med. J.* **3**, 371–375.

Coleman, D. V., Daniel, R. A., Gardner, S. D., Field, A. M., and Gibson, P. E. (1977a). *Lancet* **2**, 709–710.

Coleman, D. V., Russell, W. J. I., Hodgson, J., Tun Pe., and Mowbray, J. F. (1977b). *J. Clin. Pathol.* **30**, 1015–1020.

Coleman, D. V., Wolfendale, M. R., Daniel, R. A., Dhanjal, N. K., Gardner, S. D., Gibson, P. E., and Field, A. M. (1980). *J. Infect. Dis.* **142**, 1–8.

Collins, D. N., and Gilbert, E. F. (1977). *Lab. Invest.* **36**, 91–99.

Cossart, Y. E., Field, A. M., Hargreaves, F. D., and Porter, A. A. (1971). *Microbios* **3**, 5–14.

Cossart, Y. E., Field, A. M., Cant, B., and Widdows, D. (1975). *Lancet* **1**, 72.

Coulepis, A. G., Locarnini, S. A., Lehman, N. I., and Gust, I. D. (1980). *J. Infect. Dis.* **141**, 151–156.

Coursaget, P., Maupas, P., Levin, P., and Barin, F. (1979). *Lancet* **2**, 92.

Cross, G. F., Waugh, M., Ferris, A. A., Gust, I. D., and Kaldor, J. (1971). *Aust. J. Exp. Biol. Med. Sci.* **49**, 1–9.

Cross, G., Forsyth, J., Greenberg, H., Harrison, J., Irving, L., Luke, R., Moore, B., and Schnagl, R. (1979). *Med. J. Aust.* **1**, 56–57.

Cruickshank, J. G., Bedson, H. S., and Watson, D. H. (1966). *Lancet* **2**, 527–530.

Cubitt, W. D., McSwiggan, D. A., and Moore, W. (1979). *J. Clin. Pathol.* **32**, 786–793.

Cubitt, W. D., McSwiggan, D. A., and Arstall, S. (1980). *J. Clin. Pathol.* **33**, 1095–1098.

Cunliffe, H. R., Blackwell, J. H., and Walker, J. S. (1979). *Appl. Environ. Microbiol.* **37**, 1044–1046.

Cunningham, W. P., and Crane, F. L. (1966). *Exp. Cell Res.* **44**, 31–45.

Dalen, A. (1978). *Acta Pathol. Microbiol. Scand., Sect. B* **86**, 249–251.

Dalton, A. J. (1972). *J. Natl. Cancer Inst.* **48**, 1095–1099.

Dalton, A. J. (1975). *J. Natl. Cancer Inst.* **54**, 1137–1148.

Dalton, A. J., and Haguenau, F., eds. (1973). "Ultrastructure of Animal Viruses and Bacteriophages: An Atlas." Academic Press, New York.

Dalton, A. J., Hellman, A., Kalter, S. S., and Helmke, R. J. (1974). *J. Natl. Cancer Inst.* **52**, 1379–1381.

Dane, D. S., Cameron, C. H., and Briggs, M. (1970). *Lancet* **1**, 695–698.

Dayan, A. D., and Almeida, J. D. (1975). *In* "Viral Diseases of the Central Nervous System" (L. S. Illis, ed.), pp. 32–55. Baillière, London.

Dayan, A. D., and Cumings, J. N. (1969). *Arch. Dis. Child.* **44**, 187–196.

Dayan, A. D., and Stokes, M. I. (1973). *Lancet* **1**, 177–179.

De Harven, E. (1974). *Adv. Virus Res.* **19**, 221–264.

De Reuck, J., De Coster, W., Otte, G., and Vander Eecken, H. (1976). *J. Neurol.* **213**, 179–188.

Derrick, K. S. (1973). *Virology* **56**, 652–653.

Dienstag, J. L., Krugman, S., Wong, D. C., and Purcell, R. H. (1976a). *Infect. Immun.* **14**, 1000–1003.

Dienstag, J. L., Wagner, J. A., Purcell, R. H., London, W. T., Lorenz, D. E. (1976b). *Nature (London)* **264**, 260–261.

Dirksen, E. R., and Levy, J. A. (1977). *J. Natl. Cancer Inst.* **59**, 1187–1192.

Dmochowski, L., Grey, C. E., Sykes, J. A., Dreyer, D. A., Langford, P., Jesse, R. H., MacComb, W. S., and Ballantyne, A. J. (1964). *Tex. Rep. Biol. Med.* **22**, 454–491.

Doane, F. W. (1974). *In* "Viral Immunodiagnosis" (E. Kurstak and R. Morisset, eds.), pp. 237–255. Academic Press, New York.

Doane, F. W., and Anderson, N. (1977). *In* "Comparative Diagnosis of Viral Diseases. Human and Related Viruses" (E. Kurstak and C. Kurstak, eds.), Vol. II, Part B, pp. 506–539. Academic Press, New York.

Doane, F. W., Anderson, N., Chatiyanonda, K., Bannatyne, R. M., and McLean, D. M. (1967). *Lancet* **2**, 751–753.

Donelli, G., Araco, M., and Ruggeri, F. (1979). *Ann. Ist. Super Sanita* **15**, 801–832.

Dougherty, R. M., and DiStefano, H. S. (1974). *Proc. Soc. Exp. Biol. Med.* **146**, 481–487.

Douglas, S. D., Glade, P. R., Fudenberg, H., and Hirschhorn, K. (1969). *J. Virol.* **3**, 520–524.

Dourmashkin, R. R., Davies, H. A., Smith, H., and Bird, R. G. (1980). *Lancet* **2**, 971–972.

Drury, P. (1980). *Can. J. Med. Technol.* **42**, 80–82.

Dubois-Dalcq, M., Schumacher, G., and Sever, J. L. (1973). *Lancet* **2**, 1408–1411.

Eady, R. A. J., and Odland, G. F. (1975). *Br. J. Dermatol.* **93**, 165–173.

Earnshaw, W. C., King, J., and Eiserling, F. A. (1978). *J. Mol. Biol.* **122**, 247–253.

Edwards, E. A., Valters, W. A., Boehm, L. G., and Rosenbaum, M. J. (1975). *J. Immunol. Methods* **8**, 159–168.

Ellis, D. S., Simpson, D. I. H., Francis, D. P., Knobloch, J., Bowen, E. T. W., Lolik, P., and Deng, I. M. (1978). *J. Clin. Pathol.* **31**, 201–208.

Emmons, R. W., and Riggs, J. L. (1977). *Methods Virol.* **6**, 1–28.

England, J. J., and Reed, P. E. (1980). *Cornell Vet.* **70**, 125–135.

Epstein, M. A., and Achong, B. G. (1968). *J. Natl. Cancer Inst.* **40**, 593–607.

Epstein, M. A., Achong, B. G., and Barr, Y. M. (1964). *Lancet* **1**, 702–703.

Feinstone, S. M., Kapikian, A. Z., and Purcell, R. H. (1973). *Science* **182**, 1026–1028.

Feinstone, S. M., Kapikian, A. Z., Gerin, J. L., and Purcell, R. H. (1974). *J. Virol.* **13**, 1412–1414.

Field, A. M., Gardner, S. D., Goodbody, R. A., and Woodhouse, M. A. (1974). *J. Clin. Pathol.* **27**, 341–347.

Flewett, T. H. (1973). *Postgrad. Med. J.* **49**, 398–400.

Flewett, T. H. (1979). *In* "Virus Diseases" (R. B. Heath, ed.), pp. 45–58. Pitman, London.

Flewett, T. H., and Apostolov, K. (1967). *J. Gen. Virol.* **1**, 297–304.

Flewett, T. H., Bryden, A. S., and Davies, H. (1973). *Lancet* **2**, 1497.

Flower, A. J. E., Banatvala, J. E., and Chrystie, I. L. (1977). *Br. Med. J.* **2**, 220–223.

Floyd, R. (1979). *Appl. Environ. Microbiol.* **38**, 980–986.

Gajdusek, D. C., and Gibbs, C. J. (1976). *N. Engl. J. Med.* **294**, 553.

Galasso, G. J. (1967). *Proc. Soc. Exp. Biol. Med.* **124**, 43–48.

Gallagher, R. E., and Gallo, R. C. (1975). *Science* **187**, 350–353.

Gardner, P. S., and McQuillin, J. (1980). "Rapid Virus Diagnosis: Application of Immunofluorescence," 2nd Ed. Butterworth, London.

Gardner, S. D. (1977). *Comp. Diagn. Viral Dis.* **1**, 41–84.

Gardner, S. D., Field, A. M., Coleman, D. V., and Hulme, B. (1971). *Lancet* **1**, 1253–1257.

Gerber, M. A., Schaffner, F., and Paronetto, F. (1972). *Proc. Soc. Exp. Biol. Med.* **140**, 1334–1339.

Gerber, M. A., Hadziyannis, S., Vissoulis, C., Schaffner, F., Paronetto, F., and Popper, H. (1974). *Am. J. Pathol.* **75**, 489–502.

Gerber, M. A., Shah, K. V., Thung, S. N., and Zu Rhein, G. (1980). *Am. J. Clin. Pathol.* **73**, 794–797.

Gibson, P. E., Field, A. M., Gardner, S. D., and Coleman, D. V. (1981). *J. Clin. Pathol.* **34**, 674–679.

Giraldo, G., Beth, E., Coeur, P., Vogel, C. L., and Dhru, D. S. (1972). *J. Natl. Cancer Inst.* **49**, 1495–1507.

Glauert, A. M. (1975). "Practical Methods in Electron Microscopy" (A. M. Glauert, ed.), Vol. 3, pt. I. Elsevier, Amsterdam.

Gmelin, K., Kommerell, B., Waldherr, R., and Ehrlich, B. V. (1980). *J. Med. Virol.* **5**, 317–322.

Goffe, A. P., Almeida, J. D., and Brown, F. (1966). *Lancet* **2**, 607–609.

Gonda, M. A., Fine, D. L., and Gregg, M. (1978). *Arch. Virol.* **56**, 297–307.

Goodfellow, A., and Calvert, H. (1979). *Br. J. Dermatol.* **101**, 341–344.

Graham, J. L., and Jaeger, R. F. (1968). *Appl. Microbiol.* **16**, 177.

Gravelle, C. R., Hornbeck, C. L., Maynard, J. E., Schable, C. A., Cook, E. H., and Bradley, D. W. (1975). *J. Infect. Dis.* **131**, 167–171.

Gray, A., Francis, R. J., and Scholtz, C. L. (1980). *Lancet* **2**, 152.

Gray, E. W., Angus, K. W., and Snodgrass, D. R. (1980). *J. Gen. Virol.* **49**, 71–82.

Green, N. M. (1969). *Adv. Immunol.* **11**, 1–30.

Green, R. J. L., Webb, J. N., and Maxwell, M. H. (1979). *J. Pathol.* **129**, 9–12.

Greenberg, H. B., Wyatt, R. G., Valdesuso, J., Kalica, A. R., London, W. T., Chanock, R. M., and Kapikian, A. Z. (1978). *J. Med. Virol.* **2**, 97–108.

Greenberg, H. B., Wyatt, R. G., and Kapikian, A. Z. (1979). *Lancet* **1**, 55.

Greenberg, H. B., Valdesuso, J. R., Kalica, A. R., Wyatt, R. G., McAuliffe, V. J., Kapikian, A. Z., and Chanock, R. M. (1981). *J. Virol.* **27**, 994–999.

Gregory, D. W., and Pirie, B. J. S. (1972). *Proc. Eur. Congr. Electron Microsc., 5th, Inst. Phys. London,* pp. 234–235.

Grimaud, J.-A., Peyrol, S., Vitvitski, L., Chevalier-Queyron, P., and Trepo, C. (1980). *N. Engl. J. Med.* **303**, 818–819.

Grohmann, G. S., Greenberg, H. B., Welch, B. M., and Murphy, A. M. (1980). *J. Med. Virol.* **6**, 11–19.

Gyorkey, F., Min, K.-W., Gyorkey, P., and Melnick, J. L. (1974). *N. Engl. J. Med.* **290**, 1488–1489.

Gyorkey, F., Cabral, G. A., Gyorkey, P. K., Uribe-Botero, G., Dressman, G. R., and Melnick, J. L. (1978). *Intervirology* **10**, 69–77.

Hadler, N. M., and Dourmashkin, R. R. (1975). *J. Immunol. Methods* **7**, 85–90.

Hallauer, C., Kronauer, G., and Siegl, G. (1971). *Arch. Gesamte Virusforsch.* **35**, 80–90.

Hanks, C. T., Fischman, S. L., and Guzman, M. N. (1972). *Oral Surg. Oral Med. Oral Pathol.* **33**, 934–943.

Hansen, B. L., Hansen, G. N., and Vestergaard, B. F. (1979). *J. Histochem. Cytochem.* **27**, 1455–1461.

Hantz, O., Vitvitski, L., and Trepo, C. (1980). *J. Med. Virol.* **5**, 73–86.

Harkness, J. W., Scott, A. C., and Hebert, C. N. (1977). *Br. Vet. J.* **133**, 81–87.

Harland, W. A., Adams, J. H., and McSeveney, D. (1967). *Lancet* **2**, 581–582.

Hashida, Y., and Yunis, E. J. (1970). *Am. J. Clin. Pathol.* **53**, 537–543.

Hashimoto, K., and Thompson, D. F. (1970). *Arch. Dermatol.* **101**, 565–577.

Hayano, M., Sung, J. H., and Mastri, A. R. (1976). *J. Neuropathol. Exp. Neurol.* **35**, 287–294.

Henle, W., Hummeler, K., and Henle, G. (1966). *J. Bacteriol.* **92**, 269–271.

Henry, C., Hartsock, R. J., Kirk, Z., and Behrer, R. (1978). *Am. J. Clin. Pathol.* **69**, 435–439.

Herndon, R. M., and Rubinstein, L. J. (1968). *Neurology* **18**, 8–18.

Herndon, R. M., Johnson, R. T., Davis, L. E., and Descalzi, L. R. (1974). *Arch. Neurol.* **30**, 475–479.

Hillman, E. A., Charamella, L. J., Temple, M. J., and Elser, J. E. (1977). *Cancer Res.* **37**, 4546–4558.

Hirschman, R. J., Shulman, N. R., Barker, L. F., and Smith, K. O. (1969). *J. Am. Med. Assoc.* **208**, 1667–1670.

Hoenig, E. M., Margolis, G., and Kilham, L. (1974). *Am. J. Pathol.* **75**, 375–394.

Hogan, T. F., Borden, E. C., McBain, J. A., Padgett, B. L., and Walker, D. L. (1980a). *Ann. Intern. Med.* **92**, 373–378.

Hogan, T. F., Padgett, B. L., Walker, D. L., Borden, E. C., and McBain, J. A. (1980b). *J. Clin. Microbiol.* **11**, 178–183.

Hollinger, F. B., Bradley, D. W., Maynard, J. E., Dreesman, C. R., and Melnick, J. L. (1975). *J. Immunol.* **115**, 1464–1466.

Holmes, I. H. (1979). *Prog. Med. Virol.* **25**, 1–36.

Holmes, I. H., Wark, M. C., and Warburton, M. F. (1969). *Virology* **37**, 15–25.

Holmes, I. H., Ruck, B. J., Bishop, R. F., and Davidson, G. P. (1975). *J. Virol.* **16**, 937–943.

Hoofnagle, J. H., Gerety, R. J., and Barker, L. F. (1973). *Lancet* **2**, 869–873.

Horne, R. W. (1967). *Methods Virol.* **3**, 521–574.

Horne, R. W. (1975). *Adv. Opt. Electron Microsc.* **6**, 227–274.

Horne, R. W., and Wildy, P. (1963). *Adv. Virus Res.* **10**, 101–170.

Horta-Barbosa, L., Fucillo, D. A., London, W. T., Jabbour, J. T., Zeman, W., and Sever, J. L. (1969). *Proc. Soc. Exp. Biol. Med.* **132**, 272–277.

Hosaka, Y. (1968). *Virology* **35**, 445–457.

Howatson, A. F., Nagai, M., and Zu Rhein, G. M. (1965). *Can. Med. Assoc. J.* **93**, 379–386.

Howe, C., Bächi, T., and Hsu, K. C. (1974). *In* "Viral Immunodiagnosis" (E. Kurstak and R. Morisset, eds.), pp. 215–234. Academic Press, New York.

Hsiung, G.-D., Fong, C. K. Y., and August, M. J. (1979). *Prog. Med. Virol.* **25**, 133–159.

Hsu, S.-M., and Ree, H. J. (1980). *Am. J. Clin. Pathol.* **74**, 32–40.

Huang, S.-N. (1971). *Am. J. Pathol.* **64**, 483–500.

Huang, S.-N., and Groh, V. (1973a). *Lab. Invest.* **29**, 353–366.

Huang, S.-N., and Groh, V. (1973b). *Lab. Invest.* **29**, 743–750.

Huang, S.-N., and Neurath, A. R. (1979). *Lab. Invest.* **40**, 1–17.

Huang, S.-N., Millman, I., O'Connell, A., Aronoff, A., Gault, H., and Blumberg, B. S. (1972). *Am. J. Pathol.* **67**, 453–470.

Huang, S.-N., Minassian, H., and More, J. D. (1976). *Lab. Invest.* **35**, 383–390.

Hughes, J. H., Gnau, J. M., Hilty, M. D., Chema, S., Ottolenghi, A. C., and Hamparian, V. V. (1977). *J. Med. Microbiol.* **10**, 203–212.

Hughes, J. H., Mitchell, M., and Hamparian, V. V. (1979). *Arch. Virol.* **61**, 313–319.

Hummeler, K., Henle, G., and Henle, W. (1966). *J. Bacteriol.* **91**, 1366–1368.

Imamura, M., Phillips, P. E., and Mellors, R. C. (1976). *Am. J. Pathol.* **83**, 383–394.

Iwasaki, Y., and Koprowski, H. (1974). *Lab. Invest.* **31**, 187–196.

Jenis, E. H., Knieser, M. R., Rothouse, P. A., Jenson, G. E., and Scott, R. M. (1973). *Arch. Pathol.* **95**, 81–89.

Jensen, E. M., Korol, W., Dittmar, S. L., and Medrek, T. J. (1967). *J. Natl. Cancer Inst.* **39**, 745–754.

Johansson, M. E., Uhnoo, I., Kidd, A. H., Madeley, C. R., and Wadell, G. (1980). *J. Clin. Microbiol.* **12**, 95–100.

Johnson, K. M., Webb, P. A., Lange, J. V., and Murphy, F. A. (1977). *Lancet* **1**, 569–571.

Johnson, K. P., Swoveland, P. T., and Emmons, R. W. (1980). *J. Am. Med. Assoc.* **244**, 41–43.

Joncas, J., Berthiaume, L., and Pavilanis, V. (1969a). *Virology* **38**, 493–496.

Joncas, J. H., Berthiaume, L., Williams, R., Beaudry, P., and Pavilanis, V. (1969b). *Lancet* **1**, 956–959.

Joncas, J., Berthiaume, L., McLaughlin, B., and Granger-Julien, M. (1975). *Can. J. Public Health* **66**, 44.

Juneau, M. L. (1979). *Can. J. Med. Technol.* **41**, 53–57.

Kapikian, A. Z., Almeida, J. D., and Stott, E. J. (1972a). *J. Virol.* **10**, 142–146.

Kapikian, A. Z., Wyatt, R. G., Dolin, R., Thornhill, T. S., Kalica, A. R., and Chanock, R. M. (1972b). *J. Virol.* **10**, 1075–1081.

Kapikian, A. Z., Gerin, J. L., Wyatt, R. G., Thornhill, T. S., and Chanock, R. M. (1973). *Proc. Soc. Exp. Biol. Med.* **142**, 874–877.

Kapikian, A. Z., Kim, H. W., Wyatt, R. G., Rodriguez, W. J., Ross, S., Cline, W. L., Parrott, R. H., and Chanock, R. M. (1974). *Science* **185**, 1049–1053.

Katagiri, S., Hinuma, Y., and Ishida, N. (1967). *Virology* **32**, 337–343.

Katayama, I., Uehara, H., Gleser, R. A., and Weintraub, L. (1974). *Am. J. Clin. Pathol.* **61**, 540–548.

Katz, M., Oyanagi, S., and Koprowski, H. (1969). *Nature (London)* **222**, 888–890.

Kelen, A. E., and McLeod, D. A. (1974). *In* "Viral Immunodiagnosis" (E. Kurstak and R. Morisset, eds.), pp. 257–275. Academic Press, New York.

Kelen, A. E., Hathaway, A. E., and McLeod, D. A. (1971). *Can. J. Microbiol.* **17**, 993–1000.

Kellenberger, E., and Arber, W. (1957). *Virology* **3**, 245–255.

Kimura, A., Tosaka, K., and Nakao, T. (1972). *Arch. Gesamte Virusforsch.* **36**, 1–12.

Kimura, A., Tosaka, K., and Nakao, T. (1975). *Arch. Virol.* **47**, 295–307.

Kirk, J., and Hutchinson, W. M. (1978). *Neuropathol. Appl. Neurobiol.* **4**, 343–356.

Kistler, G. S., Sonnabend, W., and Krech, U. (1973). *Pathol. Microbiol.* **39**, 313–318.

Kogasaka, R., Sakuma, Y., Chiba, S., Akihara, M., Horino, K., and Nakao, T. (1980). *J. Med. Virol.* **5**, 151–160.

Konn, M., Yohn, D. S., Hinuma, Y., Yamaguchi, J., and Grace, J. T. (1969). *Cancer (Philadelphia)* **23**, 990–997.

Koprowski, H., Barbanti-Brodano, G., and Katz, M. (1970). *Nature (London)* **225**, 1045–1047.

Krohn, K., Finlayson, N. D. C., Jokelainen, P. T., Anderson, K. E., and Prince, A. M. (1970). *Lancet* **2**, 379–383.

Krzyzek, R. A., Watts, S. L., Anderson, D. L., Faras, A. J., and Pass, F. (1980). *J. Virol.* **36**, 236–244.

Kumanishi, T., and Hirano, A. (1978). *J. Neuropathol. Exp. Neurol.* **37**, 790–795.

Kurstak, E., and Kurstak, C. (1974). *In* "Viral Immunodiagnosis" (E. Kurstak and R. Morisset, eds.), pp. 3–30. Academic Press, New York.

Kurtz, J., and Lee, T. (1978). *Med. Microbiol. Immunol.* **166**, 227–230.

Kurtz, J. B., Lee, T. W., and Pickering, D. (1977). *J. Clin. Pathol.* **30**, 948–952.

Kurtz, J. B., Lee, T. W., Craig, J. W., and Reed, S. E. (1979). *J. Med. Virol.* **3**, 221–230.

Lack, E. E., Vawter, G. F., Smith, H. G., Healy, G. B., Lancaster, W. D., and Jenson, A. B. (1980). *Lancet* **2**, 592.

Lamontagne, L., Marsolais, G., Marois, P., and Assaf, R. (1980). *Can. J. Microbiol.* **26**, 261–264.

Laverty, C. R., Russell, P., Hills, E., and Booth, N. (1978). *Acta Cytol.* **22**, 195–201.

Lecatsas, G., and Boes, E. (1979). *Lancet* **2**, 533–534.

Lecatsas, G., and Prozesky, O. W. (1975). *Arch. Virol.* **47**, 393–397.

Lecatsas, G., Prozesky, O. W., Van Wyk, J. V., and Els, H. J. (1973). *Nature (London)* **241**, 343–344.

Lecatsas, G., Crew-Brown, H., Boes, E., Ackthun, I., and Pienaar, J. (1978). *Lancet* **2**, 433–434.

Lee, F. K., Nahmias, A. J., and Stagno, S. (1978). *N. Engl. J. Med.* **299**, 1266–1270.

Lehrich, J. R., and Arnason, B. G. W. (1978). *In* "Handbook of Clinical Neurology" (P. J. Vinken, C. W. Bruyn, and H. L. Klawans, eds.), Vol. 34, pp. 435–450. Elsevier, Amsterdam.

Lemercier, G., Mattei, X., Rey, M., and Collomb, H. (1970). *Bull. Soc. Med. Afr. Noire Lang. Fr.* **15**, 454–458.

Lewis, H. M., Parry, J. V., Davies, H. A., Parry, R. P., Mott, A., Dourmashkin, R. R., Sanderson, P. J., Tyrrell, D. A. J., and Valman, H. B. (1979). *Arch. Dis. Child.* **54**, 339–346.

Lhermitte, F., Escourolle, R., Cathala, F., Hauw, J.-J., and Marteau, R. (1973). *Rev. Neurol. Paris* **129**, 3–19.

Lipman, M., Andrews, L., Niederman, J., and Miller, G. (1975). *J. Infect. Dis.* **132**, 520–523.

Lo, R. C. F., and Howatson, A. F. (1978). *Cancer Res.* **38**, 932–938.

Locarnini, S. A., Ferris, A. A., Stott, A. C., and Gust, I. D. (1974). *Lancet* **2**, 1007.

Locarnini, S. A., Ferris, A. A., Lehmann, W. I., and Gust, I. D. (1977). *Intervirology* **8**, 309–318.

Locarnini, S. A., Garland, S. M., Lehmann, N. I., Pringle, R. C., and Gust, I. D. (1978). *J. Clin. Microbiol.* **8**, 277–282.

Lord, M. A., Itzhaki, R. F., and Sutton, R. N. P. (1980). *Lancet* **2**, 92.

Love, D. N., and Sabine, M. (1975). *Arch. Virol.* **48**, 213–228.

Luton, P. (1973). *J. Clin. Pathol.* **26**, 914–917.

Macaya, G., Visona, K. A., and Villarejos, V. M. (1979). *J. Med. Virol.* **4**, 291–301.

Maciejewski, W., Dabrowski, J., Jablonska, S., and Jakubowicz, K. (1973). *Dermatologica* **146**, 141–148.

McCormick, W. F., Rodnitsky, R. L., Schochet, S. S., and McKee, A. P. (1969). *Arch. Neurol.* **21**, 559–570.

McIntosh, K., Dees, J. H., Becker, W. B., Kapikian, A. Z., and Chanock, R. M. (1967). *Proc. Natl. Acad. Sci. U.S.A.* **57**, 933–940.

McSwiggan, D. A., Cubitt, W. D., and Moore, W. (1978). *Lancet* **1**, 1215.

Macrae, A. D., Field, A. M., McDonald, J. R., Meurisse, E. V., and Porter, A. A. (1969). *Lancet* **2**, 313–316.

Madeley, C. R. (1972). "Virus Morphology." Churchill, London.

Madeley, C. R. (1979). *J. Infect. Dis.* **139**, 519–523.

Madeley, C. R., and Cosgrove, B. P. (1975). *Lancet* **2**, 124.

Madeley, C. R., and Cosgrove, B. P. (1976). *Lancet* **1**, 199–200.

Maiztegui, J. I., Laguens, R. P., Cossio, P. M., Casanova, M. B., De La Vega, M. T., Ritacco, V., Segal, A., Fernández, N. J., and Arana, R. M. (1975). *J. Infect. Dis.* **132**, 35–43.

Mak, T. W., Manaster, J., Howatson, A. F., McCulloch, E. A., and Till, J. E. (1974). *Proc. Natl. Acad. Sci. U.S.A.* **71**, 4336–4340.

Mathan, M., Mathan, V. I., Swaminathan, S. P., Yesudoss, S., and Baker, S. J. (1975). *Lancet* **1**, 1068–1069.

Mathiesen, L. R., Feinstone, S. M., Wong, D. C., Skinhoej, P., and Purcell, R. H. (1978). *J. Clin. Microbiol.* **7**, 184–193.

Matsumara, T., Stollar, V., and Schlesinger, R. W. (1971). *Virology* **46**, 344–355.

Maynard, J. E., Lorenz, D., Bradley, D. W., Feinstone, S. M., Krushak, D. H., Barker, L. F., and Purcell, R. H. (1975). *Am. J. Med. Sci.* **270**, 81–85.

Melnick, J. L., Bunting, H., Banfield, W. G., Strauss, M. J., and Gaylord, W. H. (1952). *Ann. N.Y. Acad. Sci.* **54**, 1214–1225.

Middleton, P. J., Szymanski, M. T., Abbott, G. D., Bortolussi, R., and Hamilton, J. R. (1974). *Lancet* **1**, 1241–1244.

Middleton, P. J., Holdaway, M. D., Petric, M., Szymanski, M. T., and Tam, J. S. (1977). *Infect. Immun.* **16**, 439–444.

Milne, R. G., and Luisoni, E. (1977). *Methods Virol.* **6**, 265–281.

Minnich, L., and Ray, C. G. (1980). *J. Clin. Microbiol.* **12**, 391–394.

Mirra, S. S., and Takei, Y. (1976). *Prog. Neuropathol.* **3**, 69–88.

Miyamoto, K., Morgan, C., Hsu, K. C., and Hampar, B. (1971). *J. Natl. Cancer Inst.* **46**, 629–646.

Mohanty, S. B. (1975). *Am. J. Vet. Res.* **36**, 319–321.

Monroe, J. H., and Brandt, P. M. (1970). *Appl. Microbiol.* **20**, 259–262.

Montplaisir, S., Belloncik, S., Leduc, N. P., Onji, P. A., Martineau, B., and Kurstak, E. (1972). *J. Infect. Dis.* **125**, 533–538.

Moodie, J. W., Stannard, L. M., and Kipps, A. (1974). *J. Clin. Pathol.* **27**, 693–697.

Moore, D. H., Sarkar, N. H., Kelly, C. E., Pillsbury, N., and Charney, J. (1969). *Tex. Rep. Biol. Med.* **27**, 1027–1039.

Moore, G. E., Gerner, R. E., and Franklin, H. A. (1967). *J. Am. Med. Assoc.* **199**, 519–524.

Morecki, R., and Zimmerman, H. M. (1969). *Arch. Neurol.* **20**, 599–604.

Mori, Y., Ogata, S., Ata, S., and Nakano, Y. (1980). *Lancet* **2**, 318.

Morin, C., and Meisels, A. (1980). *Acta Cytol.* **24**, 82–84.

Moritsugu, Y., Dienstag, J. L., Valdesuso, J., Wong, D. C., Wagner, J., Routenberg, J. A., and Purcell, R. H. (1976). *Infect. Immun.* **13**, 898–908.

Moses, H. L., Glade, P. R., Kasel, J. A., Rosenthal, A. S., Hirshaut, Y., and Chessin, L. N. (1968). *Proc. Natl. Acad. Sci. U.S.A.* **60**, 489–496.

Murphy, A. M., Grohmann, G. S., Christopher, P. J., Lopez, W. A., Davey, G. R., and Millsom, R. H. (1979). *Med. J. Aust.* **2**, 329–333.

Murphy, F. A., and Whitfield, S. G. (1975). *Bull. W.H.O.* **52**, 409–419.

Murphy, F. A., Webb, P. A., Johnson, K. M., Whitfield, S. G., and Chappell, W. A. (1970). *J. Virol.* **6**, 507–518.

Murphy, F. A., Van Der Groen, G., Whitfield, S. G., and Lange, J. V. (1978). *Proc. Int. Colloq. Ebola Virus Infect. Other Haemorrh. Fevers, Antwerp, Dec. 6–8, 1977*, pp. 61–84.

Murti, K. G., Bean, W. J., and Webster, R. G. (1980). *Virology* **104**, 224–229.

Nagler, F. P. O., and Rake, G. (1948). *J. Bacteriol.* **55**, 45–51.

Narang, H. K. (1975). *Acta Neuropathol.* **32**, 163–168.

Narang, H. K., and Codd, A. A. (1980a). *Lancet* **1**, 1192–1193.

Narang, H. K., and Codd, A. A. (1980b). *Lancet* **2**, 703.

Narang, H. K., and Field, E. J. (1973). *Acta Neuropathol.* **25**, 281–290.

Narayan, O., Penney, J. B., Johnson, R. T., Herndon, R. M., and Weiner, L. P. (1973). *N. Engl. J. Med.* **289**, 1278–1282.

Nelson, J. M., Barker, L. F., and Danovitch, S. H. (1970). *Lancet* **2**, 773–774.

Nermut, M. V. (1972). *J. Microsc. (Oxford)* **96**, 351–362.

Nermut, M. V. (1977). *In* "Principles and Techniques of Electron Microscopy" (M. A. Hayat, ed.), Vol. 7, pp. 79–117. Van Nostrand-Reinhold, Princeton, New Jersey.

Nettleton, P. F., and Rweyemamu, M. M. (1980). *Arch. Virol.* **64**, 359–374.

Newman, J. T., and Smith, K. O. (1972). *Infect. Immun.* **5**, 961–967.

Nicolaieff, A., Obert, G., and Van Regenmortel, M. H. V. (1980). *J. Clin. Microbiol.* **12**, 101–104.

Norrby, E., Marusyk, H., and Örvell, C. (1970). *J. Virol.* **6**, 237–242.

Nowosławski, A., Brzosko, W. J., Madalinski, K., and Krawczynski, K. (1970). *Lancet* **1**, 494–498.

Ogilvie, M. M. (1970). *J. Hyg.* **68**, 479–490.

Omata, M., Choong-Tsek, L., Ashcavai, M., and Peters, R. L. (1980). *Am. J. Clin. Pathol.* **73**, 626–632.

Oriel, J. D., and Almeida, J. D. (1970). *Br. J. Vener. Dis.* **46**, 37–42.

Oshiro, L. S., Cremer, N. E., Norris, F. H., and Lennette, E. H. (1976). *Neurology* **26**, 57–60.

Oyanagi, S., Ter Meulen, V., Müller, D., Katz, M., and Koprowski, H. (1970). *J. Virol.* **6**, 370–379.

Padgett, B. L., Walker, D. L., Zu Rhein, G. M., and Eckroade, R. J. (1971). *Lancet* **1**, 1257–1260.

Padgett, B. L., Walker, D. L., Zu Rhein, G. M., Hodach, A. E., and Chou, S. M. (1976). *J. Infect. Dis.* **133**, 686–690.

Palmer, E. L., Martin, M. L., and Murphy, F. A. (1977). *J. Gen. Virol.* **35**, 403–414.

Panem, S., Prochownik, E. V., Reale, F. R., and Kirsten, W. H. (1975). *Science* **189**, 297–299.

Paradis, N., Martineau, B., and Côté, J. R. (1969). *Union Med. Can.* **98**, 1287–1290.

Parrino, T. A., Schreiber, D. S., Trier, J. S., Kapikian, A. Z., and Blacklow, N. R. (1977). *N. Engl. J. Med.* **297**, 86–89.

Pathak, S., and Webb, H. E. (1976). *Lancet* **2**, 311.

Patterson, S. (1975). *J. Immunol. Methods* **9**, 115–122.

Pattison, J. R., Jones, S. E., Hodgson, J., Davis, L. R., White, J. M., Stroud, C. E., and Murtaza, L. (1981). *Lancet* **1**, 664–665.

Paver, W. K., Caul, E. O., Ashley, C. R., and Clarke, S. K. R. (1973). *Lancet* **1**, 237–239.

Paver, W. K., Caul, E. O., and Clarke, S. K. R. (1974). *J. Gen. Virol.* **22**, 447–450.

Paver, W. K., Caul, E. O., and Clarke, S. K. R. (1975). *Lancet* **1**, 232.

Pavilanis, V., Joncas, J. H., Skvorc, R., Berthiaume, L., Blanc, M. F., Martineau, B., and Podoski, M. O. (1971). *Union Med. Can.* **100**, 2195–2202.

Peters, D., Nielsen, G., and Bayer, M. E. (1962). *Dtsch. Med. Wochenschr.* **87**, 2240–2246.

Pinkerton, H., and Carroll, S. (1971). *Am. J. Pathol.* **65**, 543–548.

Polson, A., and Stannard, L. (1970). *Virology* **40**, 781–791.

Portnoy, B. L., Conklin, R. H., Menn, M., Olarte, J., and Du Pont, H. L. (1977). *J. Lab. Clin. Med.* **89**, 560–563.

Praetorius-Clausen, F., and Willis, J. M. (1971). *Scand. J. Dent. Res.* **79**, 362–365.

Prineas, J. (1972). *Science* **178**, 760–763.

Provost, P. J., Ittensohn, O. L., Villarejos, V. M., and Hilleman, M. R. (1975a). *Proc. Soc. Exp. Biol. Med.* **148**, 962–969.

Provost, P. J., Wolanski, B. S., Miller, W. J., Ittensohn, O. L., McAleer, W. J., and Hilleman, M. R. (1975b). *Proc. Soc. Exp. Biol. Med.* **148**, 532–539.

Purcell, R. H., Wong, D. C., Moritsugu, Y., Dienstag, J. L., Routenberg, J. A., and Boggs, J. D. (1976). *J. Immunol.* **116**, 349–356.

Rabin, E. R., and Jenson, A. B. (1967). *Prog. Med. Virol.* **9**, 392–450.

Reese, J. M., Reissig, M., Daniel, R. W., and Shah, K. V. (1975). *Infect. Immun.* **11**, 1375–1381.

REPORT (1979). "Classification and Nomenclature of Viruses: Third Report of the International Committee on Taxonomy of Viruses." Karger, Basel.

Reuss, K., Plescher, I. C., Hülser, D., and Herzberg, K. (1967). *Zentralbl. Bakteriol. Parasitenkd. Infektionskr. Hyg. Abt. 1 Orig. Reihe A* **203**, 47–58.

Roberts, J. L., Curry, A., and Paver, W. K. (1980). *Lancet* **2**, 151–152.

Ronald, W. P., Schroeder, B., Tremaine, J. H., and Paliwal, Y. C. (1977). *Virology* **76,** 416–419.

Roos, C. M., Feltkamp-Vroom, T. M., and Helder, A. W. (1976). *J. Pathol.* **118,** 1–8.

Ross, C. A. C. (1973). *Postgrad. Med. J.* **49,** 401–402.

Rossi, G. L., Luginbühl, H., and Probst, D. (1970). *Virchows. Arch. Abt. A Pathol. Anat.* **350,** 216–224.

Roy, S., and Wolman, L. (1969). *J. Clin. Pathol.* **22,** 51–59.

Sabel, F. L., Hellman, A., and McDade, J. J. (1969). *Appl. Microbiol.* **17,** 645–646.

Saif, L. J., Bohl, E. H., Kohler, E. M., and Hughes, J. H. (1977). *Am. J. Vet. Res.* **38,** 13–20.

Saitanu, K., and Lund, E. (1975). *Appl. Microbiol.* **29,** 571–574.

Sangar, D. V., Rowlands, D. J., Smale, C. J., and Brown, F. (1973). *J. Gen. Virol.* **21,** 399–406.

Sarkar, N. H., and Moore, D. H. (1972). *J. Natl. Cancer Inst.* **48,** 1051–1058.

Sarkar, N. H., Manthey, W. J., and Sheffield, J. B. (1975). *Cancer Res.* **35,** 740–749.

Sarkkinen, H. K., Halonen, P. E., and Arstila, P. P. (1979). *J. Med. Virol.* **4,** 255–260.

Schnagl, R. D., Holmes, I. H., and Mackay-Scollay, E. M. (1978). *Med. J. Aust.* **1,** 307–309.

Schulman, A. N., Dienstag, J. L., Jackson, D. R., Hoofnagle, J. H., Gerety, R. J., Purcell, R. H., and Barker, L. F. (1976). *J. Infect. Dis.* **134,** 80–84.

Schürch, W., and Fukuda, T. (1974). *Pathol. Microbiol.* **40,** 89–99.

Schwerdt, P. R., Schwerdt, C. E., Silverman, L., and Rubinstein, L. J. (1966). *Virology* **29,** 511–514.

Scotto, J., Stralin, H., and Caroli, J. (1970). *C. R. Acad. Sci.* **271,** 1603–1604.

Seigneurin, J. M., Scherrer, R., Baccard-Longere, M., Genoulaz, O., Feynerol. C., and Gout, J. P. (1979). *Biomedicine* **31,** 99–104.

Shah, K. V., Hudson, C., Valis, J., and Strandberg, J. D. (1976). *Proc. Soc. Exp. Biol. Med.* **153,** 180–186.

Shimizu, Y. K., Feinstone, S. M., Purcell, R. H., Alter, H. J., and London, W. T. (1979). *Science* **205,** 197–200.

Shipman, C., Van der Weide, G. C., and Ma, B. I. (1969). *Virology* **38,** 707–710.

Shneerson, J. M., Mortimer, P. P., and Vandervelde, E. M. (1980). *Br. Med. J.* **280,** 1580.

Shukla, D. D., and Gough, K. H. (1979). *J. Gen. Virol.* **45,** 533–536.

Singer, I. I. (1974). *In* "Viral Immunodiagnosis" (E. Kurstak and R. Morisset, eds.), pp. 101–123. Academic Press, New York.

Sisson, S. P., and Vernier, R. L. (1980). *J. Histochem. Cytochem.* **28,** 441–452.

Smit, J. W., Meijer, C. J. L. M., Decary, F., and Feltkamp-Vroom, T. M. (1974). *J. Immunol. Methods* **6,** 93–98.

Smith, J. W., Pinkel, D., and Dabrowski, S. (1969). *Cancer (Philadelphia)* **24,** 527–531.

Smith, K. O., and Melnick, J. L. (1962). *Virology* **17,** 480–490.

Solaas, M. H. (1978). *Acta Pathol. Microbiol. Scand. Sect B* **86,** 125–129.

Spadbrow, P. B., and Francis, J. (1969). *Vet. Rec.* **84,** 244–246.

Spence, L., Fauvel, M., Bouchard, S., Babiuk, L., and Saunders, J. R. (1975). *Lancet* **2,** 322.

Spoendlin, H., and Kistler, G. (1978). *Arch. Otol. Rhinol. Laryngol.* **218,** 289–292.

Stannard, L. M., Moodie, J., Keen, G. A., and Kipps, A. (1973). *J. Clin. Pathol.* **26,** 209–216.

Stannard, L. M., Almeida, J. D., and Tedder, R. S. (1980). *J. Med. Virol.* **6,** 153–164.

Steel, C. M., and Edmond, E. (1971). *J. Natl. Cancer Inst.* **47,** 1193–1201.

Stein, O., Fainaru, M., and Stein, Y. (1971). *Lancet* **2,** 90–91.

Strauss, M. J., Shaw, E. W., Bunting, H., and Melnick, J. L. (1949). *Proc. Soc. Exp. Biol. Med.* **72**, 46–50.

Sutton, J. S., and Burnett, J. W. (1969). *J. Ultrastruct. Res.* **26**, 177–196.

Suzuki, H., Konno, T., Kutsuzawa, T., Imai, A., Tazawa, F., Ishida, N., Katsushima, N., and Sakamoto, M. (1979). *J. Med. Virol.* **4**, 321–326.

Svehag, S. E., and Bloth, B. (1967). *Virology* **31**, 676–687.

Swoveland, P. T., and Johnson, K. P. (1979). *J. Infect. Dis.* **140**, 758–764.

Taber, L. H., Mirkovic, R. R., Adam, V., Ellis, S. S., Yow, M. D., and Melnick, J. L. (1973). *Intervirology* **1**, 127–134.

Tabor, E., Mitchell, F. D., Goudeau, A. M., and Gerety, R. J. (1979). *J. Med. Virol.* **4**, 161–169.

Takamiya, H., Batsford, S., Gelderblom, H., and Vogt, A. (1979). *J. Bacteriol.* **140**, 261–266.

Takeda, M. (1969). *Acta Cytol.* **13**, 206–209.

Tanaka, R., Iwasaki, Y., and Koprowski, H. (1974). *Lancet* **1**, 1236–1237.

Tanaka, R., Iwasaki, Y., and Koprowski, H. (1975a). *Am. J. Pathol.* **79**, 335–346.

Tanaka, R., Iwasaki, Y., and Koprowski, H. (1975b). *Am. J. Pathol.* **81**, 467–478.

Taniguchi, K., Urasawa, S., and Urasawa, T. (1979). *J. Clin. Microbiol.* **10**, 730–736.

Tawara, J., Ichikawa, H., Akatsuka, K., Kumon, H., Fujio, K., and Shishido, A. (1976). *Jpn. J. Med. Sci. Biol.* **29**, 99–103.

Ter Meulen, V., Koprowski, H., Iwasaki, Y., Käckell, Y. M., and Müller, D. (1972). *Lancet* **2**, 1–5.

Tevethia, M. J., and Tevethia, S. S. (1976). *Virology* **69**, 474–489.

Thomson, R. O., Walker, P. D., Batty, I., and Baillie, A. (1967). *Nature (London)* **215**, 393–394.

Thornhill, T. S., Wyatt, R. G., Kalica, A. R., Dolin, R., Chanock, R. M., and Kapikian, A. Z. (1977). *J. Infect. Dis.* **135**, 20–27.

Tikhonenko, A. S. (1970). "Ultrastructure of Bacterial Viruses" (B. Haigh, trans.). Plenum, New York.

Tischer, I., Rasch, R., and Tochtermann, G. (1974). *Zentralbl. Bakteriol. Parasitenkd. Infektionskr. Hyg. Abt. 1 Orig. Reihe A* **226**, 153–167.

Torikai, K., Ito, M., Jordan, L. E., and Mayor, H. D. (1970). *J. Virol.* **6**, 363–369.

Torrella, F., and Morita, R. Y. (1979). *Appl. Environ. Microbiol.* **37**, 774–778.

Trepo, C. G., Zuckerman, A. J., Bird, R. G., and Prince, A. M. (1974). *J. Clin. Pathol.* **27**, 863–868.

Tsiquaye, K. N., Bird, R. G., Tovey, G., Wyke, R. J., Williams, R., and Zuckerman, A. J. (1980). *J. Med. Virol.* **5**, 63–71.

Tyrrell, D. A. J., and Almeida, J. D. (1967). *Arch. Gesamte Virusforsch.* **22**, 417–425.

Valentine, R. C., and Pereira, H. G. (1965). *J. Mol. Biol.* **13**, 13–20.

Vallat, J. M., Vital, C., Amouretti, M., Faucher, P., and Julien, J. (1977). *Rev. Neurol. Paris* **133**, 637–646.

Valters, W. A., Boehm, L. G., Edwards, E. A., and Rosenbaum, M. J. (1975). *J. Clin. Microbiol.* **1**, 472–475.

Van Rooyen, C. E., and Scott, G. D. (1948). *Can. J. Public Health* **39**, 467–477.

Van Wyk, C. W., Staz, J., and Farman, A. G. (1977). *J. Oral Pathol.* **6**, 14–24.

Vernon, M. L., Horta-Barbosa, L., Fucillo, D. A., Sever, J. L., Baringer, J. R., and Birnbaum, G. (1970). *Lancet* **1**, 964–966.

Vernon, S. K., Lawrence, W. C., Cohen, G. H., Durso, M., and Rubin, B. A. (1976). *J. Gen. Virol.* **31**, 183–191.

Viac, J., Schmitt, D., and Thivolet, J. (1978). *J. Invest. Dermatol.* **70**, 263–266.

Viloria, J. E., and Garcia, J. H. (1976). *Beitr. Pathol.* **157**, 14–22.
Vuillaume, M., and De Thé, G. (1973). *J. Natl. Cancer Inst.* **51**, 67–80.
Waldeck, W., and Sauer, G. (1977). *Nature (London)* **269**, 171–173.
Warnaar, S. O., Te Velde, J., Van Muijen, G., Prins, F., Van Der Loo, E. M., Koopmans-Broekhuizen, N., and De Man, J. C. H. (1976). *Mol. Biol. Rep.* **3**, 1–8.
Watanabe, H., and Holmes, I. H. (1977). *J. Clin. Microbiol.* **6**, 319–324.
Watanabe, I., and Okazaki, H. (1973). *Lancet* **2**, 569–570.
Watson, D. H., and Wildy, P. (1963). *Virology* **21**, 100–111.
Webb, M. J. W. (1973). *J. Microsc. (Oxford)* **98**, 109–111.
Weiner, L. P., Herndon, R. M., Narayan, O., Johnson, R. T., Shah, K., Rubinstein, L. J., Preziosi, T. J., and Conley, F. K. (1972). *N. Engl. J. Med.* **286**, 385–390.
Weintraub, M., Ragetli, H. W. J., and Townsley, P. M. (1962). *Virology* **17**, 196–198.
Wendelschafer-Crabb, G., Erlandsen, S. L., and Walker, D. H. (1976). *J. Histochem. Cytochem.* **24**, 517–526.
Whitby, H. J., and Rodgers, F. G. (1980). *J. Clin. Pathol.* **33**, 484–487.
WHO Scientific Working Group (1980). *Bull. W.H.O.* **58**, 183–198.
Williams, M. G., Almeida, J. D., and Howatson, A. F. (1962). *Arch. Dermatol.* **86**, 290–297.
Winn, W. C., Monath, T. P., Murphy, F. A., and Whitfield, S. G. (1975). *Arch. Pathol.* **99**, 599–604.
Wolanski, B., and Maramorosch, K. (1970). *Virology* **42**, 319–327.
Woode, G. N., Bridger, J. C., Jones, J. M., Flewett, T. H., Bryden, A. S., Davies, H. A., and White, G. B. B. (1976). *Infect. Immun.* **14**, 804–810.
Wrigley, N. G. (1968). *J. Ultrastruct. Res.* **24**, 454–464.
Yabe, Y., and Sadakane, H. (1975). *J. Invest. Dermatol.* **65**, 324–330.
Yolken, R. H., and Stopa, P. J. (1979). *J. Clin. Microbiol.* **10**, 703–707.
Yolken, R. H., Wyatt, R. G., Kalica, A. R., Kim, H. W., Brandt, C. D., Parrott, R. H., Kapikian, A. Z., and Chanock, R. M. (1977). *Infect. Immun.* **16**, 467–470.
Yoshizawa, H., Akahane, Y., Itoh, Y., Iwakiri, S., Kitajima, K., Morita, M., Tanaka, A., Nojiri, T., Shimizu, M., Miyakawa, Y., and Mayumi, M. (1980). *Gastroenterology* **79**, 512–520.
Zanem-Lim, O. G. (1976). *Lancet* **1**, 18–20.
Zeve, V. H., Lucas, L. S., and Manaker, R. A. (1966). *J. Natl. Cancer Inst.* **37**, 761–773.
Zissis, G., and Lambert, J. P. (1978). *Lancet* **1**, 38–39.
Zissis, G., Lambert, J. P., and De Kegel, D. (1978). *J. Clin. Pathol.* **31**, 175–178.
Zuckerman, A. J., Taylor, P. E., and Almeida, J. D. (1970). *Br. Med. J.* **1**, 262–264.
Zu Rhein, G. M., and Chou, S.-M. (1965). *Science* **148**, 1477–1479.

ADVANCES IN VIRUS RESEARCH, VOL. 27

STUDIES OF JAPANESE ENCEPHALITIS IN CHINA

C. H. Huang

Institute of Virology
Chinese Academy of Medical Sciences
Beijing, People's Republic of China

I. Introduction

Japanese encephalitis (JE) is a mosquito-borne disease caused by a member of the group B arboviruses. It is classified under the family Togaviridae, genus *Flavivirus*. The disease was first recognized as an encephalitis in Japan in 1871 and the virus was first isolated from a fatal case in Japan in 1934.

Japanese encephalitis is a disease of the Far East, extending from maritime Siberia to eastern India, Sri Lanka, part of the Indonesian archipelago, Borneo, and the Phillipines. In China the disease is very widespread. Practically all provinces except the two most western ones, Xinjiang and Xizang, have the disease. Before the liberation in 1949, very little study of JE had been done in China. Huang and Liu (1940) and Chu *et al.* (1940) reported clinical cases with serological study. The virus was first isolated in China by Yen (1941). After 1949 large numbers of epidemiological and ecological studies were carried out. Antimosquito campaigns and development of a vaccine for the control of the disease form the main subjects of the study. The morbidity rate in Beijing in the years after 1949 was 15–25/100,000 population and in recent years it has dropped to 2.5/100,000. The mortality rate was about 40–50% in previous years and in recent years it has been around 10%. Studies on the pathogenicity and pathogenesis of JE virus have also been done. The following is a review of the studies done in China.

II. Pathogenic Properties of the Virus

A. *Existence of Virus Strains with Different Peripheral Pathogenicity*

The existence of strains of JE virus with different peripheral virulence was first demonstrated by Huang (1957a,b; Huang and Tai, 1958). Different strains of virus isolated from human cases and from mosquitoes were compared as to their pathogenicity by intracerebral (ic) and subcutaneous (sc) routes of inoculation in 3-week-old mice. If only strains with low passage history in mouse brain are included in the analysis (since we have found that repeated passages of virus intracerebrally will lead to a decrease in peripheral pathogenicity), the results show that while there is not much difference in the LD_{50} among these strains when the virus is introduced by the ic route, there is, however, greater difference in the LD_{50} if the virus is inoculated subcutaneously. The subcutaneous virus titer in negative logs for virus strains isolated from 16 human fatal cases varied from 4.0 to 8.1; and 21 virus strains from different species of mosquitoes varied from 2.2 to 6.5. All these findings indicate that there are strains of JE virus with different peripheral pathogenicity. The implication of this finding for the study of ecology will be presented in Section IV.

B. Existence of Virus Strains with Different Neurovirulence

In addition to finding virus strains with different peripheral pathogenicity, Wang (unpublished data, 1975) from our laboratory also demonstrated the existence of virus strains with different neurovirulence. Wang compared 31 strains of virus originating from *Culex tritaeniorrhynchus* and isolated directly in baby hamster kidney cell culture with respect to the titer of virus obtained in tissue culture contrasted with that obtained from intracerebral LD_{50} of 3-week-old mice. The results (Table I) indicate some difference in the neurovirulence among these virus strains.

C. Relationship between the Subcutaneous Pathogenicity of the Virus and the Peripheral Multiplication

Huang *et al.* (1958) compared 18 strains of virus with different peripheral pathogenicity and the duration of viremia after infection in mice. The results (Table II) show that strains with high peripheral pathogenicity when inoculated subcutaneously in 3-week-old mice had a longer duration of viremia and those with low peripheral pathogenicity had a shorter duration of viremia. As shown in Table II, for the P1 strain, which has the highest peripheral pathogenicity (1 log unit as determined by the difference of ic and sc titer), the duration of viremia was 6 days; for strains with a peripheral virulence of 1.9–2.4 logs,

TABLE I

NEUROVIRULENCE OF VIRUS ISOLATED FROM *C. tritaeniorrhynchus* DIRECTLY FROM TISSUE CULTURE

Difference in log of hk and ic titer[a]	Number of strains	%
1	6	19.3
1.17–2.0	20	64.5
2.17–3.0	3	9.7
3.17	2	6.3

[a] hk, Titrated in baby hamster kidney tissue culture; ic, intracerebral, titrated in mouse brain.

TABLE II

RELATIONSHIP OF PERIPHERAL PATHOGENICITY TO DURATION OF VIREMIA AMONG
DIFFERENT STRAINS OF VIRUS

Virus strain	Difference of ic and subcutaneous viral titer in log	Duration of viremia at day						
		1	2	3	4	5	6	7
P1	1.0	5/5[a]	5/5	5/5	4/5	3/5	2/5	0/5
C. pipiens	1.8	5/5	5/5	1/4	2/5	2/5	0/4	—
Anhui	1.9	4/4	4/4	2/4	4/4	2/4	0/4	—
Reho1	2.0	5/5	5/5	5/5	5/5	1/5	0/5	0/5
Reho2	2.0	5/5	5/5	5/5	5/5	2/5	0/4	0/5
Chengdu1	2.2	5/5	5/5	5/5	4/5	2/5	0/5	0/5
Chengdu5	2.4	5/5	5/5	5/5	2/5	0/5	2/5	0/5
Chengdu7	2.6	5/5	5/5	5/5	1/5	0/5	0/5	0/5
Chengdu2	3.6	5/5	5/5	5/5	4/5	0/5	1/5	0/5
Peng46	3.9	5/5	5/5	0/5	0/4	0/5	0/5	0/4
Shi	4.0	5/5	5/5	3/5	1/5	0/5	1/5	0/5
Chengdu3	4.2	5/5	5/5	5/5	2/5	0/5	0/5	0/5
Chengdu4	4.8	5/5	5/5	5/5	2/5	0/5	0/5	0/5
Chengdu6	4.8	5/5	5/5	4/5	5/5	0/5	0/5	0/5
Xi An	5.4	5/5	3/5	0/4	1/5	0/4	0/5	0/5
M47	5.6	5/5	3/5	0/4	0/5	0/4	0/5	—
C. fatigans	6.3	1/5	4/5	1/5	1/5	1/5	—	—
Nakayama	6.8	5/5	2/5	0/5	0/5	0/5	0/5	—

[a] Number of mice which died/number of mice inoculated.

the duration of viremia was around 5 days; when the peripheral virulence was between 2.6 and 4.8 logs, the duration of viremia was only 4 days and dropped to 2 days when the peripheral virulence was between 5.4 and 6.8 logs. The duration of viremia reflects the index of peripheral multiplication.

D. Relationship between the Peripheral Pathogenicity and Stability to Heat

In comparing the P1 strain, which has the highest peripheral pathogenicity of 1 log unit, with the Nakayama strain, which has the lowest peripheral pathogenicity of 6.8 log units, Huang (1957a) found that the P1 strain is much more stable to heat. From this and the above studies the following relationship may be deduced. A virus strain which has a higher subcutaneous pathogenicity multiplies better in peripheral tissue and is more stable to heat with a longer duration of viremia than strains with lower subcutaneous pathogenicity.

E. Hemagglutinin and Hemolytic Activities among Strains with Different Virulence

Zhu *et al.* (1978) found that JE virus possessed hemolytic activity and there is a good correlation between hemagglutinin (HA) and hemolytic activity. Further studies of seven strains with high neurovirulence on the HA property all show high titer of HA. However, when the attenuated Xi-An strain, which shows very low neurovirulence (0.15 log) was studied, the HA titer was very low. The authors did not mention the relationship of HA and the peripheral pathogenicity among different strains. However, the ic and sc LD_{50} in 3-week-old mice of the above virus strains have been previously studied by others and when these properties are correlated with HA and hemolytic activities, no significant difference of HA or hemolytic activity among strains with different peripheral pathogenicity was found as long as the virus was still neurovirulent (Table III). The study of Zhu *et al.* (1978) also indicates that passage of virus in cell culture leads to a decrease in HA titer and passage in the brain of mice leads to reversal of neurovirulence and HA and hemolytic activities. After seven ic passages of an attenuated Xi-An (HK_{100}-12-1-7) strain in infant mice, the LD_{50} in infant mice increased from 5 logs to 8.5 logs and the HA titer increased from 1:5 to 1:5120. From the above studies it might be deduced that the HA and hemolytic properties are related to neurovirulence and have no relationship to peripheral pathogenicity.

TABLE III

HA AND HEMOLYTIC ACTIVITY IN VIRUS STRAINS WITH DIFFERENT PERIPHERAL PATHOGENICITY AND NEUROVIRULENCE

	P1	K.S.S.	L.L.$_{17}$	Nakayama	SA$_{14}$	SA$_{14}$ attenuated 12-7-7
Intracerebral titer LD_{50}	9.1	9.5	8.8	8.5	6.5	0.15
Subcutaneous titer LD_{50}	8.1	5.8	4.4	1.1	3.0	0
HA titer	1:2560	1:5120	1:2560	1:5260	1:5120	1:5
Hemolytic activity	3.25	4.98	4.28	5.25	4.76	0.001

III. Inapparent Infection and Pathogenesis of the Disease

Japanese encephalitis, like many other infectious diseases, has both clinical and inapparent infections. The ratio between clinical and inapparent infections varies in different localities and in different years. It has been estimated to be 1:1000–1:2000 in Beijing in 1957–1959 (Sung, unpublished). Clinical cases also have varied manifestations. Encephalitis is a typical form with different degrees of severity. With the improvement of laboratory diagnosis, many mild cases of meningitic form with no encephalitic symptoms are quite common.

In China, the mechanism of inapparent infection and the pathogenesis of the disease have been investigated.

A. The Inapparent Infection

Studies of virus factors and host factors in mice as causes of inapparent infection in JE by Huang (1957a,b) indicate that both virus and host factors play a part in determining whether the infection is apparent or inapparent. The existence of strains with low peripheral pathogenicity and with lower neurovirulence is sufficient to explain the occurrence of inapparent infection. However, the isolation of low virulence virus strains is too few to explain the great number of inapparent infections in man. This may be due to the difficulty of isolating the low virulence viruses. The question is whether high and intermediate virulence strains which have been isolated from human cases and vector mosquitoes can also lead to inapparent infection. It is a well-known fact that in a given strain of pathogen of any infectious disease there is a close correlation between the amount of pathogen introduced into the host and the result of infection. A subthreshold dose of infective agent is considered to be one of the factors leading to inapparent infection. The study of Wang and Yen from our laboratory (Huang, 1957a) on the influence of environmental temperature on the amount of virus in the body of infected mosquitoes showed that infected mosquitoes kept at 26–31°C, corresponding to the epidemic environmental temperature in Beijing, in a majority of cases have a higher virus content and greater power to transmit the disease than those kept at 18–22°C, which corresponds to the pre- and postepidemic environmental temperature. Although the experiments were carried out in mice, it is reasonable to deduce that mosquitoes infected with either high or intermediate virulence strains of virus, especially those infected with intermediate virulence strains, in the pre- and postepidemic periods will carry a low virus content, so that when bitten by these mosquitoes, susceptible persons would be inoculated with a subthreshold dose resulting in inapparent infection.

In a study of host factors, Huang (1957b) found two forms of nonspecific resistance of the host to infection with the JE virus: one depends on the constitution of the host and the other on the condition of the host at the time of infection. Studies on the influence of age and nutrition of mice on the outcome of infection with JE virus show that there is an increase of resistance to peripheral inoculation as the animal grows older or fatter. Since brain is the target organ of JE virus infection, existence of a constitutional barrier to the progression of the virus from the site of inoculation to the central nervous system, as suggested by Sabin (1941) and Olitsky et al. (1936) in their studies on other neurotropic viruses, also holds true in JE virus infection. Although the increase in resistance to infection with the JE virus as the animal grows older is clearly demonstrated in mice, it is important to see whether or not the same type of mechanism of resistance to infection with JE virus also exists in man. The incidence of the disease in man decreases with advancing age (Huang et al., 1951), but this can be explained by the increase in the percentage of specific immunity in the population with increase in age resulting from inapparent and apparent infections. This has been demonstrated by Zhang et al. (1964) of our laboratory. From epidemiological studies, Huang et al. (1951), however, found no increase of nonspecific resistance to JE virus infection in man with increase of age as demonstrated in mice. In 1949 the morbidity rate in Beijing for 20- to 30-year-old males was 45.1/100,000, which is much higher than for males aged 10–15 and 15–20, who had a morbidity rate of 17.4 and 13.4, respectively. There was no such increase in females aged 20–30, who had a morbidity rate of only 6.3. Moreover, there was no such increase in males aged 20–30 in the next year, 1950: the morbidity rate dropped to 24.3, and in females of the same age, the morbidity rate was 14.4. The morbidity rate was 22.2 for males aged 13–20 and 7.6 for males aged 30–40 in 1950. The marked increase of the morbidity rate in males aged 20–30 in the year 1949 is attributed to the fact that soldiers brought up in a nonepidemic area came to the epidemic area, Beijing, for the first time. If only soldiers coming from and brought up in nonepidemic areas are calculated, the morbidity rate is as high as 245.5, which is even higher than that for the male age group of 1–5 brought up in Beijing: 71.5/100,000. However, Chan and Hu (1955) reported that in autopsy cases of JE there is a higher incidence of cysticercosis of brain, 33% as compared with the low incidence of 0.014–1.24 in cases of death due to other diseases. This finding suggests that under normal condition the blood–brain barrier in man to a certain extent prevents the passage of JE virus into the brain, but there is no increase of nonspecific resistance with increase in age as seen in mice.

B. Relation of the Peripheral Multiplication of JE Virus to the Pathogenesis of the Infection

In neurotropic virus infection, the main problem in pathogenesis is whether the virus transported into the central nervous system (CNS) is from the circulating blood, passing through the blood–brain barrier (Hurst, 1950; Albrecht, 1957), or whether the infection is spread via the olfactory nerve or other peripheral nerves (Sabin and Olitsky, 1938; Hurst, 1936). Huang and Wong (1963) tackled this problem by studying the dynamics of virus multiplication in different organs and tissues after infecting mice with large and small doses of the high and low peripheral pathogenic strains and also the dynamics of virus multiplication of high virulence strains in two different age groups of mice. The results of the study clearly indicated that after peripheral inoculation of the virus, two phases of virus multiplication occur. The early phase takes place in the peripheral tissues and the second phase takes place in the brain. As shown in Table IV no virus was detected at 6 hours and at 1 day in all tissues tested when only $10^{1.5}$ LD_{50} of the highly virulent P1 strain was inoculated into the pads of 3-week-old mice. Virus was first detected in all peripheral tissues tested on the second day after inoculation and lasted 4 days in the blood or blood-rich organ, liver, while in the heart, kidney, and tissue of inoculation site, virus can be detected even at 11 days after inoculation. The termination of viremia

TABLE IV

Virus Titer in Different Tissues after Inoculation of $10^{1.5}$ LD_{50} of the P1 Strain into the Pads of 3-Week-Old Mice

Tissue tested	Virus titer (log LD_{50}) at intervals after inoculation at day											
	0.25	1	2	3	4	5	6	7	8	9	10	11
Blood	0	0	1.8	2.2	1.8	0.6	0	0	0	0	0	0
Liver	0	0	1.6	2.2	1.2	0.6	0	0	0	0	0	0
Heart	0	0	1.5	2.0	1.5	2.4	2.5	2.2	2.4	2.0	0	2.0
Kidney	0	0	1.2	1.8	2.5	3.4	3.2	2.4	0	0	0.8	0.6
Tissue of inoculation site	0	0	1.2	0.6	0.6	1.7	2.0	1.4	1.3	0	0	1.8
Brain	0	0	0	0	0	3.6	4.2	5.0	7.5	7.5	6.2	8.2

is probably related to the production of specific antibodies. Multiplication of virus in the brain was first detected 5 days after inoculation and the titer increased progressively until death of the animal. When large doses were given iv or sc, two phases of multiplication were also demonstrated, although not as clearly. The involvement of brain took place earlier, 2 days after infection. The studies suggest that, in view of the short interval it takes for the virus to get into the brain, the virus is probably transported into the CNS from the circulating blood, passing through the blood–brain barrier. Further studies of virus multiplication in the aorta indicate a greater multiplication as shown by a higher titer in this tissue than in the other tissues tested. This would suggest that the virus multiplies in the blood vessels of the brain and penetrates into the brain. Studies on the dynamics of virus multiplication with the low peripheral pathogenic Nakayama strain indicate that the development of disease or inapparent infection depends on the dose of infection. High doses of 6.6 logs of virus lead to two phases of virus multiplication, as seen in small doses of 1.5 logs of the P1 strain; smaller doses of 5.5 logs of the Nakayama virus strain, when given subcutaneously, showed multiplication of the virus at the site of inoculation starting 3 days after inoculation, with a maximum virus titer of 2.7 logs at 6 days. No virus was detected in the opposite site of inoculation, blood or brain. The multiplication of virus only at the local site would explain the inapparent infection with the development of specific immunity.

C. Cell-Mediated Immunity in JE Virus Infection and Its Role in Apparent and Inapparent Infection

Japanese encephalitis is caused by a budding virus and cell-mediated immunity (CMI) is expected. However, the role CMI plays in resistance and recovery of JE virus infection has not been clarified. There have been some studies in Japan (Yasui *et al.*, 1976) and in India (Kelkar and Banerjee, 1976) which demonstrated some protection in JE virus infection in mice by previous passive transfer of immune cells. In studies *in vitro*, Zhang and An (1981), using the macrophage inhibition test, found the development of macrophage inhibition after infection with a small dose of P1 strain of JE virus in mice. These studies suggest that there is development of CMI after JE virus infection, but what role the CMI plays in the inapparent infection and pathogenesis of clinical disease remains unsolved. Jia and Huang (1981) studied the dynamics of the CMI response in mice after infection with different virulence strains of JE virus. By using the macrophage inhibition test as a criterion for evaluation of CMI, comparison was made between

different virus virulence strains infected with different doses and introduced by different routes. Results indicate that in the primary phase of
infection, the rate of appearance of macrophage inhibition depends on
the amount and the virulence property of the virus innoculated. As
shown in Table V, with high peripheral virulence strain P1 inhibition
developed on the first day after infection with medium and large doses
and on the third day when a very small dose of 3.3 LD_{50} was given
intraperitoneally. In the case of low peripheral pathogenic strains,
Nakayama or the neuroattenuated 2-8 strain, the appearance of macrophage inhibition occurred 3 days after infection. In the secondary
phase of infection, macrophage inhibition disappears 1–2 days after
infection with large and medium doses of high peripheral pathogenic
P1 strain and 3–4 days when large and medium doses of low peripheral
pathogenic Nakayama strain are used. This has been found to be related to the death of the animal, which occurred about 2 days after the
disappearance of macrophage inhibition. In further studies on 60 mice
showing encephalitis after infection, none showed detectable macrophage inhibition. It is therefore suggested that in the secondary
phase of infection when virus multiplication has overcome the specific
CMI, virus then invades the sensitized T cells, resulting in immunosuppression and death of the animal. In the case of small doses of
virus given intraperitoneally, the appearance of macrophage inhibition
persists for about 2 weeks and the animal survives, suggesting the
possible role of CMI in the control of a small dose of infection when
given peripherally but not when given intracerebrally.

IV. Ecology of JE Virus

Ecology of a virus involves the study not only of where and how the
virus resides and circulates but also of the influence of the environment
on changes in virus properties; especially important are where and how
strains with different virulence emerge. Virus ecologists have generally paid attention to the study of where and how the virus resides and
circulates and few have been interested in the emergence of strains
with different virulence. There have been extensive studies on the ecology of JE in Japan but many problems remain unsolved.

In China special emphasis was given to the study of the ecology of JE
virus, which causes widespread disease. Besides studies on the transmission vectors, the amplifier host, and the overwintering of the virus,
special attention was devoted to the study of where and how strains
with different virulence might possibly emerge.

TABLE V

CMI Response of Mice after Infection by Different Routes with Different Doses of JE Virus with Different Virulence

Strain	Route of infection	Dosage	Degree of macrophage inhibition (days after infection)												
			1	2	3	4	5	6	7	8	9	12	15	18	21
P1	ip	3.3 LD$_{50}$	−		+++			+++			+++	++++	+++	+	−
		3.5×10^4 LD$_{50}$	+	+	−	−	−								
		10.4×10^7 LD$_{50}$	++	−	−										
	ic	1 LD$_{50}$	+	++	+++	++	++	−	−						
		31.6×10^6 LD$_{50}$	++	−	−	−									
Nakayama	ip	1.6 LD$_{50}$	−	−	++	++	++	+++	+	−	++	+			
		52.5×10^3 LD$_{50}$	−	−	++	+	++++	+	−	−	−	−	−		
		10.4×10^7 LD$_{50}$	−	−	++++	+	++++	−							
2-8	ip	10^7 TCID$_{50}$	−	−	++	++++	++	++	++		+	−			
	ic	3.3×10^6 TCID$_{50}$	−	−	−	+	++++	+	+		+	−			

A. The Transmission Vector

Extensive studies on the ecology of JE virus vector mosquitoes have been carried out by Scherer et al. (1959) between 1952 and 1958 and later by Wada and his group in Japan and Okuno et al. (1971, 1973) in Taiwan. In mainland China Culex tritaeniorrhynchus is also considered to be the main vector of JE. From the epidemiological studies of JE, Huang et al. (1951) demonstrated that the infection was transmitted mainly outdoors and after sunset. The reasons are (1) the rarity of two cases occurring in a family, in spite of the fact that more than two susceptible children in one family is quite common at that time; (2) deductions from the study of the morbidity rate of different age groups in Beijing in 1950. Although in general the morbidity rate decreases with increased age as a result of accumulated specific resistance due to inapparent and apparent infections, there is a marked increase in the morbidity rate in the 3- to 6-year-old age group. The morbidity rate is 36.3 for the 0–1 age group, 34.6 for 1–2, 31.4 for 2–3, 56.8 for 3–4, 79.9 for 4–5, 65.1 for 5–6, and 38.4 for the 6–7 age group. Theoretically at the age of 1–3 when the maternal antibodies are lost, children should be most susceptible; but actually the morbidity rate is significantly lower than it is for the higher 3- to 6-year-old age group. This suggests that the majority of infections could not take place indoors at night by the bite of Culex pipiens var. pallens. The reason why children between 3 and 6 years of age have a higher morbidity rate is probably because most parents allow their children beginning at the age of 3–4 to play outdoors, especially after sunset, where they are exposed to the bite of C. tritaeniorrhynchus and Aedes chemulpoensis, two prevalent mosquitoes which usually bite human beings outdoors.

The antimosquito campaign carried out extensively in 1952 has since eradicated the A. chemulpoensis mosquitoes (Section V). There was a drop in the morbidity rate from 25.8 in 1951 to 13.4 in 1952, suggesting the possible role of this species of mosquito as one of the vectors. However, from A. chemulpoensis, which was shown to be able to transmit the disease to mice after artificial infection (Huang and Zheng, 1951), no virus had been isolated. Moreover, after the eradication of this mosquito in Beijing, epidemics of JE still occurred in subsequent years, although on a smaller scale. Thus, C. tritaeniorrhynchus has been considered to be the main vector. The fact that this mosquito has never been found in the two most western provinces, where no JE has been reported, further supports the view that this species of mosquito is the main vector of JE in China.

Virus has been isolated from C. pipiens var. pallens but the epidemiological study indicates that it could not be the main vector. The role of *Aedes Albopictus,* which is prevalent in South China, as a vector has to be considered. Seven strains of JE virus have been isolated from larvae of this mosquito species (Wu and Wu, 1957).

B. Isolation of Virus

JE virus has been isolated from different species of mosquito in different zones. In the tropical zone, virus has been isolated from *Anopheles barbirostris, Anopheles hyrcanus, Culex epidesmus, Culex bitaeniorrhynchus, Culex vishnui, Culex gelidus,* and *Culex fuscocephala* in India, Sarawak, and Indonesia. In the subtropical zone, virus has been isolated from *C. tritaeniorrhynchus, C. gelidus,* and *C. fuscocephala* in Thailand [summarized by Dhanda and Kaul (1979) and Pant (1979) in W.H.O. Interregional Meeting on JE, New Delhi, 1979]. In the temperate zone, virus has been isolated chiefly from *C. tritaeniorrhynchus* in Japan (Scherer *et al.,* 1959) and has occasionally been isolated from *C. pipiens* (Buescher *et al.,* 1959). In the Republic of Korea, virus has been isolated from *C. tritaeniorrhynchus* and *Aedes vexans.* In China, Taiwan province, virus has been isolated from *C. tritaeniorrhynchus* (Wang *et al.,* 1962) and *C. vishnui* (Okuno *et al.,* 1971). In mainland China virus has been isolated from *C. tritaeniorrhynchus, C. pipiens var. pallens, Culex fatigans, A. hyrcanus, Aedes albopictus,* and *Armegeres obturbans.*

Isolation of virus from other countries gives no information on the virulence of the virus. Studies on where virus strains with different virulence occur is of great ecological importance. Table VI summarizes the virulence of virus strains isolated from humans, pigs, and different species of mosquito from different part of China. From Table VI one can see a very good correlation of the virulence of strains isolated from *C. tritaeniorrhynchus* in pigs and in humans. Both high and intermediate virulent strains of virus were isolated. This gives further evidence that *C. tritaeniorrhynchus* is the main transmission vector and that the pig is an amplifier host.

The transmission vector is usually correlated with apparent infection. However, the ratio of apparent to inapparent infection is about $1:1000$ in Beijing. Inapparent infection is undoubtedly also transmitted by mosquitoes. However, only three strains of JE virus with low virulence have been isolated and they were isolated from *C. pipiens var. pallens* and *C. fatigans.* The small number of low virulence strains isolated may be related to the difficulty of isolating the virus by ic inoculation of 3-week-old mice. The finding of virus strains with differ-

TABLE VI

COMPARISON OF THE PERIPHERAL PATHOGENICITY OF JE VIRUS STRAINS ISOLATED FROM HUMANS, PIGS, AND MOSQUITOES[a]

Virus isolated from	Total number	High virulence		Intermediate		Low virulence	
		1.0–2.0	2.1–3.0	3.1–4.0	4.1–5.0	5.1–6.0	6.1–7.0
Human	16	5	2	3	6		
Pig	24	8	6	6	4		
Mosquito	60						
C. tritaeniorrhynchus	46	24	20	7	6	2	1
C. pipiens var. pallens	9	22	17	4	3		
C. fatigans	2	2	2	1	2	1	
A. albopictus	2			1	1	1	1
A. obturbans	1		1	1			

[a] Virus titrated in 3-week-old mice.

ent neurovirulence, isolated directly in cell culture from *C. tritaenior-rhynchus,* as shown in Table I, suggests that there might exist in other species of mosquito virus strains with similar or even lower or no neurovirulence. A more sensitive method of virus isolation, e.g., the use of cell culture derived from *A. Albopictus* clone C6/36, which has proved to be more sensitive for virus isolation than the suckling mouse brain (Igarashi *et al.,* 1981), will probably settle this question. The difficulty of isolating low virulence virus is also reflected by the study of W. K. Müller (personal communication, 1981) on the isolation of low virulence tick-borne encephalitis virus. He found that positive isolation of strains with low virulence required at least 10 serial passages in suckling mouse brain.

C. The Vertebrate Host–The Amplifier Host

The role of vertebrates in the ecology of JE virus has also been studied in China. Serological surveys of different species of mammals and birds were carried out in Beijing between 1951 and 1954. The following animals show a very high percentage of seropositive neutralizing antibodies: pigs 100%, horses 94%, donkeys 94%, cattle 92%, and dogs 66%. Among the birds studied, seropositive antibodies were detected only in ducks (21.8%) and chickens (3.7%). No neutralizing antibodies were detected in 56 sparrows, 35 pigeons, 27 magpies, 30 crows, 15 swifts, 11 warbles, 10 house mice, and 34 rats. The absence of infection of JE virus in sparrows is different from the studies of Monath's group (Bowen *et al.,* 1980) on St. Louis encephalitis virus.

Among the animals infected with the virus, only horses and donkeys develop encephalitis. Epidemics of JE in horses occurred in horse breeding areas. There is no encephalitis in pigs but susceptible pregnant pigs when infected often have stillbirths.

Among the vertebrate hosts, the pig has been considered as the amplifier host (Wang *et al.,* 1958a,b). In 1955 Wang *et al.* of our laboratory placed 32 pigs (5 months old) in two districts of Beijing between June and October. Blood was collected weekly, virus isolation was attempted, and neutralizing antibodies were tested. After the epidemic was over all pigs got inapparent infections as demonstrated by positive neutralization tests. From the isolation of the virus, it was found that pigs were infected 2–3 weeks earlier than humans, as shown in Fig. 1.

D. Overwintering of the Virus and Transovarian Transmission

Overwintering of the JE virus in the temperate zone has been the subject of much speculation and study in Japan, Korea, and China. The

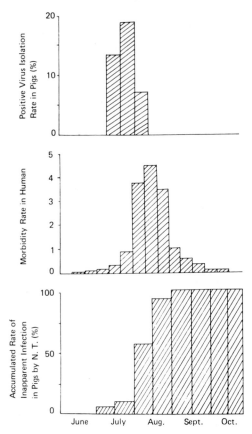

FIG. 1.   Time relationship of JE infection in pigs to the morbidity rate of humans in Beijing in 1955.

possibility that warm-blooded vertebrates carry the virus over the winter appears unlikely because of the ready development of both humoral and cellular immunity. No persistent infection of JE virus has ever been encountered in warm-blooded vertebrate hosts. The possibility of carry over in hibernating animals such as bats has been considered by Sulkin *et al.* (1970), who isolated virus from bats caught in epidemic areas and not in nonepidemic areas in Japan. However, extensive study by the department of microbiology of Shanxi Medical College, China, on bats caught in Shanxi showed no evidence of JE virus infection either by isolation of virus or detection of antibodies. It is difficult to imagine how the bat became infected before hibernation since no virus has ever been isolated from mosquitoes caught after the end of an epidemic. If the infection took place during the epidemic season, specific antibodies would develop and virus would disappear from the circulation.

Although Mifune (1965) was able to show that experimentally infected *C. tritaeniorrhynchus* could infect pigs after hibernation, all efforts to isolate the virus from naturally hibernating *C. tritaeniorrhynchus* adults, since they start appearing in March, have been negative (Buescher *et al.*, 1959; Hayashi *et al.*, 1973; Shichijo *et al.*, 1968; Wang, 1980). Moreover, overwintering mosquitoes usually do not suck blood before hibernation. These studies do not favor the hypothesis of overwintering of the virus in hibernating adult mosquitoes.

Transovarian transmission of JE virus by mosquitoes was one of the hypotheses offered to explain how the virus survives between epidemics. Japanese workers in 1938 were the first to report the successful experimental transovarian transmission of *C. pipiens var. pallens*. Because of failure to repeat the experiment, coincident infection of tested mice by virus-carrying mosquitoes was suspected and most investigators tended to feel that transovarian transmission of JE virus does not occur. In China many experiments were carried out to prove the possibility of transovarian transmission. In 1954 and 1955, 33 transovarian transmission experiments were carried out in our laboratory (Wang *et al.*, 1958a,b). Adult *C. pipiens var. pallens* reared from larvae were artificially infected by feeding them infected mouse brain suspension. Infected mosquitoes were allowed to lay eggs and reared to adult mosquitoes. Infant mice were exposed to these mosquitoes. Results showed that 2 of 63 batches of mosquitoes transmit the disease to infant mice. Virus so isolated was proved to be JE virus.

In 1956 Wang *et al.* (1958a,b) carried out another experiment. From June to October mosquito eggs were collected and reared in the laboratory. Adult mosquitoes so reared were directly tested for the presence of virus. Altogether 60 batches from 2500 *C. tritaeniorrhynchus* were directly inoculated intracerebrally in mice with negative results. Two pigs and five ducks (serologically negative to JE virus) isolated in screened rooms and cages were exposed to the bite of adult *C. pipiens var. pallens* repeatedly as soon as adult mosquitoes were reared. Altogether 2000 mosquitoes were allowed to bite the pigs and 700 mosquitoes were allowed to bite the ducks. No virus was isolated from blood collected 2–3 days after the bite, but one duck developed neutralizing antibodies, with a titer of 214 in October, 899 in November, and 2050 in December.

Other evidence for transovarian transmission includes the following: (1) In 1953 one strain of JE virus was isolated from adult male *C.*

pipiens var. pallens in Nanking. Since male mosquitoes do not suck blood, the transovarian origin of the virus has to be considered; (2) two strains of JE virus were isolated from larvae of *C. pipiens var. pallens* reared from eggs collected in Shenyang, one in 1953 and the other in 1954 (personal communication); (3) two strains of JE virus were isolated directly from larvae of *C. pipiens var. pallens* caught in 1953 in Xi-An (personal communication); (4) one strain of JE virus was isolated from *C. pipiens var. pallens* reared from larvae collected in Nanking in 1954 (personal communication); and (5) seven strains of JE virus were isolated from *A. albopictus* reared from larvae collected in Fukien between June 20 and 27, 1955 by Wu and Wu (1957).

The above studies together with the recent positive transovarian transmission of JE virus by Rosen *et al.* (1978, 1981) support the hypothesis that transovarian transmission of JE virus by mosquitoes is probably the best explanation of how this virus survives between epidemics in the temperate zone.

E. Studies on the Variation of JE Virus—Ecological Considerations

Experimental studies on the variation of JE virus might help explain how strains of JE virus with different virulence emerged.

1. Variation in Peripheral Pathogenicity

Huang (1964) reported his study on the variation of peripheral pathogenicity of JE virus in the most susceptible animal, the white mouse. The virus used is a high peripheral pathogenic strain P1. Passage of the virus was carried out by intracerebral and subcutaneous routes separately, using mice of different age groups. Harvested viral material for passage from either route of inoculated mice was always obtained from the brain. Figure 2 shows the result of passage intracerebrally in 7-day-old mice while Fig. 3 shows the result of passage in 3-week-old mice. Results indicate that after cerebral passages the virulence (of the virus), which showed no marked change when titrated by the intracerebral route in 3-week-old mice, decreased as the number of passages increased when titrated by the subcutaneous route. Definite evidence of a decrease in peripheral pathogenicity of the virus was detected at about 130 passages when the virus was passed in 7-day-old mice and 100 passages in the case of 3-week-old mice. When adult 5-week-old (Fig. 4) and 1-year-old mice (Fig. 5) were used for intracerebral passages, definite evidence of a decrease in peripheral pathogenicity of the virus started very much earlier, at about 10 passages. The magnitude of the decrease in subcutaneous virulence was also different when virus was passed in mice of different ages. The

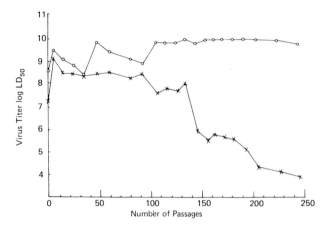

FIG. 2.   Change of virulence of P1 strain after ic passages in 7-day-old mice: ○——○, ic titration; ×——×, sc titration.

magnitude of the decrease in subcutaneous virulence when the virus was passed in 7-day-old and 3-week-old mice is smaller (4 logs) than in 5-week-old mice (6.5 logs) and 1-year-old mice (8 logs). This magnitude of decrease in 7-day-old and 3-week-old mice was reached after more than 200 passages and maintained at the same level on further passages, while with passage in adult mice the highest magnitude was attained in 50–117 passages in 5-week-old mice and in 1-year-old mice. From the above findings, it can be seen that there seem to be three periods of change in virus virulence in the course of cerebral passages: (1) a latent period—a period before evidence of change in

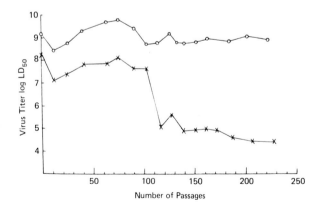

FIG. 3.   Change of virulence of P1 strain after ic passages in 3-week-old mice: ○——○, ic titration; ×——×, sc titration.

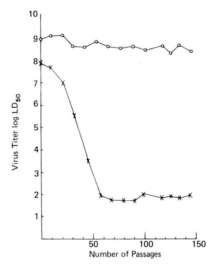

FIG. 4. Change of virulence of P1 strain after ic passages in 5-week-old mice: O——O, ic titration; ×——×, sc titration.

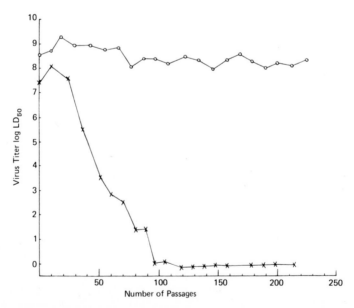

FIG. 5. Change of virulence of P1 strain after ic passages in 1-year-old mice: O——O, ic titration; ×——×, sc titration.

virulence can be detected; (2) a variation period—a period in which a progressive decrease in virulence was noted; and (3) a maintainence period—a period in which the decrease in virulence had reached its maximum and was maintained at the same level.

The decrease in the peripheral virulence of the same virus strain also occurred when it was passed in Ehrlich ascites tumor cells (Hsu, unpublished data) and in chorioallantoic membranes of chick embryos (Jen, unpublished data). The decrease in peripheral pathogenicity occurred earlier than in mouse brain passages. In the study on the attenuation of JE virus, Yu *et al.* (1962) used a low peripheral pathogenic strain, SA_{14}, and carried passages in baby hamster kidney cell culture. They found a drop not only in peripheral pathogenicity but also in neurovirulence. At the first tissue culture passage, the virus titer was $10^{-6.5}$ when titrated intracerebrally in 3-week-old mice and $10^{-3.0}$ when titrated subcutaneously. At the twenty-seventh passage, the ic titer dropped to $10^{-1.35}$ and the subcutaneous titer dropped to less than $10^{-0.6}$.

Contrary to the above findings by cerebral passages, there was no decrease in peripheral virulence when the virus was passed by the subcutaneous route even after 190 and 269 passages in 3-week-old and 7-day-old mice, respectively, as shown in Figs. 6 and 7. By subcutaneous passage the virus invaded not only the peripheral tissues but also the brain. It is interesting to note that although the virus also entered the brain tissue, no drop in peripheral pathogenicity occurred, in comparison with passage of the virus only in the brain. This suggests that multiplication of the virus in the peripheral tissue is the crucial factor in maintaining and possibly in increasing the peripheral virulence if a low peripheral pathogenic strain is used for the passage. This was proved by passing subcutaneously a low peripheral pathogenic virulence variant obtained from the fiftieth cerebral passage of P1 high

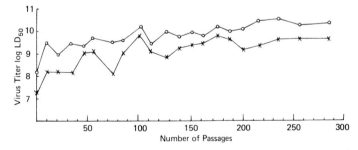

FIG. 6.   Change of virulence of P1 strain after sc passages in 7-day-old mice: ○——○, ic titration; ×——×, sc titration.

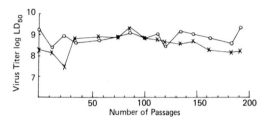

Fig. 7. Change of virulence of P1 strain after sc passages in 3-week-old mice: ○——○, ic titration; ×——×, sc titration.

virulence strain in 1-year-old mice. As shown in Fig. 8, the subcutaneous virulence gradually increased as the number of the passage increased and by the one hundred and sixteenth passage the subcutaneous virus titer rose from 10^{-3} to $10^{-7.1}$.

From the above studies, it can be seen that peripheral pathogenicity of JE virus can be decreased or increased under different environments in which the virus replicates. Passages of virus in mouse brain, chorioallantoic membrane of chick embryo, Ehrlich ascites tumor cells, or baby hamster kidney cell culture result in a decrease of peripheral

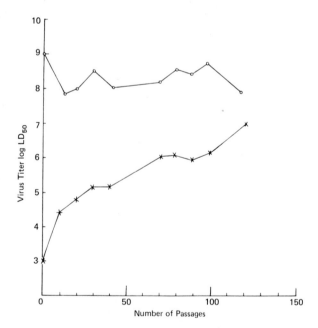

Fig. 8. Subcutaneous passages of low peripheral pathogenic variant in 3-week-old mice: ○——○, ic titration; ×——×, sc titration.

pathogenicity, while subcutaneous passages in mice result in an increase of peripheral virulence.

2. Variation in Neurovirulence

In studies on the attenuation of JE virus, Li *et al.* (1962) passed the virus in suspended tissue culture of chick embryo and found no drop of neurovirulence even after 230 passages with P3 strain. On the other hand, a rapid decrease of neurovirulence was found when a low peripheral pathogenic strain, SA_{14}, was passed in baby hamster kidney cell culture. The titer when titrated in 3-week-old mice intracerebrally was $10^{-6.5}$ at the first passage, and dropped to $10^{-4.5}$ at the sixteenth passage and $10^{-1.77}$ at the twentieth passage. Further passages of up to 100 generations showed no further drop (Li *et al.*, 1966). However, when the one hundredth passaged virus was plaqued, three of nine plaques showed very low neurovirulence (0–0.5 log), and the rest varied between 1.83 and 5.5 logs, although the titer as titrated in hamster kidney cell culture was as high as 7.0–8.5 logs. When the low neurovirulence plaques 9-4-2 and 12-1-1 were passed intracerebrally in 3-week-old mice, the neurovirulence increased very rapidly after a few passages, as shown in Table VII.

From the above studies, it can be seen that the neurovirulence strain can be obtained very rapidly by passage of low neurovirulence virus in the brain of mice, but loss of neurovirulence results from passage in baby hamster kidney cell culture but not in suspended chick embryo tissue culture. The inability to obtain change of neurovirulence of JE virus in suspended chick embryo tissue culture may be due either to

TABLE VII

REVERSION OF NEUROVIRULENCE BY INTRACEREBRAL PASSAGE IN MICE

Plaque	Number of ic passage	Intracerebral LD_{50} (log)	$TCID_{50}$ (log)
9-4-2	0	<0.50	7.5
	1	≤1.83	6.0
	3	7.17	6.5
	7	8.33	—
	10	8.50	—
12-1-1	0	0	7.5
	1	2.5	6.5
	3	6.0	7.0
	7	6.5	—
	10	8.5	9.0

cells not multiplying or to the fact that the drop of neurovirulence requires young developing epithelial cells, not fibroblasts.

F. Hypothetical Considerations of the Origin of JE Virus Strains with Different Virulence

Based on studies on the pathogenic properties of different virulence strains of JE virus, on the variation of virus virulence in mice and tissue culture, and on the ecology of JE virus, a hypothetical sketch on the possible origin of emergence of virus strains with different virulence is deduced, as shown in Fig. 9.

Figure 9 indicates a mosquito–pig, duck–mosquito cycle for the origin of high virulence virus strains. The reason for this deduction is based on the study of Huang (1964) that high peripheral pathogenic virus strains can be obtained only when the virus is passed subcutaneously in white mice. Since the pig has been considered as an amplifier host, it is deduced that a mosquito–pig–mosquito cycle will lead to an increase of peripheral pathogenicity of the virus. However, whether a mosquito–pig–mosquito cycle will eventually lead to loss of neurovirulence requires further study by repeated passage of virus in a mosquito–pig–mosquito cycle. Studies on the passage of JE virus SA_{14} strain in baby hamster kidney cell culture (Yu et al., 1962) indicate that neurovirulence dropped quite rapidly after 20 passages. This finding suggests that subcutaneous passage in the pig might eventually lead to a decrease or loss of neurovirulence, since virus does not invade the CNS after peripheral infection of the pig. Li et al. (1966) found a rapid increase of neurovirulence after a few intracerebral passages in mice. So, besides the mosquito–pig–mosquito cycle to increase periph-

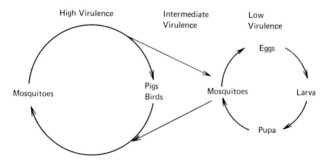

FIG. 9. Hypothetical sketch of the possible origin of high- and low-virulence strains of JE virus.

eral pathogenicity, passage of the virus once in a while in some animal brain might be necessary to maintain neurovirulence. In an unpublished study, we found that infant ducks when infected subcutaneously will develop encephalitis; the rate of inapparent infection in ducks is as high as 21.8% in Beijing. Thus, a mosquito–duck–mosquito cycle must also be considered. The high peripheral pathogenic strain has been shown to be more resistant to heat than the low peripheral pathogenic strain and since birds have a body temperature of 41–42°C, while pigs have a body temperature of 39–39.6°C, it is considered that a passage in infant ducks will not only retain the property of neurovirulence but will also select out strains with high virulence.

The origin of low virulence virus strains, as sketched in Fig. 9, is a mosquito–egg–larva–pupa–mosquito cycle. The basis for this conclusion is that a drop in peripheral pathogenicity of JE virus took place when the virus was passed intracerebrally in white mice, in the chorioallantoic membrane of developing chick embryo, in Ehrlich ascites tumor cells, or in baby hamster kidney cell culture. The passage of the virus in baby hamster tissue culture also leads to a loss of neurovirulence. All these conditions have one feature in common, i.e., young developing cells or brain cells with a longer life. Another point in common is that all these cells are related to epithelial cells. Brain cells and the outer layer of skin are all derived from the same outer layer in the embryonic stage. It is interesting to see that a drop in neurovirulence of the virus occurs in baby hamster kidney cell culture and not in chick embryo tissue suspended culture which contains primarily nonmultiplying fibroblasts. Because of this it is necessary to look for young developing cells, preferably of epithelial cell origin, to discern where the virus circulates. The larva and pupa stages of the mosquito meet this requirement well. Increasing evidence of transovarian transmission of JE virus has accumulated. This further supports the above hypothesis. The intermediate virulence strain will emerge when the high virulence strain passes into the mosquito transovarian cycle or when the low virulence strain passes into the vertebrate cycle.

V. Control of JE in China

In the years after 1949, the morbidity rate (cases/100,000) for JE in Beijing was between 15 and 25. In recent years, the morbidity rate in Beijing has been reduced to around 2.5, in spite of the fact that rice fields, a good breeding place for the vector *C. tritaeniorrhynchus*, have been greatly increased. The reduction in the morbidity rate is attributed primarily to the effect of mass immunization of children under 10 and partly to the antimosquito campaign.

A. Antimosquito Campaign

The antimosquito campaign was started in 1950 and was nationwide. It was directed by the Bureau of Prevention, Ministry of Health, sponsored by the local health authority, and carried out by the people. After 2 years a slogan for the antimosquito campaign had been formulated: prevent early, small, and thorough, i.e., start the campaign early, kill the mosquitoes' larva and pupa while they are small, and emphasize thoroughness in prevention.

The results of the antimosquito campaign are summarized in Table VIII. It can be seen that the morbidity rate dropped from 25.8/100,000 population in 1951 to 13.4 in 1952. That the drop is not due to differences in the severity of epidemics in different years is shown by the fact that no significant drop was seen in the suburbs, where the antimosquito campaign is much more difficult to carry out. The result of the antimosquito campaign has been the eradication of *A. chemulpoensis,* which was commonly seen in Beijing before the campaign. Transmission experiments showed that *A. chemulpoensis* can be readily infected after taking an infective meal from an infected mouse. Furthermore, effective transmission from mice to mice has been carried out through the bite of *A. chemulpoensis* (Huang *et al.,* 1950–51). Whether or not the drop of the morbidity rate in 1952 is related to the eradication of *A. chemulpoensis* is difficult to determine, but has to be considered. One fact is certain: after the eradication of *A. chemulpoensis* in Beijing the morbidity rate remained around 15 for quite a few years, indicating that *C. tritaeniorrhynchus* is the main vector.

The antimosquito campaign also reduced *C. pipiens var. pallens* to a

TABLE VIII

Morbidity Rate[a] of JE in Beijing (1951–1954)

Year	City proper	Suburb	Total
1951	29.9	17.2	25.8
1952	16.1	13.0	13.4
1953	17.3	17.9	17.0
1954	16.4	19.8	17.0

[a] Cases/100,000.

great extent, but what role it plays in the transmission of the disease is uncertain. In 1951 Huang and Zheng isolated one strain of JE virus from 1 of 43 lots of 371 *C. pipiens var. pallens,* but it was very difficult to isolate the virus from the same species of mosquito thereafter in Beijing. The isolation rate was calculated to be 1 : 10,181 in Beijing while in Xi-An it was 1 : 661 (Wang, 1980).

The importance of an early antimosquito campaign in the control of epidemics has been reported by Zhang (1978). From Table IX it can be seen that when the antimosquito campaign was carried out in the first generation of the mosquito in June, more than 2 months before the occurrence of maximum cases, the morbidity rate was 5, when it was carried out in July the morbidity rate was 44, and when it was carried out in August the morbidity rate was 75.7.

B. Inactivated Tissue Culture Virus Vaccine

Inactivated mouse brain vaccine was first prepared and used on a small scale in the 1950s. Because of a possible allergic reaction with that vaccine, inactivated baby hamster kidney cell culture liquid vaccine was developed and has been used since 1967 on a massive scale among children between 1 and 10 years of age. The efficacy of the vaccine is shown in Tables X and XI. Table X is the result of a study on the effectiveness of inactivated tissue culture vaccine observed in the Huiyang health station of Guangdong province (Ren, 1978). Vaccinated children received two doses of inactivated tissue culture vaccine in the first year in 1973 and one reenforced dose was given in 1974. Table X indicates a protective rate of 95% even 5 years after primary immunization. Table XI shows the rate of immunization of JE-inactivated tissue culture vaccine and the morbidity rate carried out in the Lu-ta area of Liaoning province in 1974 (Zhang, 1978). Results show that the morbidity rate decreased as the immunization rate increased.

TABLE IX

ANTIMOSQUITO CAMPAIGN AND MORBIDITY RATE OF JE IN CHAO YANG AREA IN 1974

	Campaign carried out in		
	June	July	August
Number of productive brigade	24	64	129
Number of cases	1	44	106
Morbidity rate (cases/100,000)	5	44	75.7

TABLE X

EFFECTIVENESS OF INACTIVATED HKTC VACCINE[a]

	Vaccinated group			Unvaccinated group			
Year	Number of children	Number of cases	MR[b]	Number of children	Number of cases	MR[b]	Protection rate (%)
1973	72,309	4	5.5	3526	5	141.8	96.1
1974	71,070	4	5.6	3521	2	56.7	90.7
1975	71,066	1	1.4	3519	0	0	
1976	71,065	3	4.2	3519	3	85.3	95.1
1977	71,062	1	1.4	3516	1	28.4	95.0

[a] Five-year observation (Ren, 1978).
[b] Morbidity rate (cases/100,000).

C. Development of Live Attenuated Vaccine

Li and Yu (1959) found no loss of neurovirulence of JE virus P3 strain even after 110 passages in chopped pieces of chick embryo without head and spinal column suspended tissue culture. However, when the SA_{14} strain was passed in baby hamster kidney cell culture, a remarkable drop of neurovirulence was found after 20 passages, as stated in Section IV,E. The virus at the one hundredth passage was then plaqued. Low neurovirulent plaque 12-1-7 was selected (Li et al., 1966). By further plaquing and selection, an attenuated 5-3 strain was selected for trial in humans and its biological properties were studied (Yu et al., 1973). Another attenuated strain, 2-8, also originating from 12-1-7, was attenuated by ultraviolet light radiation and plaque selection by Chen and Wang (1974). The prepared vaccine was also given to 8000 children from nonepidemic areas with a 50% positive seroconversion rate (Chen, unpublished data). Comparison of one dose of live

TABLE XI

RATE OF IMMUNIZATION OF INACTIVATED HKTC VACCINE AND MORBIDITY RATE IN THE LU-TA AREA IN 1974[a]

District	Population of 1–10 year olds	Number of children immunized	Percentage immunized	Morbidity rate for 1–10 year olds
Zhuang Ho	237,457	33,637	14.2	35.0
Fu Hsien	187,414	161,491	86.7	6.5
Xin Chin	135,754	120,839	89.0	3.6

[a] From Zhang (1978).

attenuated 5-3 strain vaccine with tissue culture-inactivated vaccine (two doses for the first year and one reenforced dose in the second year) in children brought up in epidemic areas was made (Ren, 1978). No difference in the protective rate was found between these two types of vaccine after a 5-year observation. To improve the immunogenicity of the attenuated 5-3 strain, Yu *et al.* (1981) passed the virus five times subcutaneously in infant mice and obtained a highly attenuated stable virus mutant 14-2 strain after further plaquing and selection. This strain was found to multiply better and has a better immunogenicity than the original 5-3 strain when tested in mice.

D. Safety, Epidemiological, and Serological Evaluation of Attenuated 2-8 Strain Vaccine after Immunization of Horses

Tissue culture vaccine prepared from attenuated 2-8 strain was given a field trial in horses (Huang *et al.*, 1974). The vaccine was inoculated intracerebrally with 158,100 $TCID_{50}$ to 5-year-old horses with negative neutralizing antibodies to JE virus. There was only a mild febrile reaction for a short time. Only one horse showed transient encephalitic symptoms. The animal recovered the next day with no residual signs. After preliminary studies no reaction of any kind was found when the vaccine was given intramuscularly. The positive seroconversion rate was as high as 84.9–86.6%. A field trial to evaluate the efficacy was made. Altogether 126,039 horses were immunized im with 1 ml of the vaccine prepared in hamster kidney cell culture with a titer of 10^7–$10^{8.5}$ $TCID_{50}$/ml and compared with 40,674 unvaccinated controls. The morbidity rate in the vaccinated group was 23/100,000 and 172.1 in the unvaccinated control, with a protective rate of 86.7% calculated after the epidemic.

TABLE XII

PREVENTION OF STILLBIRTH AFTER IMMUNIZATION OF 5-3 AND 14-2 STRAINS OF LIVE VIRUS VACCINE

Group	Number of sows	Total number of births	Number of stillbirths	Rate of stillbirth (%)
14-2 Vaccine	60	739	59	7.98
Control	20	181	100	55.25
5-3 Vaccine	127	1364	111	8.10
Control	62	455	215	47.30

E. Prevention of Stillbirths in Swine with 5-3 and 14-2 Attenuated Live Vaccine

Ao *et al.* (1981) compared the prevention of stillbirths in swine vaccinated with 5-3 and 14-2 attenuated strains. Live vaccine were all prepared in hamster kidney cell culture with a titer of $10^{6.5}$–$10^{7.5}$ $TCID_{50}$ and given peripherally to sows. The incidence of still birth due to infection with JE virus was significantly lower in the vaccinated sows than in the unvaccinated ones. Only 7.98% of newborn piglets from sows immunized with 14-2 strains were stillborn as comparad with 55.25% in the control unvaccinated group. No significant difference was seen between the two attenuated virus vaccines in the prevention of stillbirth in sows as shown in Table XII.

REFERENCES

Albrecht, P. (1957). *Acta Virol.* **1**, 188.
Ao, J., Yu, Y. X., Wu, P. F., Zhang, G. M., Yao, L. T., and Zhu, Q. C. (1981).*Acta Microbiol. Sin.* **21**(2), 174.
Bowen, G. S., Monath, T. P., Kemp, G. E., Kerschner, J. H., and Kirk, L. J. (1980).*Am. J. Trop. Med. Hyg.* **29**, 1411.
Buescher, E. L., Scherer, W. F., Rosenberg, M. Z., Gresser, I., Hardy, J. L., and Bullock, H. R. (1959). *Am. J. Trop. Med. Hyg.* **8**, 651.
Chan, S., and Hu, C. H. (1955). *Chin. J. Pathol.* (in Chinese) **1**, 53.
Chen, B. C., and Wang, Y. M. (1974). *Acta Microbiol. Sin.* **14**(2), 176.
Chu, F. T., Wu, J. P., and Teng, C. H. (1940). *Chin. Med. J.* **58**, 68.
Hayashi, K., Shichijo, A., Mifune, K., and Matsuo, S. (1973). *Trop. Med.* **15**(4), 214.
Huang, C. H. (1957a). *Acta Virol.* **1**, 36.
Huang, C. H. (1957b). *Acta Virol.* **1**, 83.
Huang, C. H. (1964). *Acta Microbiol. Sin.* **10**(1), 1.
Huang, C. H., and Liu, S. H. (1940). *Chin. Med. J.* **58**, 427.
Huang, C. H., and Tai, Y. (1958). *Acta Microbiol. Sin.* **6**, 42.
Huang, C. H., and Wong, C. (1963). *Acta Virol.* **7**, 322.
Huang, C. H., and Zheng, Y. K. (1951). *Chin. Med. J.* (in Chinese) **37**(4), 296.
Huang, C. H., Feng, L. C., and Jen, K. H. (1950–51).*Peking Nat. Hist. Bull.* **19**(2–3), 249.
Huang, C. H., Sung, K., and Tien, F. T. (1951). *Chin. Med. J.* (in Chinese) **37**(4), 253.
Huang, C. H., Zhou, M. X., Wang, J., and Tai, Y. (1958). *Acta Microbiol. Sin.* **6**(1), 47.
Huang, C. H., Hang, C. S., and Chen, B. C. (1974). *Acta Microbiol. Sin.* **14**(2), 185.
Hurst, E. W. (1936). *J. Pathol. Bacteriol.* **42**, 271.
Hurst, E. W. (1950). *J. Comp. Pathol.* **60**, 237.
Igarashi, A., Buel, K., Ueba, N., Yoshido, M., Sasao, F., and Fukai, K. (1981).*Int. Congr. Virol., 5th,* p. 144.
Jia, F. L., and Huang, C. H. (1981). *Acta Acad. Med. Sin.,* in press.
Kelkar, S. D., and Banerjee, K. (1976). *Indian J. Med. Res.* **64**(12), 1720.
Li, H. M., and Yu, Y. X. (1959). *Acta Microbiol. Sin.* **7**, 327.
Li, H. M., Yu, Y. X., and Ji, S. R. (1962). *Acta Microbiol. Sin.* **8**(3), 251.
Li, H. M., Yu, Y. X., Ao, J., and Fang, Z. (1966). *Acta Microbiol. Sin.* **12**(1), 41.

Mifune, K. (1965). *Endem. Dis. Bull. Nagasaki* **7**(3), 178.

Okuno, T., Tseng, P. T., Lim, S. Y., Hsu, S. T., and Huang, C. T. (1971). *Bull. W.H.O.* **44**, 599.

Okuno, T., Mitchell, C. J., Chen, P. S., Wang, J. S., and Lin, S. Y. (1973). *Bull. W.H.O.* **49**, 347.

Olitsky, P. K., Sabin, A. B., and Cox, H. R. (1936). *J. Exp. Med.* **64**, 723.

Ren, Y. L. (1978). *Biol. Prod. Com.* (in Chinese) **7**(3), 111.

Rosen, L., Tesh, R. B., Lien, J. Ch., and Cross, J. H. (1978). *Science* **199**, 909.

Rosen, L., Shroyer, D. A., and Lien, J. C. (1981). *Int. Congr. Virol., 5th* W10/04.

Sabin, A. B. (1941). *J. Pediatr.* **19**, 596.

Sabin, A. B., and Olitsky, P. K. (1938). *Proc. Soc. Exp. Biol. Med.* **38**, 595.

Scherer, W. F., Buescher, E. L., Flemings, M. B., Noguchi, A., and Scanlon, J. (1959). *Am. J. Trop. Med. Hyg.* **8**(6), 665.

Shichijo, A., Mifune, K., Hayashi, K., Wada, Y., Ito, S., Kawai, S., Miyagi, I., and Oda, T. (1968). *Trop. Med.* **10**(3), 168.

Sulkin, S. E., Allen, R., Miura, T., and Toyokawa, K. (1970). *Am. J. Trop. Med. Hyg.* **19**(1), 77.

Wang, S. P., Grayston, J. P., and Hu, S. M. K. (1962). *Am. J. Trop. Med.* **11**, 149.

Wang, Y. M. (1980). *In* "Epidemiology," Chap. 58 (in Chinese). Ren Ming Wei Sheng Chu Ban She.

Wang, Y. M., Liu, Z. W., Jen, K. H., and Zhang, X. (1958a). *Liu Xing Bing Fang Zhi Yang Jui* (in Chinese), p. 257.

Wang, Y. M., Jen, K. H., Zheng, Y. K., and Liu, Z. X. (1958b). *Liu Xing Bing Fang Zhi Yang Jui* (in Chinese), p. 197.

Wu, C. J., and Wu, S. Y. (1957). *Acta Microbiol. Sin.* **5**(1), 27.

Yasui, C., and Arai, H. (1976). *Virus* (in Japanese) **26**(3-4), 103.

Yen, C. H. (1941). *Proc. Soc. Exp. Biol. Med.* **46**(4), 609.

Yu, Y. X., Ao, J., Lei, W. X., and Li, H. M. (1962). *Acta Microbiol. Sin.* **8**(3), 260.

Yu, Y. X., Ao, J., Zhu, Y. G., Fang, T., Huang, N. J., Liu, L. H., Wu, P. F., and Lis, H. M. (1973). *Acta Microbiol. Sin.* **13**(1), 16.

Yu, Y. X., Wu, P. F., Ao, J., Liu, L. H., and Li, H. M. (1981). *Chin. J. Microbiol. Immunol.* **1**(2), 77.

Zhang, K. X. (1978). *Liu Xing Bing Fang Zhi Yan Jiu* (in Chinese), p. 247.

Zhang, L. B., and An, H. Q. (1981). *J. Microbiol. Immunol.* (in Chinese) **1**(4), 281.

Zhang, Y. H., Jin, E. Y., and Sung, K. (1964). *Zhonghua Yixue Zazhi* (in Chinese) **50**(7), 428.

Zhu, J. H., Zeng, Y., and Qui, F. X. (1978). *Acta Microbiol. Sin.* **18**, 59.

ADVANCES IN VIRUS RESEARCH, VOL. 27

PLANTAGO AS A HOST OF ECONOMICALLY IMPORTANT VIRUSES

John Hammond*

Department of Botany and Plant Pathology
Purdue University
West Lafayette, Indiana

* Present address: USDA, Florist and Nursery Crops Laboratory, Beltsville, Maryland 20705.

I. Introduction

Weed hosts of plant viruses play an important part in the epidemiology of many diseases of economic significance. In order to control such diseases it is necessary to understand their ecology, so that suitable methods may be employed to break the cycle of infection. There is often more than one source of the virus, and there may also be a variety of vectors responsible for transmission of the virus to the crop. The importance of each factor in contributing to crop infection must therefore be assessed so that effective control measures can be developed.

The introduction of new crops may lead to the appearance of new virus diseases or to the increasing importance of previously minor problems. These considerations necessitate a better understanding of the role of weeds and wild plants in the epidemiology of currently important viruses, and also of those viruses endemic in weeds but as yet not known to cause significant crop diseases.

The wide and successful distribution of any weed, and its presence in and around varied types of crop, makes it a potential reservoir of plant viruses. Most plant species that have been thoroughly examined have been found to be naturally infected by at least one type of virus, and it seems likely that all species are potential reservoirs of some virus. Plants that grow among many other types have a greater opportunity for contact with a wide range of viruses and vectors. Hence, because of their ecology, weeds have a high potential for natural virus acquisition.

The wide distribution of successful weed species is clearly advantageous to the survival of the viruses affecting them, and in general weed plants are robust and tolerant of adverse conditions. Tolerance to virus infection is another factor that may help to make some weeds successful. As a result of the genetic diversity of a natural population, at least some individuals are more tolerant to virus infection as well as other factors. The opposite is usually true of crop plants where, because of uniformity of genetic background, if one plant is susceptible, its neighbors probably are also susceptible.

Fewer than 250 plant species are weeds of world-wide importance or of significance in more than one crop or locale. Of these, both *Plantago lanceolata* and *Plantago major* occur over large areas of the world (Holm *et al.*, 1977). Indeed, Allard (1965) has stated that *P. lanceolata* is one of the 12 most successful noncultivated colonizing species. The distribution of *Plantago* species over a diverse range of crops and climates creates a high potential for natural virus acquisition, and indeed plantains are infected naturally by a wide range of viruses. The potential importance of *Plantago* species as reservoirs of economically important viruses is greater than that of many weeds because many *Plantago* species are perennials, and thus can act as overwintering (or oversummering) sources of virus between crop plantings. In addition plantains are tolerant to infection by many viruses, frequently showing no obvious signs of infection. Several virus vector species have been found to overwinter or to reproduce on plantains, and *Plantago* species may therefore act as reservoirs of both viruses and vectors.

The viruses affecting *Plantago* are of varying importance in a wide variety of crops, while the relative importance of plantains as virus reservoirs also varies considerably from one area, crop, or season to another, and with each virus.

Summarizing, *Plantago* species are very efficient weeds, and are tolerant of adverse conditions. Being spread throughout the world and through many different crops, they have the potential for contact with a wide range of viruses, and are indeed susceptible to many of them. They are also tolerant to many of these viruses, frequently showing no obvious symptoms of infection. *Plantago* species therefore provide an excellent model to examine the importance of weeds as reservoir of plant viruses of economic importance.

II. The Occurrence and Ecology of *Plantago* Species

The genus *Plantago* is distributed almost world-wide, primarily as a result of the cosmopolitan range of a few widely introduced species. Most of the species occur almost entirely in temperate climates; in many parts of the tropics only *P. lanceolata* and *P. major* are present, and even these are absent from low-lying areas of tropical America (Good, 1964). The genus includes approximately 250 species, of which about 20 have wide geographical ranges, 9 have discontinuous ranges, about 200 are confined to one region, and 9 have very narrow ranges, limited to one or a few islands (Good, 1964).

The wide distribution of *P. lanceolata* and *P. major* is a result of their aggressive nature as weeds, and their long association with agriculture and man (Fryer and Evans, 1968). Like most *Plantago* species they are perennials, and owe their success as weeds to their efficiency as colonizing species and to their prolific seed production. Not only are they efficient colonizers, but they are able to maintain their ecological niche when succession leads to the establishment of more complex communities.

Most *Plantago* species are wind pollinated and self-compatible, although they are also commonly cross-pollinated. Self-fertility is of vital importance to the success of colonizing species, since self-fertile plants are able to produce seed when only a few or single plants are present, or pollinating insects are absent. The ability to produce seed under these conditions is advantageous for the long distance dispersal of even a single propagule, and for the colonization of any adjacent freshly disturbed habitats (Mulligan and Findlay, 1970). The seeds have a high viability, and are produced in large numbers. *Plantago major* produces an average of 13,000–15,000 seeds per plant, with large plants producing up to 30,000 seeds, while *P. lanceolata* produces about 2500 and up to 10,000 seeds per plant. In both species final germination is in the range 60–90% (Salisbury, 1961). The seed is also viable for long periods in the soil, germinating when suitable conditions arise. Chippendale and Milton (1934) estimated that *P. major* seeds survive for 50–60 years in soil, while Crocker (1938) reported 83.5% germination after burial for 20 years, and 10% after 40 years. *Plantago lanceolata* is less long-lived in the soil, with only 8% germination reported after 10 years (Salisbury, 1961). However, the large numbers of seeds shed and their longevity mean that the soil may contain very large numbers of viable seeds. Champness and Morris (1948) estimated that there may be up to 2.18 million viable seeds of *P. lanceolata* and 3.52 million of *P. major* per acre in pasture soils, while Milton (1936) found a maximum of 8.84 million seeds of *P. major* per acre in lowland grassland soil.

The seed also frequently survives passage through the gut of birds and animals. Brenchley (1920) reported that a cow was estimated to have ingested 89,000 plantain seeds with its fodder, of which 85,000 were apparently undamaged when voided in the dung, and 58% germinated. The seed mucilage is either bacteriostatic or bactericidal, remaining effective even after gut passage (Freerksen, 1950). The plants also resist adverse conditions, *P. major* being the more tolerant of waterlogging and *P. lanceolata* the more drought resistant.

The efficiency of spread of *Plantago* species is quite remarkable, the more so because its dispersal is passive. *Plantago* species have long been associated with man; seed of both *P. lanceolata* and *P. major* were found in the stomachs of Tollund and Graubelle man, preserved in Danish peat bogs from the third and fifth centuries AD (Helbaek, 1950, 1958). Both North American Indians and Maoris referred to *P. major* as "Englishman's foot," presumably because of its spread in the wake of the colonists (Leyel, 1977). As with other weeds brought by settlers from the Old World, the seeds were probably carried along with grain and seeds, since until comparatively recently equipment was not available to clean crop seed efficiently. Several *Plantago* species have a viscid seed mucilage which would facilitate attachment of seed to clothing and equipment, a further potential means of dispersal.

The introduction and spread of *P. lanceolata* and *P. major* to New Zealand (Thomson, 1922) illustrate how rapidly the two species can be spread passively. New Zealand is quite effectively isolated by distance from natural means of plant introduction. Therefore until the introductions made by settlers much of its flora was distinct from that of other areas. At the turn of this century over 75% of the plant species present were unique to New Zealand, and many more limited to Australasia (Cheeseman, 1906).

The first visit of Western man to New Zealand was that of Captain Cook in 1769. Cook is known to have left animals and to have planted European seed on a second visit in 1773. Several hundred years prior to Cook's visit Polynesians and Melanesians are known to have carried plants and seeds to New Zealand. However, plants with European centers of diversity, such as *P. lanceolata* and *P. major,* seem more likely to have been introduced after Cook's initial visit. Occasional and scattered settlement by sealers and whalers may be the cause of some of the introductions to New Zealand, but many of the plants now naturalized there were probably introduced only after regular settlement began in 1839. *Plantago lanceolata* and *P. major* may have been among the earlier introductions, probably among seed brought for cultivation, or among the bedding and belongings of the immigrants. *Plantago major* was first recorded there in 1839, at which time it was apparently as widespread as in England, spreading beyond the regions known to have been visited by Europeans (Bidwill, 1841). *Plantago lanceolata* was first recorded in New Zealand in 1864 by Hooker (1867). By 1922 it was one of the most common of the introduced plants (Thomson, 1922). Thus at least *P. major* had become very widely spread within 70 years of its probable date of introduction.

Man is also responsible for dispersal by means other than contaminated crop seed. Both *P. lanceolata* and *P. major* were valued as herbal remedies (Gerarde, 1636) and were carried to the colonies for that reason by some settlers (Haughton, 1978). Salisbury (1961) collected dry dust from church floors, presumably brought in on boots as mud, and found viable seed of plantains and other species. Seed of *Plantago lagopus* and *Plantago indica* was found in malt refuse from the brewing industry, which was once used as a fertilizer; and seed of *Plantago virginica* and *Plantago varia* was found in wool imported to Tweedside (Ridley, 1930). *Plantago major* seed was once a major constituent of bird seed, being very palatable to birds, and was traded for that purpose. However, the seeds are frequently voided without being digested, and viable seeds of several *Plantago* species have been found in the feces of many birds and animals (Brenchley, 1920; Ridley, 1930; Salisbury, 1961).

Birds and animals probably also cause spread because the viscid seed mucilage readily sticks to feathers, fur, and farm equipment (Ridley, 1930). Guppy (1906) had difficulty in setting up experiments to test seed viability after gut passage in a canary because the bird always carries seed on its feathers after feeding. The viscid mucilage is usually dry, but becomes very sticky when wet, and may be the major natural means of seed dispersal.

Russell and Musil (1961) suggest that seed may be spread by becoming attached to dry leaves and carried by the wind. Ridley (1930) says that the seeds are probably also blown by themselves in the dry state, aided by the long weak peduncles that bear the capsules. However, the evidence he cites, of plants growing naturally on walls and rooftops, may indicate dispersal by birds.

III. *Plantago* SPECIES IN AGRICULTURAL SITUATIONS

Plantains cause a variety of problems in agriculture, but also have some commercial usefulness. They occur as weeds in a wide variety of crops and situations, competing for resources with the crop, and sometimes causing contamination of the desired product. This is most common in crops grown for seed. Seed may be dispersed in the field by unintentional sowing with the crop, and by the attachment of seed to animals or equipment entering the field. Several species are grown commercially for a variety of purposes, but only on a moderate scale.

A. As Weeds within Crops

More than 50 countries report *P. lanceolata* and *P. major* as important weeds in crops as diverse as alfalfa, coffee, rice, cereals, onions, cotton, and small fruits, and also in pastures and temperate orchards. *Plantago* species are seldom regarded as the major weed of a particular crop, but the sheer number of places, crops, and cultural systems in which they prosper makes it necessary to regard them as important world weeds (Holm *et al.*, 1977). They are a common weed of irrigated crops in Rhodesia (Salisbury, 1961), and are especially common in those crops originally grown in Europe and now grown throughout the world. This is largely through contamination of crop seed, especially in the small-seeded legumes (Holm *et al.*, 1977). Most of the serious problems caused by plantains are in cereals (Holm *et al.*, 1977) and in grasslands. The often abundant foliage may cause spoilage in hay, as the leaves do not dry readily, and therefore mold easily (Gill and Vear, 1966). Plantains are particularly common in the arable fields of sugarbeet-producing areas, while in Hawaii *P. lanceolata* is the dominant weed of dry zone pastures at 1000–1200 m, where it persists and increases even during periods of drought (Holm *et al.*, 1977). It is a common and troublesome weed of grasslands in the midwestern and eastern states of North America (Reed, 1970).

Competition between plantains and the crop may be for light, space, or nutrients, which may be especially important where water is a limiting factor. *Plantago lanceolata* is a good source of calcium, chlorine, phosphorus, potassium, sodium, and cobalt, and is sometimes used in herbage mixtures to supplement the mineral supply for livestock for this reason (Holm *et al.*, 1977). This suggests that plantains can deprive crop plants of these elements by more efficiently sequestering them, as does the requirement for heavy fertilization with high phosphorus when *Plantago psyllium* is grown commercially (Wilson, 1945).

Plantago lanceolata rapidly reproduces vegetatively in open areas, forming dense swards where its expansion is limited by competition. It will thus readily compete for space in a crop situation, and may become established and spread before the crop canopy closes, reducing the favorability of the area to the weed. Maximum germination of *P. lanceolata* coincides in temperate climates with the main spring and fall planting times of crops. Thus the necessity for an open area during the transition from reliance on seed reserves to photosynthetic assimilates (Holm *et al.*, 1977) is met, and a high proportion of seedlings may become established.

Plantago major grows best in arable land, and cannot become established unless the surrounding vegetation is short. However, it can survive and set seed in closed communities (as when the crop canopy closes) provided that there are temporary open spaces for the seedlings to become established, as in newly sown field (Sagar and Harper, 1964).

Germination of both *P. lanceolata* and *P. major* seems best when the seeds are close to the surface, and when open sites are available in the herbage (Holm *et al.*, 1977). These conditions are admirably met in a seedbed, the preparation of which will undoubtedly bring some buried seed to a more favorable position near the surface.

B. As a Problem in Seed Cleaning in Specific Crops

Even with the advent of seed cleaning methods much more advanced than a simple winnow, *Plantago* seeds are still very difficult to separate from the seed of some crop plants due to similarities of size, shape, and weight. This is particularly true of the small seeded legumes (the clovers, sanfoin, and lucerne) and timothy grass (Salisbury, 1961; Gill and Vear, 1966; Fryer and Evans, 1968; Holm *et al.*, 1977). Problems of weed contamination of seed lots are much more prevalent where farmers save their own seed, and do not have it properly cleaned. Farm-saved seed is also more likely to contain diseased seed, which will often renew infection in the new crop. The combination of poorer quality seed, early onset of disease, and sowing weeds with the crop may lead to greater losses in these crops as the weeds will be in a favorable position to take advantage of poor growth in the affected crop.

C. Commercial Usage

Several species are grown for a variety of reasons, but none are major crops, and they are restricted to a few locations and small areas.

1. Forage Mixtures

Plantago lanceolata is still used occasionally in herbage strips in pastures, or in general grazing mixtures for variety, and as a supplementary source of minerals (Gill and Vear, 1966; Holm *et al.*, 1977). Sheep find the plants particularly palatable and will often bite them off at the crown (Salisbury, 1961).

2. Herbal Remedies

Gerarde (1636) claimed many virtues for all parts of the plants, especially of *P. major*. This superiority was also recognized in Hooper's Medical Dictionary (Grant, 1848), which reported the retention of *P. major* in the *materia medica* of the Edinburgh College "until of late years." Here the leaves are named as the pharmaceutical part of the plant, having "refrigerant, attentuating, astringent and diuretic" properties (Grant, 1848). *Plantago major* was used by many in colonial times as a home remedy for a wide variety of ills including deep cuts and snake and scorpion bites, and also as a green vegetable (Haughton, 1978).

Plantago lanceolata is still grown on a field scale in Holland for medicinal purposes, demonstrating a continuing demand for such herbal remedies. However, the major cropping of *Plantago* species is for psyllium seed from *P. psyllium, Plantago arenaria,* and *Plantago ovata.* It is used because of its indigestibility and bulking properties in certain cases of chronic constipation (Wilson, 1945) or because the mucilage acts as a mild laxative comparable to agar or mineral oil (Hill, 1952). The mucilage extracted from the seed is employed as a bandoline and a cloth-sizing material (Wilson, 1945). *Plantago psyllium* and *P. arenaria* are grown in several European countries, especially in southeastern France (Wilson, 1945), while *P. ovata* is grown primarily in India (Hill, 1952).

3. Bird Seed

Some *Plantago* seed, primarily that of *P. major,* is still raised for sale as bird seed.

IV. *Plantago* Species as Virus Host

There are many reports in the literature of weeds harboring viruses and mycoplasma-like organisms, and of their role as reservoirs for infection of crop plants. Several reports are of natural infection of *Plantago* species, and other authors have experimentally inoculated *Plantago* species with a number of viruses (e.g., Hollings, 1959). In order to assess the importance of infected weed plants in the ecology of the crop diseases, it is also necessary to understand the means of transmission and hence whether the virus is likely to be transmitted from weed to crop, and whether such instances are of major or minor importance.

A. The Number and Type of Viruses and Mycoplasma-like Organisms Infecting Plantago Species

At least 26 viruses and 3 mycoplasma-like organisms are known to infect one or more *Plantago* species naturally, and a further 13 viruses and at least 1 mycoplasma-like organism have been shown capable of infecting plantains under experimental conditions (Table I). The viruses infecting *Plantago* represent a wide variety of types, from at least 19 different groups, demonstrating the range of susceptibility of *Plantago* species to viral infection. However, there are few reports of severe symptoms of infection in any *Plantago* species. In many cases the infections are systemic and symptomless, or produce very mild or indistinct symptoms. The reports of natural infection come from many parts of the world (Table I), reflecting the distribution of *Plantago* species. The fact that the majority of reports concern *P. lanceolata* or *P. major* emphasizes the spread of these two species throughout the world, and their common presence as weeds in many crops.

TABLE I

THE VIRUSES INFECTING *Plantago* SPECIES AND THEIR GEOGRAPHICAL OCCURRENCE

Virus group	Virus name	*Plantago* species affected	Type of infection	Country of occurrence	Reference
Carlavirus	Plantain virus 8(PIV8)	*P. lanceolata*	N[a]	Great Britain	Hammond (1981)
Closterovirus	Beet yellows virus (BYV)	*P. major*	N	Germany	Beiss (1956)
		P. major ⎫ *P. media* ⎬ *P. erecta* ⎭	N	Ukraine (USSR)	Goryshin (1959)
		P. insularis ⎫ *P. lanceolata* ⎬	N	Belgium	Roland and Tahon (1961)
		P. lanceolata ⎫ *P. psyllium* ⎬ *P. racemosa* ⎭	N	Germany	Weisner (1962)
		P. major	N	Belgium	Jadot (1974)
		P. erecta ⎫ *P. insularis* ⎬	E		Bennett and Costa (1954)
Potexvirus	Hydrangea ringspot virus (HRSV)	*P. lanceolata*	E		Hollings (1958, 1959)
	Plantago severe mottle virus (PSMV)	*P. major*	N	Canada	Rowhani and Peterson (1980)
	Plantain virus X (PIVX)	*P. lanceolata*	N	Great Britain	Hammond and Hull (1981)
		P. arenaria ⎫ *P. coronopus* ⎪ *P. indica* ⎬ *P. major* ⎪ *P. rugelii* ⎪ *P. serpentina* ⎭	E		
Potyvirus	Potato virus X (PVX)	*P. lanceolata*	E		Hollings (1959)
	Carnation vein mottle virus (CVMV)	*P. lanceolata*	E		Hollings (1959)
	Potato virus Y (PVY)	*P. lanceolata* ⎫ *P. major* ⎬	N	USSR	Akhatova *et al.* (1979)

Virus	Plantago sp.		Country	Reference
Plantain virus 7 (PIV7)	*P. lanceolata*	N	Great Britain	Hammond (1981)
Tobacco etch virus (TEV)	*P. lanceolata*	E		Hollings (1959)
Turnip mosaic virus (TuMV)	*P. depressa*	N	China	Wang et al. (1965)
(Anemone mosaic)[b]	*P. lanceolata*	E		Hollings (1959)
Ribgrass mosaic virus (RMV)	*P. lanceolata* ⎫ *P. major*	N	United States	Holmes (1941)
	P. major	N	United States	Valleau and Johnson (1943)
	P. lanceolata ⎫ *P. major* *P. rugelii*	N	United States	Holmes (1950)
	P. lanceolata	N	Italy	Sibilia (1948)
	P. lanceolata	N	USSR	Goldin (1953)
	P. lanceolata ⎫	N	Scotland	Harrison (1956)
	P. major	N	Bulgaria	Kovachevsky (1963)
	P. media	N	Yugoslavia	Miličič (1968)
	P. lanceolata	N	Yugoslavia	Juretić et al. (1969)
	P. asiatica	N	Armenia (USSR)	Kostin and Volkov (1976)
	P. media	N	Armenia (USSR)	Tamrazyan et al. (1977)
	P. lanceolata ⎫ *P. major* *P. media*	N	Czechoslovakia	Prochazkova (1977)
	P. major	N	USSR	Polák and Chlumská (1980)
	P. media	N	United States	Chessin et al. (1980)
	P. lanceolata	N	Great Britain	Hammond (1981)
Tobacco mosaic virus (TMV)	*P. lanceolata* ⎫ *P. rugelii*	N	United States	Boyle and Wharton (1957)
	P. lanceolata *P. major* *P. rugelii*	E		Holmes (1938)
	P. lanceolata ⎫ *P. rugelii*	E		McKinney (1952)
(2 strains)	*P. lanceolata*	E		Hollings (1959)
(Tomato strain)	*P. major*	N	USSR	Blaszcsak (1976)

Tobamovirus

(continued)

TABLE I (Continued)

Virus group	Virus name	Plantago species affected	Type of infection	Country of occurrence	Reference
Tobravirus	Tobacco rattle virus (TRV)	P. lanceolata ⎫ P. major ⎭	N	Czechoslovakia	Polák et al. (1968)
		P. major	N	Germany	Uschraweit and Valentin (1956)
		P. major	N	Denmark	Kristensen and Engsbro (1966)
		P. indica ⎫ P. lanceolata ⎬ P. psyllium ⎭ P. sempervirens	E		Schmelzer (1957)
		Plantago sp.	E	United States	Komuro et al. (1970)
	(Aster ringspot)	P. virginica	N	United States	Anderson (1959)
Caulimovirus	Plantain virus 4 (PIV4)	P. lanceolata ⎫ P. major ⎭	N	Great Britain	Hammond (1981)
Cucumovirus	Cucumber mosaic virus (CMV)	P. major	N	Great Britain	Tomlinson et al. (1970)
		P. major	N	United States	Dodds and Taylor (1980)
		P. lanceolata	N	France	Quiot et al. (1979)
Geminivirus	Beet curly top virus (BCV)	P. erecta	N	United States ⎫	Severin (1934)
		P. major	E		Giddings (1938)
		P. erecta	E		Giddings (1944)
		P. major	E	⎭	
Ilarvirus	Prune dwarf virus (PDV)	P. virginica	E		Fulton (1957)
	Tobacco streak virus (TSV)	P. major	N	United States	Fulton (1948)
	(Colorado red node)	P. lanceolata	E		Thomas (1949)
Luteovirus	Beet mild yellows virus (BMYV)	P. lanceolata ⎫ P. major ⎭	N	Belgium	Roland et al. (1968)
Nepovirus	Arabis mosaic virus (ArMV)	P. major	E		Lister and Murant (1967)
		P. sempervirens	E		Schmelzer (1963b)

Virus group	Virus	Plantago species	N/E	Country	Reference
	(Raspberry yellow dwarf)	*Plantago* spp.	N	Great Britain	Harrison (1958)
	Cherry leaf roll virus (CLRV)	*P. major*, *P. lanceolata*	E		Cropley (1961)
	Elm mottle virus (EMV)	*P. major*, *P. sempervirens*	E		Schmelzer (1966)
		P. lanceolata	E		Schmelzer (1969)
	Raspberry ringspot virus (RRV) (Raspberry leaf curl)	*P. lanceolata*, *P. major*	E		Cadman et al. (1954)
	Tobacco ringspot virus (TRSV)	*P. lanceolata*	N	United States	Rush (1970)
		P. major	N	United States	Uyemoto (1975)
		P. lanceolata	E		Price (1940)
		P. lanceolata	E		Hollings (1959)
	Tomato blackring virus (TBRV)	*P. lanceolata*, *P. psyllium*, *P. sempervirens*	E		Schmelzer (1963c)
	(Beet ringspot) (Celery yellow vein)	*P. lanceolata*	E		Hollings (1959)
	Tomato ringspot virus (TomRSV)	*Plantago* sp., *P. lanceolata*	N	United States	Frazier et al. (1961)
		P. major	N	United States	Uyemoto (1975)
Tobacco necrosis virus group	Tobacco necrosis virus (TNV)	*P. major*	E		Smith (1937)
		P. major	E		Roland (1957)
	(3 strains)	*P. lanceolata*	E		Hollings (1959)
	Plantain virus 5 (PlV5)	*P. lanceolata*	N		Hammond (1981)
Tomato spotted wilt virus group	Tomato spotted wilt virus (TSWV)	*P. major*	E	Great Britain	Smith (1932)
Tombusvirus	Carnation mottle virus (CarMV)	*P. lanceolata*	E		Hollings (1959)
	Plantain virus 6 (PlV6)	*P. lanceolata*	N	Great Britain	Hammond (1981)
	Tomato bushy stunt virus (TBSV)	*P. major*	N	Italy	Lovisolo et al. (1965)
		P. lanceolata	E		Hollings (1959)

(continued)

TABLE I (*Continued*)

Virus group	Virus name	Plantago species affected	Type of infection	Country of occurrence	Reference[a]
Tymovirus	Plantago mottle virus (PlMV)	*P. major*	N	United States	Granett (1973)
Unclassified isometric viruses	Broad bean wilt virus (BBWV)	*P. lanceolata*	N	Australia	Taylor *et al.* (1968)
		P. lanceolata	N	United States	Uyemoto and Provvidenti (1974)
		P. coronopus	N	Poland	Schmelzer *et al.* (1975)
		P. lanceolata	N	Poland	Schmelzer (1975)
		P. lanceolata	N	Argentina	Gracia and Feldman (1976)
		P. lanceolata	E		Schumann (1963)
		P. lanceolata ⎫ *P. psyllium* ⎭	E		Schmelzer (1960)
Rhabdovirus	Dodder latent virus (DLV)	*P. major*	E		Bennett (1944)
	Plantain virus A (PlVA)	*P. lanceolata*	N	Great Britain	Hitchborn *et al.* (1966)
		P. lanceolata	N	Great Britain	Hammond (1981)
	Plantain virus B (PlVB)	*P. lanceolata*	N	Great Britain	Hitchborn *et al.* (1966)
Other bacilliform viruses	Alfalfa mosaic virus (AMV)	*P. lanceolata*	N	Poland	Ksiązek (1976)
		P. lanceolata	E		Hollings (1959)
		P. psyllium ⎫ *P. sempervirens* ⎭	E		Schmelzer (1963a)
Unidentified virus	?	*P. major*	E	Canada	Bezner and Schmelzer (1972)
Mycoplasma-like organisms	Aster yellows	*P. major*	N	United States	MacClement and Richards (1956)
		P. major	N	United States	Severin (1929)
		P. major	N		Linn (1940)
		P. major	E		Kunkel (1928)
		P. major	E		Frazier and Thomas (1953)
	Cloverwitches broom (Strawberry green petal) ⎫	*P. major*	E		Frazier and Posnette (1957)
	Kok-saghyz yellows	*P. major*	N	USSR	Ryjkoff (1943)
	Plantago yellows	*P. media*	N	Lithuania (USSR)	Staniulis and Genyte (1976)

[a] N, Natural infection; E, experimental infection.

[b] Names in parentheses are those used in the reference cited.

B. The Known Modes of Transmission

The methods of spread among the viruses infecting plantains (Table I) vary considerably. Many are readily mechanically transmissible, but this is often unimportant in natural situations, where transmission may be most frequently via a specific vector.

The members of the carlavirus, potyvirus, caulimovirus, cucumovirus, and luteovirus groups, some of the rhabdoviruses, and broad bean wilt virus (BBWV) and alfalfa mosaic virus (AMV) are most commonly transmitted by various species of aphids, and indeed the luteoviruses are not known to be transmissible otherwise.

The geminiviruses are transmitted either by whiteflies or by leafhoppers. Beet curly top virus (BCV), the only geminivirus so far recorded in *Plantago* species, is transmitted by leafhoppers.

The tymoviruses and some tombusviruses are transmitted mainly by leaf-eating beetles, while other tombusviruses may have fungal vectors as do tobacco necrosis virus (TNV) and some of the potexviruses. Species of thrips are the only known vectors of tomato spotted wilt virus (TSWV).

There are several reports of various insects transmitting TMV, but such infections are probably due to mechanical damage, and insect transmission of TMV may not be common in the field.

Other types of vectors include the nematodes, the trichodorid species that transmit the tobraviruses, and the genera *Longidorus* and *Xiphinema* that are vectors of the nepoviruses. The nepoviruses are also frequently transmitted via the seed and pollen of infected plants, as are the ilarviruses, which have no known vectors. Dodder latent virus (DLV) and several other viruses are capable of being transmitted by several *Cuscuta* species, but this is unlikely to be important in the field.

Mechanical transmission appears to be the normal means of infection with the tobamoviruses and the potexviruses. Infective virus in plant debris in the soil can be a potential source of infection, and transmission also occurs readily via animals, men, or equipment passing through the crop. Pruning tools have been shown to be the means of spread within crops of a variety of mechanically transmissible viruses.

C. A Survey of the Viruses Endemic to Plantago Species in Great Britain

Over a period of 3 years *Plantago* species, mainly *P. lanceolata* and *P. major,* were collected from nine areas of Great Britain and examined for virus infection (Hammond, 1981). The plants were mainly chosen at random without regard to symptoms, and were collected primarily from agricultural land and nearby roadside verges. They were tested for virus infection by inoculation to a range of indicator species, and by electron microscopy of negatively stained epidermal dip preparations (Hitchborn and Hills, 1965). Where virus infection was detected by either method, extended host range studies were made to try to identify the virus, together with purification and characterization where possible. A total of eight distinct transmissible viruses were detected. Sometimes virus-like isometric particles were observed by electron microscopy, with no evidence of transmissibility even to healthy plants of *P. lanceolata* or *P. major*. Of the eight viruses detected, two were identified as previously described viruses known to infect *Plantago* species (Hammond and Hull, 1981). These were ribgrass mosaic virus (RMV; Holmes, 1941) and plantain virus A (PlVA; Hitchborn *et al.*, 1966). A third, plantain virus X (PlVX), was shown to be a potexvirus (Hammond and Hull, 1981) while the five others were not fully characterized and were provisionally named plantain viruses 4–8 (PlV4–PlV8; Hammond, 1981). They were tentatively assigned to the following groups: plantain virus 4 (PlV4), caulimovirus; plantain virus 5 (PlV5), tobacco necrosis virus group; plantain virus 6 (P1V6), tombusvirus; plantain virus 7 (PlV7), potyvirus; and plantain virus 8 (PlV8), carlavirus (Hammond, 1981).

1. The Frequency of Infection

Of 144 plants collected, 92 (nearly 64%) were infected, including 88 of 130 *P. lanceolata*, 4 of 10 *P. major*, and none of 4 *P. maritima*. Such a high percentage is remarkable considering that the majority of these plants bore no obvious symptoms of infection and were collected at random from the general population. The most commonly detected virus was PlVX, found in 51 of 130 plants of *P. lanceolata* (Hammond and Hull, 1981) from 8 of the 9 areas of Great Britain where plants were collected (Hammond, 1981). RMV was detected in 28 *P. lanceolata* from 5 areas; PlVA from 13 *P. lanceolata* from 4 areas; PlV4 from 11 *P. lanceolata* and 3 *P. major* from 5 areas; PlV5 from a single plant of *P. lanceolata;* PlV6 also from a single *P. lanceolata;* PlV7 from 8 *P. lanceolata* from 4 areas; and PlV8 from a single *P. lanceolata* (Hammond, 1981).

2. The Frequency of Multiple Infections

Multiple infections were common, and as many as four distinct viruses were detected in a single plant. Because of the precautions taken to prevent cross-infection in the glasshouse, this probably represents the situation in nature. At least 2 viruses were detected in each of 23 plants of *P. lanceolata*. These figures do not include any of the virus-like particles observed by electron microscopy but not transmitted to other plants.

V. The Importance of the Viruses Naturally Infecting *Plantago*

The viruses identified as occurring naturally in *Plantago* are of varying agricultural importance, and the significance of a particular virus may also vary widely from area to area, depending upon the crops grown there. Each of the viruses will be discussed in terms of the diseases caused in agricultural and horticultural crops, and the possible importance of *Plantago* species, among other known sources of the virus, for crop infection.

A. Carlaviruses

Plantain Virus 8 (PlV8)

This virus was found only in a single plant of *P. lanceolata,* and infected only *P. lanceolata* and cucumber plants out of 18 species tested by mechanical inoculation (Hammond, 1981). PlV8 is presumed to be aphid transmitted, but no vector was identified. No obvious symptoms were observed in infected plants, and no agricultural importance is known.

B. Closteroviruses

Beet Yellows Virus (BYV)

Beet yellows is one of the major diseases of sugar beet in most of the growing areas worldwide, and may cause severe losses in both sugar crops and those grown for seed. Losses may also be significant in table beet, Swiss chard, and spinach (Bennett, 1960). Hull (1953) estimated that as much as half of the potential sugar yield was lost in infected crops, both through loss of root weight and the lower percentage of sugar. When seed crops are infected in the early stages of development, they may yield only 40–60% of their potential (Hull, 1952).

Secondary effects of BYV infection may involve augmentation of other pathogens. Bennett (1960) reports synergism of effects between BYV and beet curly top virus (BCV), and more severe effects when BYV-infected plants are also infected by leafspot, caused by *Cercospora beticola* (Sacc.). The susceptibility to leafspot is increased in the virus-infected plants. Infected spinach may be more susceptible to some vascular diseases and root-rot organisms (Bennett, 1960).

There are several references to the natural infection of several *Plantago* species (Table I), and also to some of the several aphid vectors of BYV overwintering on *P. lanceolata* and *P. major* (Heathcote *et al.*, 1965; Jadot *et al.*, 1969). One of the aphid species found overwintering on plantains was *Myzus persicae*, the most important vector of BYV in England (Heathcote and Cockbain, 1964).

Most of the weed hosts of BYV are annuals, and do not act as overwintering sources of virus for the aphid vectors (Bennett, 1960). Plantains are perennial weeds and hosts to the vectors as well as to the virus, and thus may be more important than some of the more commonly infected annual weeds such as chickweed (*Stellaria media*) and groundsel (*Senecio vulgaris*).

There are other sources of considerable importance, such as beet and mangold clamps, wild beet, and beet left in the field. Such unharvested beet may survive to harbor virus and aphids, especially in crops of barley or alfalfa, which are frequently planted following beet crops (Bennett, 1960). The relative importance of the various sources, and of the several aphid vectors, may vary from area to area and from season to season (Bennett, 1960; Heathcote *et al.*, 1965), and plantains may be locally important virus reservoirs.

C. Potexviruses

1. Plantain Virus X (PlVX)

Although PlVX is widespread in *P. lanceolata* in Great Britain, and readily mechanically transmitted to a range of test plants, there is no evidence at present of any agricultural problems due to PlVX (Hammond and Hull, 1981).

2. Plantago Severe Mottle Virus (PSMV)

Like PlVX, PSMV has an experimental host range from several families and is mechanically transmissible, but is not known to be of any agricultural significance (Rowhani and Peterson, 1980).

D. Potyviruses

1. Potato Virus Y (PVY)

This virus is one of the major problems in potato crops around the world, either alone or in combination with other potato viruses (see

Smith, 1972). The common symptoms are "leaf drop streak" or veinal necrosis. In combination with potato virus X (PVX) it causes rugose mosaic, a disease more destructive than that caused by either virus alone. Economically important diseases are also caused by PVY infection in peppers (*Capsicum* spp.), tobacco, and tomato among other crops (Edwardson, 1974). Losses in peppers may be total in severe virus epiphytotics (Villalón, 1981).

The major means of introduction of PVY to a crop is by one of several aphid species, while transmission within the crop may be either by further aphid transmission or by plants rubbing together. At least 16 aphid species have been reported to be capable of acting as vectors (Kennedy *et al.*, 1962). The virus is stylet borne, and *Myzus persicae* may be the most efficient vector (Smith, 1972).

Plantago species are among several weeds reported as reservoirs of PVY in Russia (Akhatova *et al.*, 1979), but more common sources of infection may be volunteer potato plants and infected tubers of "seed" potato. Seed potatoes are often raised in areas away from the major ware crop, and are usually more carefully examined and rogued than ware crops. In these areas volunteer plants will be less common, and weed sources, including *Plantago,* may be more important than in the major growing districts. The aphid vectors of BYV reported to overwinter on plantains (Jadot *et al.*, 1969) are also vectors of PVY (Kennedy *et al.*, 1962).

2. Turnip Mosaic Virus (TuMV)

Also known as cabbage black ringspot virus, TuMV causes diseases in most cruciferous crops and has a wide host range, infecting species in at least 20 families (Tomlinson, 1970). The effects in the field may be quite severe, causing losses of over 50% of both leaf and root fresh weight in swedes (Tomlinson and Ward, 1978). Losses in white cabbage may be increased by storage, as TuMV is one of the causes of internal necrosis of cabbage, which lowers the quality of the cabbage and may lead to the rejection of samples for processing (Walkey and Webb, 1978).

TuMV has a large number of aphid vectors (Kennedy *et al.*, 1962), of which *M. persicae* and *Brevicoryne brassicae* are most important (Tomlinson, 1970). Like PVY, TuMV is stylet borne and nonpersistent, and also readily mechanically transmitted. Cruciferous crops are grown throughout the year in many locations, and probably provide ample reservoirs for both virus and vectors. Cruciferous weeds may be more important than *Plantago,* but *Plantago depressa* has been reported as a natural oversummering host of TuMV in China (Wang *et al.*, 1965).

3. Plantain Virus 7 (PlV7)

Known naturally to infect only *P. lanceolata*, PlV7 was not transmitted mechanically even to other *P. lanceolata*, although it was transmitted to *P. lanceolata* and *Nicotiana benthamiana* by *Myzus* sp. and *M. persicae* (Hammond, 1981). No agricultural significance is known.

E. Tobamoviruses

1. Tobacco Mosaic Virus (TMV)

One of the best known plant viruses, TMV, has a very wide host range, many strains, and no known vector. It is often hard to tell whether reports are of TMV itself, or of related viruses. Thus some reports of TMV in *Plantago* may be of ribgrass mosaic virus (RMV), now usually recognized as a separate virus. However, the report of Blaszczak (1976) is specific in describing a virus isolated from *P. major* as a tomato strain of TMV. TMV causes many diseases in many crops, among the better known and more economically important of which are the mosaics of tobacco and tomato.

There are many sources of TMV in dry and living plant tissue, and *Plantago* species are not likely to be an important source, especially as most strains of TMV are localized in *Plantago* (Holmes, 1941). Plantains may, however, be the origin of local infections.

2. Ribgrass Mosaic Virus (RMV)

This virus was originally isolated from *P. lanceolata* and *P. major* in the United States, where it was the cause of outbreaks of mosaic in tobacco (Holmes, 1941; Valleau and Johnson, 1943). It has also been reported to be the cause of internal browning of tomatoes, where *P. rugelii* was also implicated (Holmes, 1950). Ribgrass mosaic virus has since been reported from many other parts of the world (Table I) and also from other natural host species (Oshima and Harrison, 1975). Ribgrass mosaic virus differs from TMV in infecting a wide range of cruciferous species (Oshima *et al.*, 1971, 1974). No vector is known.

Ribgrass mosaic virus is probably the most widespread virus in *Plantago* species, although PlVX was found in a higher proportion of plants than RMV in the course of a survey in Great Britain (Hammond, 1981). However, PlVX has not been reported from any other areas.

Ribgrass mosaic virus is economically significant in tobacco crops, and tobacco growing areas may be where *Plantago* is most important as a reservoir of virus disease.

F. Tobraviruses

Tobacco Rattle Virus (TRV)

The major diseases caused by TRV are spraing (also called corky ringspot) and stem mottle in potatoes; yellow blotch in sugar beet, a ringspot in peppers (*Capsicum* spp.) and asters; and mottling, distortion, and stunting in several bulbous ornamentals. The rattle disease of tobacco is now of little importance (Harrison and Robinson, 1978). Spraing may render potato tubers unsaleable for human consumption, while the diseases of ornamentals such as *Tulipa, Gladiolus, Narcissus,* and *Hyacinthus* may also cause significant losses in a high-cost, high-value industry. The name beet yellow blotch describes only the foliar symptoms in sugar beet—TRV causes a reduction in the root system of infected beet, which is more fangy and has more secondary roots than that of healthy beet (Gibbs and Harrison, 1964).

Tobacco rattle virus infects over 100 species naturally, in many of which the infection is restricted to the roots, and about 400 species have been infected experimentally; it has been suggested that TRV is essentially a virus of wild plants that infects cultivated plants grown at sites where the virus is already established (Harrison and Robinson, 1978). Cooper and Harrison (1973) suggested that the nematode vectors of TRV prefer some weed species as hosts above potatoes, and that they feed most frequently on potatoes when the preferred weeds are not available.

The vectors of TRV are trichodorid nematodes of the genera *Paratrichodorus* and *Trichodorus* (Harrison and Robinson, 1978). As the vectors are of very limited mobility, introduction of the virus to a new area is likely to be in infested soil, or in infected planting material to a site where the vector is already established. Virus spread in infected weeds also shows the importance of infected seed as a reservoir of the virus in the soil during periods when nematode populations fluctuate or lose their infectivity; weed seed may maintain virus during a fallow period in which the vectors lose their infectivity in the absence of susceptible crop or weed roots to reinfect and act as fresh sources of virus (Murant and Lister, 1967). Tobacco rattle virus is occasionally transmitted in the seed of some weed species (Lister and Murant, 1967), though this is not reported in *Plantago* species.

Plantains are just one source of TRV among many. However, many of the other hosts are annuals, whereas plantains are perennials and may thus be of more importance as an overwintering source. As in other cases in which a virus has many possible weed reservoirs, the importance of each will vary from one area and one season to another, and a particular source may be locally very important. Anderson (1959) suggested that *Plantago virginica* may be an important field source for infection of peppers.

G. Caulimoviruses

Plantain Virus 4 (PlV4)

Naturally occurring in both *P. lanceolata* and *P. major* from five separate areas of Great Britain, PlV4 was sap transmissible to *P. lanceolata,* turnip, and broccoli without causing obvious symptoms (Hammond, 1981). No vector was identified, but by analogy with the definitive caulimoviruses (Shepherd, 1976), PlV4 would be aphid transmitted. No present agricultural significance is known.

H. Cucumoviruses

Cucumber Mosaic Virus (CMV)

One of the most widely spread viruses known, with a very wide host range, CMV exists as many strains, which may differ considerably from one another in host reactions. Natural infections are recorded from a range of agricultural and horticultural crops. The best known diseases are mosaic of cucumber, other cucurbit crops, and tobacco. The fruit of infected cucumbers is often misshapen (Smith, 1972). Other important crops affected are tomatoes, celery, and spinach. A severe disease caused by CMV and its associated RNA 5 has been reported, which was probably the cause of the almost total loss of the field tomato crop in the Alsace region of France in 1972 (Kaper and Waterworth, 1977). Cucumber mosaic virus also incites a variety of diseases in ornamentals including many lilies (Smith, 1972).

Cucumber mosaic virus is readily mechanically transmitted, and on occasion is transmitted through the seed of some species (Neergaard, 1977). Seed transmission in some weeds may be important in some instances (Tomlinson and Carter, 1970). Most of the initial crop infections with CMV probably originate by aphid transmission rather than by infected seed, and in-crop transmission by aphids will further spread the disease. Over 60 aphid vectors are known (Kennedy *et al.*, 1962) which vary in transmission efficiency. Cucumber mosaic virus is nonpersistent in its vectors.

Plantago major infected with CMV was found among a lettuce crop affected by the virus (Tomlinson *et al.*, 1970) and in a tobacco seedling-raising house where the crop was diseased (Dodds and Taylor, 1980) and may be a source of the virus for other crops as well. Quiot *et al.* (1979) found CMV in *P. lanceolata*. In each of these cases, however, other weeds were suggested to be of more importance than plantains as reservoirs of CMV, although there may be cases in which plantains are a major source.

I. Geminiviruses

Beet Curly Top Virus (BCV)

The disease caused by BCV in sugarbeet was described and investigated many years before the nature of the viral particle was demonstrated by Mumford (1974). The effect on beet and other crops is debilitating and often lethal. The symptoms are of yellowing, usually together with leaf curling and distortion in beet, Swiss chard, spinach, tomato, pepper, bean, cucurbits, flax, and many other species (Thomas and Mink, 1979). The virus is limited to arid and semiarid areas of western North America from Mexico to Canada, and the Eastern Mediterranean. Both virus and vector may have been brought to the United States from the Mediterranean area (Bennett and Tanrisever, 1959).

The vectors of BCV are leafhoppers, in which the virus is persistent. *Circulifer tenellus* is the only known natural vector in the United States, while *C. tenellus* and *Circulifer opacipennis* are reported as vectors in the Mediterranean (Thomas and Mink, 1979). Giddings (1938) reports that *Plantago erecta* plays an important role in the life cycle of the leafhopper in California, and that *P. erecta* is one of the most important overwintering hosts in California. In this instance the plantains grow a considerable distance from the crop, and control of the disease by control of the reservoir is impractical in terms of the area of hills that serve as the range of the vectors.

J. Ilarviruses

Tobacco Streak Virus (TSV)

Widely distributed but not common, TSV causes a systemic necrosis in tobacco, the later growth of which does not show symptoms. Other plants in which diseases caused by TSV have been reported include cotton, forage legumes, tomatoes, asparagus, peas, soybeans, potatoes, dahlias, and roses (Fulton, 1981).

The only "possible vectors" of TSV reported are thrips (*Frankliniella* sp.) in Brazil (Costa and Neto, 1976). Tobacco streak virus is readily mechanically transmissible, though unstable. It is also transmissible by dodder, and through the seed of some species (Fulton, 1971).

Fulton (1948) recovered TSV from *P. major* and four other weed species that were growing near tobacco fields that had recently become infected with TSV. Eradication of weeds for the control of TSV is probably not necessary, as TSV is of only minor importance on most hosts, at least in North America (Fulton, 1981).

K. *Luteoviruses*

Beet Mild Yellows Virus (BMYV)

First described as a separate virus by Russell (1958), BMYV is regarded by some as a strain of beet western yellows virus (BWYV), as the two cause very similar diseases (Duffus, 1973), although BMYV differs in epidemiology and host range and appears to be confined to Europe (Duffus, 1972). Beet mild yellows virus is in some years more important than BYV in England (Russell, 1958) and causes severe economic loss in sugarbeet.

Dual infections of BYV and BMYV cause significantly higher loss than BMYV alone, and BMYV-infected plants may also be more severely affected by *Alternaria* species than are BYV-infected or healthy plants (Russell, 1960). Tolerant varieties of beet lost only half as much sugar yield as did susceptible varieties (Russell, 1963).

Both *P. lanceolata* and *P. major* have been shown to be overwintering hosts of BMYV (Roland *et al.*, 1968), and *P. lanceolata* has also been shown to act as an overwintering host to some of the vector aphid species (Jadot *et al.*, 1969). Roland *et al.* (1968) found *Stellaria media* and *P. lanceolata* to be the most frequently encountered weeds infected with BMYV. Thus *P. lanceolata* may be one of the most important overwintering hosts of BMYV, despite the failure of Russell (1965) to recover either BYV or BMYV from *P. lanceolata* in England.

L. *Nepoviruses*

1. Arabis Mosaic Virus (ArMV)

Yellow dwarf of raspberry, mosaic and yellow crinkle of strawberry, stunt mottle of cucumber, chlorotic stunt of lettuce, and stunting and mosaic of celery are among the diseases caused by ArMV infection (Murant, 1970). Arabis mosaic virus is also associated with nettlehead disease of hops (Bock, 1966).

The importance of ArMV is probably greatest in the long-term high-value crops such as raspberry and hops. Infected raspberry plants produce little or no fruit (Harrison, 1958), while hop bines infected with nettlehead disease have shortened internodes and do not twist normally.

The importance of weed reservoirs of ArMV is also greatest in such high-value crops in which certified virus-free propagating material is often used to plant fresh or fallowed areas. There is little point in extra expenditure on such stock materials unless the sources of the virus are removed. Arabis mosaic virus is obviously widespread in weeds because the yellow dwarf disease frequently occurs in fields planted with raspberry for the first time (Harrison, 1958).

Arabis mosaic virus has been found in plantains from raspberry fields by Harrison (1958), and has been shown to be seed transmitted in *P. major* and other weeds (Lister and Murant, 1967). Plantains and other weed hosts should be thoroughly removed from proposed planting sites and prevented from reestablishment long enough to exceed the viability of ArMV in the nematode vector, *Xiphinema diversicaudatum,* if the virus is to be eradicated from a site.

2. *Tobacco Ringspot Virus (TRSV)*

Originally confined to North America, TRSV has now been spread to other parts of the world, probably in ornamental crops (Stace-Smith, 1970a). Among the diseases caused by TRSV are ringspot diseases of tobacco, cucumber, and many ornamentals, soybean bud blight, blueberry necrotic ringspot, and chlorotic or necrotic spotting of other crops; of these, soybean bud blight is the most damaging (Stace-Smith, 1970a). Tobacco ringspot virus is also a problem in grapevine (Gilmer *et al.,* 1970) and cherry (Uyemoto *et al.,* 1977).

The most common vector of TRSV is the nematode *Xiphinema americanum,* but there are also reports of transmission by species of thrips, spider mites, grasshoppers, and flea beetles (Stace-Smith, 1970a). The nematode is apparently not the major factor in the spread of soybean bud blight, and while seed transmission occurs in soybeans and other species, this is not the cause of major outbreaks of bud blight (Stace-Smith, 1970a).

Tobacco ringspot virus has been isolated from *P. major* in vineyards affected by severe decline (Uyemoto, 1975), and is commonly found in *P. major* in vineyards and orchards in New York state, where it is a potential source of crop infection (R. Provvidenti, personal communication). There are also many other hosts of TRSV that may be more important in other situations.

3. Tomato Ringspot Virus (TomRSV)

Many of the diseases caused by TRSV may also be caused by TomRSV. The major diseases are mosaic and ringspot of tobacco, raspberry, blackberry, and various ornamentals; rasp leaf of cherry, yellow bud mosaic of peach, yellow vein in grapevine, and stunt in *Gladiolus* (Stace-Smith, 1970b). Stem pitting of peach (Smith *et al.*, 1973) and union necrosis in apple (Parish and Converse, 1981) have been associated with TomRSV infection, as has a grapevine decline in New York which is also caused by TRSV (Uyemoto, 1975).

Soybean bud blight, the major disease caused by TRSV, may also be caused by TomRSV infection (Sinclair and Shurtleff, 1975).

Tomato ringspot virus has been isolated from *P. lanceolata* from vineyards in New York (Uyemoto, 1975) and is commonly found in both vineyards and orchards there in *P. major,* which is a potential source of infection for the crops (R. Provvidenti, personal communication). Frazier *et al.* (1961) found TomRSV common along the California coast in *Plantago* sp. and several other weed species. Tomato ringspot virus shares a common vector with TRSV, *Xiphinema americanum,* and poses a similar problem in many crops. Removal of *Plantago* and other susceptible weed species would be helpful in preventing the infection of new vineyards and orchards, as TomRSV and TRSV are common in their weed hosts.

M. Tobacco Necrosis Virus Group

Plantain Virus 5 (PlV5)

This virus was isolated only from *P. lanceolata* (Hammond, 1981) and is not known to infect any agricultural crops naturally. No vector was identified.

N. Tombusviruses

1. Tomato Bushy Stunt Virus (TBSV)

The name of TBSV describes the symptoms of the disease incited in tomato. Other diseases caused by TBSV are mottle, severe deformation, reduced growth, and sterility of globe artichoke (*Cynara scolymus*); distortion and yellow stellate spots in *Pelargonium;* asteroid mosaic in *Petunia;* fruit pitting, veinal necrosis, and stunting of cherry; and chlorotic spots and rings in carnation (Martelli *et al.,* 1971). No vector is known, but at least one strain is soil transmitted, and the vector may be a chytrid fungus (Lovisolo *et al.,* 1965).

Plantago major has been reported to be infected by TBSV, although the virus could be detected only in the roots (Lovisolo *et al.,* 1965). Smith (1957) suggested that TBSV could be harbored by some wild host plants in hedgerows. The fact that TBSV is naturally restricted to the roots of several species, and becomes systemic in few (Lovisolo *et al.,* 1965), may have hindered attempts to elucidate the epidemiology of the virus. Little is known of the importance of alternate hosts of TBSV, and hence of the relative importance of *P. major.*

2. *Plantain Virus 6 (PlV6)*

Originally isolated from *P. lanceolata,* PlV6 infected french bean and celery symptomlessly under experimental conditions (Hammond, 1981). No vector and no agricultural significance is known.

O. *Tymoviruses*

Plantago Mottle Virus (PlMV)

Isolated from *P. major* (Granett, 1972), PlMV was shortly afterward discovered infecting peas in New York state (Granett, 1973; Provvidenti and Granett, 1976). Although PlMV infects several other species experimentally (Granett, 1973), the only known disease of any agricultural importance is in pea, in which it causes systemic mottle, leaf deformation, and necrosis (Provvidenti and Granett, 1976).

No vector of PlMV was identified, but it is presumed that flea beetles or similar insects are responsible. *Plantago major* is evidently an important reservoir, as no other is known (Provvidenti and Granett, 1976).

P. *Other Isometric Viruses*

Broad Bean Wilt Virus (BBWV)

The diseases associated with BBWV infection are wilt of broad beans, streak of peas, blight of spinach, and ringspot in *Nasturtium* and *Petunia,* besides infections of a wide range of other plants (Taylor and Stubbs, 1972). Broad bean wilt virus has a number of aphid vectors of varying efficiency, of which *M. persicae* is the most efficient (Taylor and Stubbs, 1972).

Broad bean wilt virus is widespread, and natural infections of *Plantago* species have been recorded from the United States, Poland, Argentina, and Australia (Table I). However, there are also many other hosts of the virus (Taylor and Stubbs, 1972). *Plantago lanceolata* is a perennial, and thus may be more important than some of the other host plants, particularly as some of the aphid vectors of BBWV have been found to overwinter on *P. lanceolata* (Jadot *et al.*, 1969).

Q. Rhabdoviruses

1. Plantain Virus A (PlVA)

Plantain virus A is known to occur naturally only in *P. lanceolata* (Hitchborn *et al.*, 1966; Hammond, 1981), and the only other known host is *N. benthamiana* (Hammond, 1981). It was transmitted experimentally by mechanical inoculation and by *Myzus* sp., but the natural vector was not established (Hammond, 1981). No agricultural importance is known.

2. Plantain Virus B (PlVB)

Plantain virus B was reported only from *P. lanceolata*, with no transmission or vector data available (Hitchborn *et al.*, 1966). It is of no known agricultural significance.

R. Other Bacilliform Viruses

Alfalfa Mosaic Virus (AMV)

Like TMV and CMV, AMV is very common around the world and has many hosts. Alfalfa mosaic virus causes variable symptoms in alfalfa; potato calico disease and tuber necrosis; mosaic and other diseases of several leguminous crops, celery, lettuce, and many ornamentals; and severe necrosis in tomatoes, among other diseases (Jaspars and Bos, 1980).

The virus is transmitted by many species of aphid in a nonpersistent manner (Hull, 1969) and through the seed of a variety of plants, which may play a major role in crop disease epidemiology (Van Regenmortel and Pinck, 1981).

In view of the wide host range and the frequency of seed transmission, the role of plantains in the spread of AMV is probably negligible, but may be of minor importance locally.

VI. DISCUSSION AND CONCLUSIONS

Plantago species have been shown by many workers to be susceptible to a wide range of viruses (Table I). Many of these viruses have been shown to infect *Plantago* species naturally in field conditions, and plantains have been implicated as sources of infection of crop plants in several instances. The two most common species of plantain, *P. lanceolata* and *P. major,* are perennials, and several virus vectors have been found to overwinter, or to reproduce, on plantains. Thus *Plantago* is a source of both vectors and viruses affecting crop plants.

The reports of viruses in *Plantago* come from many parts of the world, reflecting the wide distribution of the genus, especially *P. lanceolata* and *P. major.* The widespread nature of plantains as successful weeds (Holm *et al.*, 1977; Allard, 1965) and the large number of viruses known to infect them make *Plantago* species an excellent model for the importance of weeds in general as reservoirs of virus diseases.

Duffus (1971) has stated that weeds serve as the main reservoir of vectors of plant viruses in many agricultural areas, and that where a weed is host of both virus and vector the virus spread may be very severe. Weeds will be most important where they sustain the vector and the virus in the absence of the crop plant—either between harvest and planting in the case of annual crops, or for longer periods in the case of perennial crops such as hops or grapes, where the land is usually fallowed before replanting. Plantains were shown to be overwintering hosts of several aphid species (Heathcote *et al.*, 1965; Jadot *et al.*, 1969). Among the viruses transmitted by one or more of these vectors are AMV, BBWV, BMYV, BYV, CMV, PVY, and TuMV (Kennedy *et al.*, 1962), all of which have been found to occur naturally in plantains (Table I). If vector and virus simultaneously overwinter in *Plantago* then the potential for severe spread (Duffus, 1971) is clear. Perennial weeds will be of most importance for virus and vector overwintering, and in long-term virus survival, those weeds in which virus is seed transmitted will be especially important. For example, Murant and Lister (1967) showed that raspberry ringspot virus (RRV) and tomato black ring virus (TBRV) were rapidly lost from soils fallowed overwinter or kept weed-free, but that when weed seeds in the soil were allowed to germinate, vector nematodes acquired the viruses from seedlings arising from infected seed. Thus it appears that seed dissemination of these viruses, rather than movement of viruliferous nematodes, is the reason for such a large proportion of the vector population in Scotland carrying RRV and TBRV (Murant and Lister, 1967).

Viruses with a wide host range have a selective advantage in situations in which some of the hosts may be annuals, or removed by cultivation, or susceptible to adverse climatic conditions. Thus some viruses such as tobacco mosaic virus and cucumber mosaic virus are found worldwide largely because of host and host–habitat diversity. Conversely, some viruses are very limited in their host ranges, and appear to be very closely adapted to their natural hosts, often causing no obvious signs of disease. Such specialized relationships tend to be geographically restricted, and are more commensal than parasitic in nature.

Virus survival is also influenced by factors such as stability, productivity, and transmissibility of the virus, and by the effects on the host plants. The survival and reproduction of the host are obviously advantageous for further virus spread. The distribution of the host is also important; in general, individuals in close proximity are more likely to become infected than those further away, whether by mechanical transmission or via a specific vector, some of which (e.g., nematodes and fungi) are of limited mobility. The wide distribution of perennial weeds such as *Plantago* species may contribute to the survival of many economically important viruses.

Weed control rather than vector control may be more beneficial in prevention of virus spread because of the importance of weeds and weed seeds in the spread and persistence of some of the nematode-transmitted viruses (Cadman, 1963). Among these viruses is ArMV, which is associated with several diseases of high-value perennial crops, including nettlehead disease of hops and yellow dwarf of raspberry. Weeds such as plantains are potential reservoirs of ArMV, and infected plantains were found near diseased raspberries by Harrison (1958), while Lister and Murant (1967) showed that ArMV can be seedborne in *P. major*.

The actual importance of *Plantago* species, or of any weeds, as reservoirs of viruses will vary from season to season, and from one location to another. Factors that affect the relative importance include cultural practices, which influence the survival of plantains and other weed sources; the numbers of vectors, the timing of their migrations, and the host plants available to them (including *Plantago*); climatic factors; previous cropping history and hence exposure of the plantain population to the viruses from the crop, rather than vice versa; the virus strain; and the proportions of each susceptible weed source actually infected. Sometimes plantains may be locally very important sources, and other times they may be negligible sources of virus for the same crop. Thus in Belgium, *P. lanceolata* may be one of the most important overwintering hosts of BMYV (Roland *et al.*, 1968) and its vectors (Jadot *et al.*, 1969), while Russell (1965) failed to recover BMYV from the same species, and *P. lanceolata* is not known to be an important host in England. Whether such differences are due to local differences in the virus, or to differing susceptibility in *P. lanceolata* is not known.

The fact remains that plantains are susceptible to a large number of plant viruses, and are potential reservoirs of many of them. Where the epidemiology of a disease suggests that weeds may be the major source of virus infection, and especially where the virus is thought to over-winter in a weed host, *Plantago* species should certainly be examined as a possible source of the virus. As infection of plantains is often also symptomless and easily overlooked on visual inspection, a detailed examination including transmission studies may be necessary.

Many plant viruses perhaps should be considered as pathogens of wild plants that become important to man when crops are grown in their vicinity (Duffus, 1971). Harrison (1964) noted that nearly all nematode-transmitted viruses and their vectors have extensive weed or wild plant host ranges that ensure their survival where immune crops are grown, and suggested that they become important when land that is already infested is brought under cultivation. Evidence support-ing this view was provided by Bennett (1952) who cites examples where the introduction of crops to areas in which they had not formerly been grown had in many instances resulted in attack by previously un-known viruses. During the 30 years after 1920, some 20 virus diseases affecting many different hosts were described as new viruses, and many are now of major concern; it is possible that many of these had previously been present in crop plants, but were undiscovered. This is most likely with those producing relatively minor disease. However, it is unlikely that most of these viruses were commonly present in crops but previously undetected. Many of these diseases were also initially of only local importance, and have apparently spread from a single area. Assuming *de novo* origin of so many viruses to be extremely unlikely, a more likely explanation is the increased dissemination of viruses that have existed as localized entities for long periods, and it is significant that many of these viruses were found in the developing agricultural areas of the world (Bennett, 1952). Bennett (1952) further suggested that the most logical original source of such viruses is native unculti-vated plants, and that these viruses had only local distribution before the era of modern agricultural expansion.

Another factor in the increased importance of virus diseases is undoubtedly the increase in monoculture and uniformity of plant varieties. The increased tendency in plant breeding toward genetic uniformity and maximization of agronomic performance has led to large areas of pure cultures grown at minimal spacing, with the same crop sometimes grown in the same field in successive years. This situation is obviously more favorable for the effective transmission of viruses by arthropod vectors than that in scattered and mixed natural communities of wild plants (Holmes, 1954). As the plant breeders develop varieties that are resistant to the currently known diseases, it is almost inevitable that new problems will arise that have been obscured by the symptoms of other diseases, or have been relatively minor. These will take on more importance as effective resistance to the current problems is introduced, and the demand for increased agricultural production continues.

With the domestication of new species and the expansion of the growing areas of those already cultivated, the balance between viruses and their natural host plants is drastically altered—plants grown in new areas come into contact with viruses to which they have no resistance, and viruses, after transfer from indigenous to cultivated hosts, are often then transported to other areas as new virus diseases (Bennett, 1952) and may attack further crops. Crop plants and weeds raised in a single area over a long period of time generally develop resistance or tolerance to the endemic viruses that enables the plants to compete more favorably with neighboring species (Holmes, 1954). Thus the weeds from which "new" viruses diseases arise are usually little affected, but the same virus may cause major damage in a crop situation because of the lack of resistance and through the close proximity of so many plants of genetically similar makeup. During the course of a few years encounter between a virus and a new host plant, the virus may adapt to the host, and a more damaging strain may arise which more readily, or more extensively invades the new host.

Distribution of the vectors of viruses endemic to wild plants may also be limited, but may be greatly changed by introduction of a new crop plant to which the vector is also suited, or by introduction of the vector to a new area (Bennett, 1952). In such instances the geographical range of the vector may be much enlarged, and the vector may come into contact with other viruses it is also able to transmit, causing increased or new crop virus diseases. Adaptation of either vector or virus to a new crop host may lead to increasingly severe outbreaks of "new" virus diseases.

The frequency of infection of plantains surveyed in Great Britain, and the frequent lack of obvious detrimental effects on the plants even when affected by as many as four viruses (Hammond, 1981), suggest that plantains have developed a high tolerance to most of their endemic viruses. This would suggest a long association with those viruses (Holmes, 1954) and their vectors (Matthews, 1970). Weeds may indeed be the preferred host of the vectors, and crop plants may be attacked only when the vectors cannot find suitable weed hosts. This was suggested in the case of TRV affecting potatoes, which may be a less attractive host to the nematodes than the weeds (Cooper and Harrison, 1973).

Control of weed plants in such situations may lead to increased vector activity on the crop plant. However, there are some situations in which control of weed hosts of virus diseases is likely to be much more effective than others, and especially with nematode-transmitted viruses. High-value horticultural crops such as raspberry, hop, and many ornamentals are particularly likely to suffer yield losses from virus infection. The fruit yield of ArMV-infected raspberry, or nettlehead-diseased hop, for example, is significantly reduced, while both flower yield and quality are lowered by virus infection of many ornamentals. One method of control is to reduce initial infection of the crop as much as possible by planting only certified virus-free material, which is usually considerably more expensive than nontested stocks. However, if weed sources of the virus are present in or around the fields the crop may not remain virus-free long enough to recoup the expense of the certified stocks. Control of the vectors is also important, and whereas fallowing or the growing of nonsusceptible crops may reduce the proportion of viruliferous nematodes, the virus may persist in weed hosts or even spread through the seed of susceptible weeds, thus increasing the availability of the virus to the vector. Integrated control is therefore necessary.

Weed control may be of some importance in control of aphid-borne viruses of annual crops as well as of nematode-transmitted diseases, but with aphid-borne viruses the reservoirs may be beyond the boundaries of the farmer trying to control the disease, and control measures within the boundaries may therefore be of only limited value. Nonpersistent viruses may be more effectively controlled by this means than the persistently transmitted viruses, but there are many factors that affect the distance a vector will fly from one plant to another, and, in general, it is only possible to say that increasing the separation of the crop from the source of the virus is likely to reduce crop infection.

In conclusion, it seems probable that there are still many viruses infecting wild plants that may eventually cause losses in economic crops (Bennett, 1952), and quite likely that *Plantago* species will be the source of some of these. An example that has occurred in recent years is that of PlMV in peas (Provvidenti and Granett, 1976). *Plantago* mottle virus was originally isolated from *P. major* (Granett, 1972) and was shortly afterward detected in a pea crop (Granett, 1973). Currently of little importance in peas, and known only in New York state (Provvidenti and Granett, 1976), PlMV may assume greater importance as the other virus diseases of pea are lessened in impact by breeding for resistance (Granett, 1973), although resistance to PlMV is already being sought (Provvidenti, 1977, 1979).

Thus *Plantago* species are already known to be hosts of a wide range of economically important viruses, and to a number of viruses that may become important in the future, and are probably also host to many other viruses that remain to be detected. It will be of interest to see if this is so with other weed plants not as yet studied in such detail.

REFERENCES

Akhatova, F. K. H., Eliseeva, Z. N., and Katin, I. A. (1979). *Vestn. Skh. Nauk. Kaz.* **4,** 36–39.
Allard, R. (1965). *In* "The Genetics of Colonizing Species" (H. Baker and G. Stebbins, eds.), pp. 49–75. Academic Press, New York.
Anderson, C. W. (1959). *Phytopathology* **49,** 97–101.
Beczner, L., and Schmelzer, K. (1972). *Acta Phytopathol. Acad. Sci. Hung.* **7,** 377–382.
Beiss, U. (1956). *Phytopathol. Z.* **27,** 83–106.
Bennett, C. W. (1944). *Phytopathology* **34,** 77–91.
Bennett, C. W. (1952). *Plant Dis. Rep. Suppl.* **211,** 43–46.
Bennett, C. W. (1960). *U.S. Dept. Agric. Tech. Bull.* No. 1218.
Bennett, C. W., and Costa, A. S. (1954). *J. Am. Soc. Sugar Beet Technol.* **8,** 230–235.
Bennett, C. W., and Tanrisever, A. (1959). *J. Am. Soc. Sugar Beet Technol.* **10,** 189–211.
Bidwill, J. C. (1841). "Rambles in New Zealand." London.
Blaszcak, W. (1976). *Rocz. Nauk Roln. Ser. E* **6,** 89–95.
Bock, K. R. (1966). *Ann. Appl. Biol.* **57,** 131–140.
Boyle, J. S., and Wharton, D. C. (1957). *Phytopathology* **47,** 199–207.
Brenchley, W. E. (1920). "Weeds of Farmland." Longmans, Green, London.
Cadman, C. H. (1963). *Ann. Rev. Phytopathol.* **1,** 143–172.
Cadman, C. H., Chambers, J., and Fisken, A. G. (1954). *Rep. Scott. Agric. Res. Inst.* for 1953–1954, p. 18.
Champness, S. S., and Morris, K. (1948). *J. Ecol.* **36,** 149–173.
Cheeseman, T. F. (1906). "Manual of the New Zealand Flora." Wellington, New Zealand.
Chessin, M., Juretić, N., Miličić, D., Perryman, J., and Giri, L. (1980). *Phytopathol. Z.* **97,** 295–301.
Chippendale, H. G., and Milton, W. E. J. (1934). *J. Ecol.* **22,** 508–531.
Cooper, J. I., and Harrison, B. D. (1973). *Ann. Appl. Biol.* **73,** 53–66.
Costa, A. S., and Neto, V. da C. L. (1976). *Congr. Soc. Bras. Fitopathol., 9th.*
Crocker, W. J. (1938). *Bot. Rev.* **4,** 235–274.
Cropley, R. (1961). *Ann. Appl. Biol.* **49,** 524–529.
Dodds, J. A., and Taylor, G. S. (1980). *Plant Dis.* **64,** 294–296.

Duffus, J. E. (1971). *Annu. Rev. Phytopathol.* **9**, 319–340.
Duffus, J. E. (1972). *CMI/AAB Descript. Plant Viruses* No. 89.
Duffus, J. E. (1973). *Adv. Virus Res.* **18**, 347–386.
Edwardson, J. R. (1974). *Fl. Agric. Exp. St. Monogr. Ser.* No. 4.
Frazier, N. W., and Posnette, A. F. (1957). *Ann. Appl. Biol.* **45**, 580–588.
Frazier, N. W., and Thomas, H. E. (1953). *Plant Dis. Rep.* **37**, 272–275.
Frazier, N. W., Yarwood, C. E., and Gold, A. H. (1961). *Plant Dis. Rep.* **45**, 649–651.
Freerksen, E. (1950). *Naturwissenschaften* **37**, 564–565.
Fryer, J. D., and Evans, D. J., eds. (1968). "Weed Control Handbook. Vol. 1. Principles," 5th Ed. Blackwell, Oxford.
Fulton, R. W. (1948). *Phytopathology* **38**, 421–428.
Fulton, R. W. (1957). *Phytopathology* **47**, 215–220.
Fulton, R. W. (1971). *CMI/AAB Descript. Plant Viruses* No. 44.
Fulton, R. W. (1981). *In* "Handbook of Plant Virus Infections" (E. Kurstak, ed.), pp. 377–413. Elsevier, Amsterdam.
Gerarde, J. (1636). "The Herball; or General Historie of Plants [New Ed.] very much enlarged and amended by Thomas Johnson." Norton & Whittakers, London.
Gibbs, A. J., and Harrison, B. D. (1964). *Plant Pathol.* **13**, 144–150.
Giddings, N. J. (1938). *J. Agric. Res.* **56**, 883–894.
Giddings, N. J. (1944). *J. Agric. Res.* **69**, 149–157.
Gill, N. T., and Vear, K. C. (1966). "Agricultural Botany," 2nd Ed. Duckworth, London.
Gilmer, R. M., Uyemoto, J. K., and Kelts, L. J. (1970). *Phytopathology* **60**, 619–627.
Goldin, M. I. (1953). *Dokl. Acad. Nauk SSSR* **88**, 933–935.
Good, R. (1964). "The Geography of the Flowering Plants," 3rd Ed. Wiley, New York.
Goryshin, V. A. (1959). *Proc. Lenin Acad. Agric. Sci. USSR* **24**, 31–36.
Gracia, O., and Feldman, J. M. (1976). *Phytopathol. Z.* **85**, 227–236.
Granett, A. L. (1972). *Phytopathology* **62**, 761 (Abstr.).
Granett, A. L. (1973). *Phytopathology* **63**, 1313–1316.
Grant, K. (1848). "Hooper's Medical Dictionary," 8th Ed. Longman, Green, London.
Guppy, H. B. (1906). "Observation of a Naturalist in the Pacific: II. Plant Distribution." MacMillan, London.
Hammond, J. (1981). *Plant Pathol.* **30**, 237–243.
Hammond, J., and Hull, R. (1981). *J. Gen. Virol.* **54**, 75–90.
Harrison, B. D. (1956). *Plant Pathol.* **5**, 147–148.
Harrison, B. D. (1958). *Ann. Appl. Biol.* **46**, 221–229.
Harrison, B. D. (1964). *In* "Plant Virology" (M. K. Corbett and H. D. Sisler, eds.), pp. 118–147. Univ. of Florida Press, Gainesville.
Harrison, B. D., and Robinson, D. J. (1978). *Adv. Virus Res.* **23**, 25–77.
Haughton, C. S. (1978). "Green Immigrants: The Plants that Transformed America." Harcourt, New York.
Heathcote, G. D., and Cockbain, A. J. (1964). *Ann. Appl. Biol.* **52**, 259–266.
Heathcote, G. D., Dunning, R. A., and Wolfe, M. D. (1965). *Plant Pathol.* **14**, 1–10.
Helbaek, H. (1950). *Arboger Nord. Oldkyndighed Hist.*
Helbaek, H. (1958). *Kuml Arbog Tusk Arkael. Selskab* 83–116.
Hill, A. F. (1952). "Economic Botany," 2nd Ed. McGraw-Hill, New York.
Hitchborn, J. H., and Hills, G. J. (1965). *Virology* **27**, 528–540.
Hitchborn, J. H., Hills, G. J., and Hull, R. (1966). *Virology* **28**, 768–772.
Hollings, M. (1958). *J. Hortic. Sci.* **33**, 275–281.
Hollings, M. (1959). *Ann. Appl. Biol.* **47**, 98–108.
Holm, L. G., Plucknett, D. L., Pancho, J. V., and Herberger, J. P. (1977). "The World's Worst Weeds: Distribution and Biology." University Press of Hawaii, Honolulu.

Holmes, F. O. (1938). *Phytopathology* **28**, 58–66.
Holmes, F. O. (1941). *Phytopathology* **31**, 1089–1098.
Holmes, F. O. (1950). *Phytopathology* **40**, 487–492.
Holmes, F. O. (1954). *Adv. Virus Res.* **2**, 1–30.
Hooker, J. D. (1867). "Handbook of the New Zealand Flora." London.
Hull, R. (1952). *J. R. Agric. Soc. Engl.* **113**, 86–102.
Hull, R. (1953). *Plant Pathol.* **2**, 39–43.
Hull, R. (1969). *Adv. Virus Res.* **15**, 365–433.
Jadot, R. (1974). *Parasitica* **30**, 37–44.
Jadot, R., Roland, G., and Riga, A. (1969). *Parasitica* **25**, 97–108.
Jaspars, E. M. J., and Bos, L. (1980). *CMI/AAB Descript. Plant Viruses* No. 229.
Juretić, N., Wrischer, M., and Polák, Z. (1969). *Biol. Plant* **11**, 284–290.
Kaper, J. M., and Waterworth, H. E. (1977). *Science* **196**, 429–431.
Kennedy, J. S., Day, M. F., and Eastop, V. F. (1962). "A Conspectus of Aphids as Vectors of Plant Viruses." Commonwealth Agricultural Bureau, Farnham Royal.
Komuro, Y., Yoshino, M., and Ichinohe, M. (1970). *Ann. Phytopathol. Soc. Jpn.* **36**, 17–26.
Kostin, U. D., and Volkov, Yu. G. (1976). *Tr. Biol. Poch. Inst. Dalnevost. Nauchn. Tsentr. Acad. Nauk SSSR* **25**(128), 205–210.
Kovachevsky, I. C. (1963). *Phytopathol. Z.* **49**, 127–146.
Kristensen, H. R., and Engsbro, B. (1966). *Tidsskr. Planteavl.* **70**, 353–379.
Książek, D. (1976). *Zesz. Probl. Postępow Nauk Roln.* **182**, 173–179.
Kunkel, L. O. (1928). *Phytopathology* **18**, 156 (Abstr.).
Leyel, C. F., ed. (1977). "A Modern Herbal, by Mrs. M. Grieve." Jonathan Cape, London.
Linn, M. B. (1940). *Bull. N.Y. Agric. Exp. Stn. (Ithaca)* **742**, 33.
Lister, R. M., and Murant, A. F. (1967). *Ann. Appl. Biol.* **59**, 49–62.
Lovisolo, O., Bode, O., and Volk, J. (1965). *Phytopathol. Z.* **53**, 323–342.
MacClement, W. D., and Richards, M. G. (1956). *Can. J. Bot.* **34**, 793–799.
McKinney, H. H. (1952). *Plant Dis. Rep.* **36**, 184–187.
Martelli, G. P., Quacquarelli, A., and Russo, M. (1971). *CMI/AAB Descript. Plant Viruses* No. 69.
Matthews, R. E. F. (1970). "Plant Virology." Academic Press, New York.
Miličić, D. (1968). *Naturwissenschaften* **55**, 90–91.
Milton, W. E. J. (1936). *Bull. Welsh Plant Breed. Stn. Ser. H* **14**, 58–86.
Mulligan, G. A., and Findlay, J. (1970). *Can. J. Bot.* **48**, 859–860.
Mumford, D. L. (1974). *Phytopathology* **64**, 136–139.
Murant, A. F. (1970). *CMI/AAB Descript. Plant Viruses* No. 16.
Murant, A. F., and Lister, R. M. (1967). *Ann. Appl. Biol.* **59**, 63–76.
Neergaard, P. (1977). "Seed Pathology," Vol. I. Wiley, New York.
Oshima, N., and Harrison, B. D. (1975). *CMI/AAB Descript. Plant Viruses* No. 152.
Oshima, N., Ohashi, Y., and Umekawa, M. (1971). *Ann. Phytopathol. Soc. Jpn.* **37**, 319–325.
Oshima, N., Ohashi, Y., and Umekawa, M. (1974). *Ann. Phytopathol. Soc. Jpn.* **40**, 243–251.
Parish, C. L., and Converse, R. H. (1981). *Plant Dis.* **65**, 261–263.
Polák, Z., and Chlumská, J. (1980). *Sb. UVTIZ Ochr. Rostl.* **16**, 101–104.
Polák, Z., Králík, O., and Čech, M. (1968). *Biol. Plant* **10**, 31–36.
Price, W. C. (1940). *Am. J. Bot.* **27**, 530–541.
Procházková, Z. (1977). *Ochr. Rostl.* **13**, 189–195.
Provvidenti, R. (1977). *Plant Dis. Rep.* **61**, 851–855.
Provvidenti, R. (1979). *J. Hered.* **70**, 350–351.

Provvidenti, R., and Granett, A. L. (1976). *Ann. Appl. Biol.* **82**, 85–89.
Quiot, J. B., Marchoux, G., Douine, L., and Vigouroux, A. (1979). *Ann. Phytopathol.* **11**, 325–348.
Reed, C. F. (1970). *U.S. Dept. Agric. Handbook* **366**, 346.
Ridley, H. N. (1930). "The Dispersal of Plants Throughout the World." Reeve, Ashford.
Roland, G. (1957). *Parasitica* **13**, 135–143.
Roland, G., and Tahon, J. (1961). *Rev. Agric. (Brussels)* **14**, 869–895.
Roland, G., Jadot, R., and Riga, A. (1968). *Parasitica* **24**, 121–128.
Rowhani, A., and Peterson, J. F. (1980). *Can. J. Plant Pathol.* **2**, 12–18.
Rush, M. C. (1970). *Phytopathology* **60**, 917–918.
Russell, G. E. (1958). *Ann. Appl. Biol.* **46**, 393–398.
Russell, G. E. (1960). *Ann. Appl. Biol.* **48**, 721–728.
Russell, G. E. (1963). *Ann. Appl. Biol.* **52**, 405–413.
Russell, G. E. (1965). *Ann. Appl. Biol.* **55**, 245–252.
Russell, P. G., and Musil, A. F. (1961). *U.S. Dept. Agric. Yearbook,* pp. 80–88.
Ryjkoff, V. L. (1943). *Dokl. Acad. Nauk SSSR NS* **41**, 94–96.
Sagar, G. R., and Harper, J. (1964). *J. Ecol.* **52**, 189–221.
Salisbury, E. (1961). "Weeds and Aliens." Collins, London.
Schmelzer, K. (1957). *Phytopathol. Z.* **30**, 281–314.
Schmelzer, K. (1960). *Z. Pflanzenkr. (Pflanzenpathol.) Pflanzenschutz* **67**, 193–210.
Schmelzer, K. (1963a). *Phytopathol. Z.* **46**, 17–52.
Schmelzer, K. (1963b). *Phytopathol. Z.* **46**, 105–138.
Schmelzer, K. (1963c). *Phytopathol. Z.* **46**, 235–268.
Schmelzer, K. (1966). *Phytopathol. Z.* **55**, 317–351.
Schmelzer, K. (1969). *Phytopathol. Z.* **64**, 39–67.
Schmelzer, K. (1975). *Zentralbl. Bakteriol. Parasitenkd. Infektionskr. Hyg. Abt. 2* **130**, 232–233.
Schmelzer, K., Gippert, R., Weissenfels, M., and Beczner, L. (1975). *Zentralbl. Bakteriol. Parasitenkd. Infektionskr. Hyg. Abt. 2* **130**, 696–703.
Schumann, K. (1963). *Phytopathol. Z.* **48**, 135–148.
Severin, H. H. P. (1929). *Hilgardia* **3**, 543–582.
Severin, H. H. P. (1934). *Hilgardia* **8**, 263–280.
Shepherd, R. J. (1976). *Adv. Virus Res.* **20**, 305–339.
Sibilia, C. (1948). *Bull. Stag. Patol. Veg.* **6**, 193–194.
Sinclair, J. B., and Shurtleff, M. C., eds. (1975). "Compendium of Soybean Diseases." American Phytopathological Society, St. Paul, Minnesota.
Smith, K. M. (1932). *Ann. Appl. Biol.* **19**, 305–330.
Smith, K. M. (1937). *Parasitology* **29**, 70–85.
Smith, K. M. (1957). "A Textbook of Plant Virus Diseases," 2nd Ed. Churchill, London.
Smith, K. M. (1972). "A Textbook of Plant Virus Diseases," 3rd Ed. Academic Press, New York.
Smith, S. H., Stouffer, R. F., and Soulen, D. M. (1973). *Phytopathology* **63**, 1404–1406.
Stace-Smith, R. (1970a). *CMI/AAB Descript. Plant Viruses* No. 17.
Stace-Smith, R. (1970b). *CMI/AAB Descript. Plant Viruses* No. 18.
Staniulis, J., and Genyte, L. (1976). *Phytopathol. Z.* **86**, 240–245.
Tamrazyan, L. G., Protsenko, A. E., and Shvedchikova, N. G. (1977). *Biol. Zh. Arm.* **30**, 29–32.
Taylor, R. H., and Stubbs, L. L. (1972). *CMI/AAB Descript. Plant Viruses* No. 81.
Taylor, R. H., Smith, P. R., Reinganum, C., and Gibbs, A. J. (1968). *Aust. J. Biol. Sci.* **21**, 929–935.

Thomas, P. E., and Mink, G. I. (1979). *CMI/AAB Descript. Plant Viruses* No. 210.

Thomas, W. D. (1949). *J. Colo.-Wyo. Acad. Sci.* **4,** 40.

Thomson, G. M. (1922). "The Naturalization of Animals and Plants in New Zealand." Cambridge Univ. Press, London and New York.

Tomlinson, J. A. (1970). *CMI/AAB Descript. Plant Viruses* No. 8.

Tomlinson, J. A., and Carter, A. L. (1970). *Ann. Appl. Biol.* **66,** 381–386.

Tomlinson, J. A., and Ward, C. M. (1978). *Ann. Appl. Biol.* **89,** 61–69.

Tomlinson, J. A., Carter, A. L., Dale, W. T., and Simpson, C. J. (1970). *Ann. Appl. Biol.* **66,** 11–16.

Uschraweit, H. A., and Valentin, H. (1956). *Nachrichtenbl. Dtsch. Pflanzenschutzdienstes (Braunschweig)* **8,** 132–133.

Uyemoto, J. K. (1975). *Plant Dis. Rep.* **59,** 98–101.

Uyemoto, J. K., and Provvidenti, R. (1974). *Phytopathology* **64,** 1547–1548.

Uyemoto, J. K., Welsh, M. F., and Williams, E. (1977). *Phytopathology* **67,** 439–441.

Valleau, W. D., and Johnson, E. M. (1943). *Phytopathology* **33,** 210–219.

Van Regenmortel, M. H. V., and Pinck, L. (1981). *In* "Handbook of Plant Virus Infections" (E. Kurstak, ed.), pp. 415–421. Elsevier, Amsterdam.

Villalón, B. (1981). *Plant Dis.* **65,** 557–562.

Walkey, D. G. A., and Webb, M. J. W. (1978). *Ann. Appl. Biol.* **89,** 435–441.

Wang, J.-L. Chai, H.-M., Niou, A.-H., Chang, L.-C., and Chang, Y.-M. (1965). *Acta Phytolac. Sin.* **4,** 355–360.

Weisner, K. (1962). *Nachrichtenbl. Dtsch. Pflanzenschutzdienst. (Berlin)* **16,** 45–53.

Wilson, C. M. (1945). "New Crops for the New World." MacMillan, New York.

ADVANCES IN VIRUS RESEARCH, VOL. 27

PENETRATION OF VIRAL GENETIC MATERIAL INTO HOST CELL

A. G. Bukrinskaya

The D. I. Ivanovsky Institute of Virology
Academy of Medical Sciences
Moscow, USSR

I. Introduction

The early events in animal virus-infected cells can be conveniently divided into two phases. During the first phase, the virus is internalized by the host cell and modified in such a way that it becomes capable of initiating infection. This first phase includes adsorption of the virus particle by the host cell plasma membrane, penetration into the cell, and uncoating, processes which are aimed at localizing the virus in the proper cellular compartment and removing its outer protective structures. Once the virus envelope and protective proteins are removed and the virus inner component, which is able to initiate infection, is liberated, the second phase, during which viral genes begin to be expressed, begins.

In contrast to the detailed knowledge available on the later phase of viral infection (e.g., transcription and replication of the viral genome and the synthesis of the viral structural components), little is known about the first, initial phase. These steps of early virus–cell interaction represent one of the more complicated and difficult fields of study in contemporary experimental virology. Although much is known about how some bacterial viruses enter the host cell and initiate infection, the precise mechanisms of their entry have not yet been fully elucidated.

The disadvantages in treating this problem involve the following: (1) the high rate of virus penetration and uncoating; (2) the short half-life of the uncoating intermediates; (3) the asynchronous character of the process; and (4) the small amount of input virus undergoing modification. Conclusions concerning many points of virus attachment, penetration, and uncoating are often based on electron microscopic observations which, however, cannot distinguish between infectious and noninfectious virus particles. In the best of circumstances, less than 20% of the virus population is infectious, and in most instances this percentage is much lower. Thus, the majority of data obtained from morphological or biochemical studies of virus penetration deal with virions which are incapable of infecting cells. In addition, at the high multiplicities of infection usually employed in morphological and biochemical experiments, only a small proportion of the virus particles undergo uncoating due to the limited capacity of the cell to uncoat virus.

The difficulties in interpreting the experimental data, on the one hand, and the seeming simplicity of the problem, on the other hand, are probably the reasons why the problem did not stimulate much interest in investigators. It is noteworthy that since the 1940s and 1950s, when the background of our knowledge concerning virus adsorption and penetration was established, very few papers concerning this problem have been published, in contrast to the wide stream of publications upon the later steps of virus reproduction.

Interest in this phase of infection has recently been revived in connection with the appearance of new experimental approaches and new aspects of the problem. Virus adsorption and cell entry became of interest not only to virologists, but also to cytologists, molecular biologists, chemists, chemotherapists, and medical practitioners. Chemists and biochemists use the virus model as a tool to study the structure of cell surface receptors and their specificity. Cytologists see in the viral models a prospective system for studying endocytosis, cell fusion from without and from within, and intracellular transport. The interest of chemotherapists and medical practitioners was stimulated by the finding that some inhibitors of virus infection act on the early steps of virus infection, preventing virus penetration and uncoating.

The present article attempts to summarize the information concerning early virus–cell interaction published during the last decade. The first part (Section II) is devoted to general problems concerning attachment, penetration, and uncoating, while the second part (Section III) is concerned with these processes in individual virus groups.

Animal viruses are usually divided into two classes, those with a lipoprotein envelope and those without. The former class contains detectable amounts of lipid (5–20% by weight) while the latter class does not. It is believed that the early virus–cell interaction for these two classes of viruses is different. An attractive role for viral envelope proteins is to effect attachment and penetration of the virus through the cell plasma membrane, thus providing efficient translocation of virus genetic material to the host cell. Surprisingly, nonenveloped viruses often follow a similar pathway in gaining entry into the cell, suggesting that the function of the virus envelope is superseded by surface proteins of these viruses. Therefore, in Section II the two classes of viruses are not described separately, and similarities in their interaction with the cell are stressed.

The complexity of the subject and the controversy surrounding it are presented fully, and the gaps in our knowledge and ignorance of many points concerning early virus–cell interaction are indicated.

II. Stages of Penetration

A. Attachment of Viruses to the Host Cell

Adsorption of viruses on the cell surface is the first step leading to infection. Adsorption is more than a nonspecific electrostatic interaction between the negatively and positively charged groups on the virus and cell surfaces as was proposed originally. This highly specific process is mediated through the recognition of specific receptors on the cell plasma membrane by defined sequences of proteins on the virus surface. So far, the proper term for this process seems to be "attachment," and for the viral surface proteins involved in it "virus attachment proteins."

1. Cellular Receptors

The process of attachment proceeds not merely through the formation of a single particle–receptor bond, but through the cumulative formation of many bonds in the same area until a sufficiently stable multivalent attachment is formed. It has been estimated that up to 3000 receptor molecules might be present in the attachment area. Such binding could occur since cellular receptors are freely mobile in the plane of the plasma membrane (Lonberg-Holm and Phillipson, 1974). Accordingly, the first unstable interaction providing reversible adsorption is followed by irreversible adsorption due to the multiple binding of a single virus particle by several receptors.

For a number of viruses, specific receptors are present on the cell surface in limited amounts. Lonberg-Holm et al. (1976) estimated that they varied from 1×10^4 to 1×10^5. Thus, they could be saturated with excess virus, and competition for binding with the cell surface could be used to construct a "receptor map" for viruses. Such a "receptor map" was constructed by Lonberg-Holm et al. for certain nonenveloped viruses. The authors defined four receptor families detected by competition for attachment to HeLa cells. The first family contained human rhinovirus type 1 and 2, the second family contained human rhinovirus type 14 (and probably other types) and Coxsackie virus type A21, the third family contained three serotypes of poliovirus, and the fourth family contained Coxsackie virus B and adenoviruses.

Cellular receptors for some viruses may be present on only a limited number of cell species and thus could define the susceptibility of cell lines to these viruses. For example, poliovirus and the other picornaviruses bind only to cells of primate origin. The receptors for other viruses are present on the surface of a wide variety of cells, and for these viruses adsorption is not a host-limiting factor. Orthomyxoviruses and paramyxoviruses bind to sialic acid-containing substances on the cell surface and, hence, have a relatively wide specificity for host receptors. Some togaviruses can bind and infect vertebrate and invertebrate cells. Much information on how viruses bind to animal cells was obtained by studying the interaction of hemagglutinating viruses with erythrocytes (reviewed by Bächi et al., 1977). The role of carbohydrates as cell receptors for orthoviruses and paramyxoviruses has been intensively studied since the 1940s and 1950s using this simple system. It has been demonstrated that the receptors for orthomyxoviruses and paramyxoviruses and for polyoma virus are sialyloligosaccharides of erythrocyte glycoproteins. Vibrio cholerae neuraminidase, which releases virtually all erythrocyte sialic acid, abolishes agglutination by these viruses. There appears to be high specificity concerning the structure to which the virus binds, since in resialylation experiments only erythrocytes resialylated with specific sialyltransferase were agglutinated with Newcastle disease and polyoma viruses. Newcastle disease virus, equine influenza virus, and polyoma virus were shown to use the same specific receptor: its sequence was identified as AcNeu α2,3-Gal and perhaps AcNeu α2,8-AcNeu. The same sialyloligosaccharides as host cell receptor determinants for Sendai virus (AcNeu α2,3-Gal β1,3-GalNac) were identified on the surface of MDBK cells (Markwell and Paulson, 1980). On the other hand, erythrocytes treated with Newcastle disease virus (NDV) retained their full capacity to be agglutinated by human influenza viruses, suggesting that they have other specific receptors. A human influenza virus (PR8) exhibited a specificity less strict than NDV, and three sialyloligosaccharides could serve as viral hemagglutinin receptors (Cahan and Paulson, 1980; Fried et al., 1981).

The general opinion is that orthomyxoviruses and paramyxoviruses bind to glycoproteins, to glycophorin in particular, the major glycoprotein of the human erythrocyte surface containing sialic acid (Steck, 1974; Paulson et al., 1979). Glycophorin is generally accepted as an erythrocyte receptor because purified glycophorin effectively inhibits hemagglutination of these viruses (Howe and Lee, 1972). The number of glycophorin molecules in human erythrocyte is about 5×10^5 per one cell. Since a maximum of 4500 virus particles can bind to each erythrocyte (Wolf et al., 1980), it follows that one virus particle binds to numerous glycophorin molecules.

For most viruses, the chemical identity of the cell receptors is not known. Their identification is complicated by the fact that the viruses could bind to nonspecific receptors, although only the specific receptors possess the capacity to pass infectious virus into the cell. The penetration into the cell and virus replication should be taken into account when the receptors are defined, a condition which is difficult to realize in many adsorption experiments. There are several examples of discrepancies in the evaluation of specific cell receptors.

  a. Receptors for Orthomyxoviruses and Paramyxoviruses. By analogy with erythrocytes, the glycoproteins were said to be specific receptors for myxoviruses and paramyxoviruses. However, the receptors on cells other than erythrocytes are not necessarily identical to the structures responsible for hemagglutination, and susceptible animal cells should differ from erythrocytes in the structure of their surface, since the virus cannot penetrate and infect erythrocytes. The possibility that sialic acid is bound to glycolipids as substrates is presently accepted. Sialic acid-containing glycolipids (gangliosides) occur naturally in the plasma membrane of most cell types. They are the receptors for cholera toxin and other bacterial toxins, interferon (Vengris et al., 1976), and some glycoprotein hormones (see Holmgren et al., 1980). There is a series of studies suggesting that gangliosides are specific receptors for orthomyxoviruses and paramyxoviruses. The first studies were made by Haywood (1974a,b, 1975a,b), who demonstrated that liposomes containing gangliosides isolated from bovine brain can inhibit hemagglutination of Sendai virus and that most gangliosides possess receptor activity and provoke the fusion between virus envelope and liposome membrane. The gangliosides active in virus binding and fusion possessed more than one molecule of sialic acid.

Later, Sharom et al. (1976) reported the binding of influenza and Sendai viruses to lecithin–cholesterol liposomes containing beef brain gangliosides. Using gangliosides spontaneously attached to plastic surfaces through strong hydrophobic bonds, Holmgren et al. (1980) showed that Sendai virus binds with high affinity to gangliotetraosylceramides with a terminal disialosyl group, i.e., to GT_{1a}, GQ_{1b}, and GP_{1c}. The presence of a sialic acid residue was an absolute requirement for binding, and the N-acetylgalactosamine residue was also essential for the receptor property. The common terminal sequence, AcNeu $\alpha2 \rightarrow$ 8 AcNeu $\alpha2 \rightarrow 3$ Gal $\beta1 \rightarrow 3$ GalNac, of these gangliosides occurs naturally in the plasma membrane of most cell types and is also present in some glycoproteins.

The role of gangliosides as virus receptors for influenza virus was also shown. Schulze (1981) demonstrated that glycolipids with a terminal galactose residue isolated from cell membranes could inhibit influenza virus binding to the cell and its infectivity. The importance of gangliosides for influenza virus adsorption and cell entry was further supported by the findings that certain gangliosides could restore the adsorption and penetration of influenza virus into the cells whose natural receptors had been destroyed by neuraminidase. Moreover, ganglioside GT_{1b} possessing terminal galactose (in accordance with the data of Schulze), when inserted into the plasma membrane, induced the accumulation of influenza virus genomes in the nuclei in amounts essentially larger than that in native cells. This suggests that some gangliosides could be involved in influenza virus penetration and transport to the nuclei (Bukrinskaya et al., 1982a; Bergelson et al., 1982).

It follows from these data that only certain gangliosides with a definite structure possess the properties of virus receptors. This could explain the failure of Wu et al. (1980) to reveal the receptor activity of gangliosides isolated from HeLa cells and rat brain, the structure of which might be inconsistent with the virus receptor activity.

Gangliosides and glycoproteins are usually regarded as playing alternative roles in virus reception, but they both may be involved in virus–cell interaction. It is likely that the virus approaching the cell initially meets the glycoprotein layer and binds to glycoproteins. However, to penetrate the plasma membrane, the virus has to interact with it more intimately, and such contact could be provoked by interaction with gangliosides. In addition to the close contact with plasma membrane lipids, the interaction with gangliosides might be essential for the fusion of virus and cell membranes (see below).

It is still not known whether orthomyxoviruses and paramyxoviruses use the same or different kinds of specific receptors. The fact that the cells can lose the ability to adsorb Sendai virus while still retaining the susceptibility to infection with influenza virus and other paramyxoviruses and that certain oligopeptides differently inhibit the fusion induced by orthomyxoviruses and paramyxoviruses (Richardson et al., 1980) suggests that receptors on the cell plasma membrane are highly specific.

b. Receptors for Polyoma Viruses. Sialic acid residues on the cell surface were shown to be nonspecific receptors providing nonspecific (noninfective) adsorption of virus particles or viral capsids on the cell surface. The material adsorbed was delivered to the lysosomes but not to the nuclei. Meanwhile, the binding to specific receptors (their nature is not yet identified) resulted in the specific transport of the virus particle to the nuclear membrane (Mackay and Consigli, 1976; Bolen and Consigli, 1979).

An important conclusion to be drawn from these data is that specific receptors on the cell surface could be involved in the penetration of the virus particle through the plasma membrane and in its intracellular transport.

c. Receptors for Alphaviruses. It has been reported that the specific receptor for Semliki Forest virus is the major histocompatibility (MHC) antigen on the surface of mouse and human cells (Helenius et al., 1978), i.e., the gene complex plays a fundamental role in certain cell–cell membrane recognition systems, in the control of some gene products, and in the susceptibility to some diseases. However, Oldstone et al. (1980) later demonstrated that MHC is hardly involved with the specific reception of at least eight of the nine viruses studied (Semliki Forest virus, Sindbis virus, Coxsackie B3, mouse hepatitis, Pichinde, vesicular stomatitis virus, herpes simplex virus, and lymphocytic choriomeningitis virus), since cells lacking this antigen supported virus replication. Only murine cytomegaloviruses did not express viral antigens and did not form infectious virus in such cells. Earlier, Haspel et al. (1977) showed that MHC could not serve as a receptor for measles and vesicular stomatitis viruses. It is possible that these glycoproteins frequently expressed on cell surfaces may nonspecifically bind virus particles and may be involved in noninfectious adsorption.

Specific receptors for alphaviruses have not yet been identified. Coombs et al. (1981) have suggested that these viruses use some nonspecific receptors, a circumstance which is in keeping with the extensive range of hosts (vertebrate and invertebrate) of alphaviruses.

There are different approaches in studying the specificity of cell surface receptors. One is the competition of viruses for binding with the cell surface to determine the existence of common receptors for different virus species (Lonberg-Holm *et al.*, 1976; Cahan and Paulson, 1980). Another approach is the destruction of native receptors by specific enzymes, with subsequent control of adsorption blockage (Oldstone *et al.*, 1980), or destruction of native receptors followed by linkage of the particular substances which are presumed to be receptors or to imitate receptors (Cahan and Paulson, 1980), or isolation from the native plasma membrane of receptor substances which inhibit adsorption (Schulze, 1981) and have high affinity for the virus attachment proteins. By these means, the protein nature of the cellular receptors for picornaviruses (reviewed by Smith, 1977) and adenoviruses (Hennache and Boulanger, 1977; Svensson *et al.*, 1981) was established.

An additional experimental strategy is the construction of model membranes (liposomes) by attaching the presumed receptors to them. Although the native membranes could contain structures which are not represented by these artificially linked membranes, or structures in other conformational states, such experiments could help in elucidation of receptor chemistry and specificity of their recognition by the virus. As described earlier, the insertion of gangliosides into preformed lipid vesicles was successfully used in the study of their receptor function (Haywood, 1974a,b, 1975a,b; Sharom *et al.*, 1976).

Adsorption of some viruses to liposomes could occur in the absence of biological receptors. Phospholipids and cholesterol in a bilayer configuration are sufficient for virus binding (Mooney *et al.*, 1975). This indicates that lipids could be of some importance in at least the first step of attachment. The different aspects of virus interaction with liposomes have been reviewed by Tiffany (1977).

2. Virus Attachment Proteins

These surface virus proteins involved in the interaction with cell membranes have been identified for most viruses. In both enveloped and nonenveloped viruses these proteins are present in multiple copies in the virus particle (Lonberg-Holm and Philipson, 1974; Meager and Hughes, 1977; Helenius and Simons, 1980). For example, there are about 240 glycoproteins in the virion of Semliki Forest virus (Helenius and Simons, 1980), 12 oligomeric fiber proteins in the virion of adenovirus, 300–450 hemagglutinin subunits in the virion of influenza virus, and 24 molecules of $\sigma 1$ protein in the virion of reovirus (Lee *et al.*, 1981).

For enveloped viruses, these proteins are mostly glycoproteins inserted in the virus envelope. With most viruses, the glycoproteins are present in the form of spikes about 7 to 10 nm in length. Usually several glycoprotein molecules constitute one spike. Rhabdovirusus and alphaviruses possess only one type of spike; orthomyxoviruses and paramyxoviruses possess two types of spikes, each one formed by a different glycoprotein species.

Glycoproteins are amphipathic molecules. They consist of an external hydrophilic part, which includes the amino terminus of the polypeptide, and a hydrophobic segment at the carboxy terminus by which the spikes are strongly associated with the lipid bilayer. The primary function of glycoproteins is known to be the efficient attachment to cell receptors. Recent information suggests another important function of these proteins connected with membrane fusion. When virions are treated with protease to remove the glycoprotein spikes, there is a dramatic decrease in infectivity, presumably due to the less efficient interaction of virions with the cell surface.

For the biological activity of most glycoproteins, posttranslational cleavage is required. The cleavage has been shown to be involved in the processing of influenza hemagglutinin, both paramyxovirus glycoproteins and togavirus glycoproteins, and envelope proteins of oncoviruses. It has been demonstrated that activation by proteolytic cleavage is an important factor in determining the host range and pathogenicity of orthomyxoviruses and paramyxoviruses (for review see Klenk *et al.*, 1980).

Oligosaccharides play an essential role in the interaction of virus glycoproteins with cell receptors. Their side chains are attached to the polypeptide molecule at special sites determined by the polypeptide backbone. The glycosylation sites can vary in different hosts. The removal of oligosaccharides from glycoproteins abolishes the ability of the virus particle to bind to the cells and erythrocytes.

Little information is available concerning the nature of attachment proteins of mammalian nonenveloped icosahedral viruses. It seems obvious that these proteins are localized in the virus outer shell. With reoviruses, this protein has been recently identified as the minor outer shell protein $\sigma 1$ (Lee *et al.*, 1981).

As a result of binding of virus to specific receptors, various biochemical reactions are induced in the host cell. These include alterations in membrane transport of cations, changes in membrane fluidity and potential (Levanon and Kohn, 1978), and chemiluminescence, which is triggered by virus glycoproteins and correlates with the generation of unstable oxygen (Peterhaus, 1980).

As indicated, the complexity in studying virus interactions with the host cell makes it imperative to devise a simple system. One approach to studying the role of viral components in this interaction is the reconstitution of viral membranes from lipids and proteins by attaching virus surface proteins to lipid carriers. It has been shown that glycoproteins of enveloped viruses could be inserted into preformed lipid vesicles, binding to the preexisting phospholipid bilayer in such a way as to regain the configuration they have in the native viral membrane.

The combination of liposomes and viral glycoproteins (so called "virosomes") provides a model system for studies of lipid–protein interaction and interaction with host cell membrane. By these means, the organization of the virus membrane of paramyxoviruses (Haywood, 1974a,b; Hosaka, 1975; Hsu et al., 1979), influenza viruses (Huang et al., 1979; Oxford et al., 1981), and Semliki Forest virus (Helenius et al., 1977), and the function of individual viral components in attachment and membrane fusion and their cooperative effects (Huang et al., 1980) have been investigated. It has been shown that in reconstituted viral membranes the density of spikes fluctuates by a factor of 4, and the spikes are less numerous than in native virus particles. This indicates that spike packing density is determined by factors other than the lipid and protein (e.g., membrane protein in the case of paramyxoviruses and orthomyxoviruses).

B. Entry of Viruses into the Host Cell

There are two traditional versions of this step of infection. According to viropexis, or endocytosis, the version proposed by Fazekas de St. Groth in 1948, attachment of virus is followed by invagination of the plasma membrane to form a vacuole which contains virus particles. The second pathway of cell entry is the fusion of the virus envelope with the host cell plasma membrane.

Reports on the mode of entry of viruses into the cell which have accumulated during the last decade differ widely. Some findings support the view that endocytosis is a general mechanism by which viruses are internalized by a cell (reviewed by Dales, 1973); other observations are compatible with the model for direct penetration of the plasma membrane, presumably involving membrane fusion (Zhdanov and Bukrinskaya, 1961; Heine and Schnaitman, 1971; Coombs et al., 1981); the third opinion is that both processes are necessary for the virus to pass into the cell, endocytosis to internalize the virus and fusion with the membrane of the intracellular vacuole to escape from the vacuole and to release viral genetic material into the cytoplasm (Helenius *et al.*, 1980a,b).

In this section, we summarize the data for and against these concepts and conclude that fusion seems to be the universal mode of entry into cells.

1. Entry by Endocytosis

Endocytosis is a general term that refers to the process by which cells entrap extracellular material within inward foldings of the plasma membrane that pinch off from the surface to form intracellular vesicles. Endocytosis was first identified in phagocytic cells (the process was called phagocytosis), but later was recognized in all animal cells.

This process could be adequately explained from the point of view of cell membrane organization. The most widely accepted model of membrane structure is the fluid mosaic model. The membrane is envisaged as a fluid lipid layer into which proteins are inserted and are free to diffuse laterally. The virus particle, with its repeating subunits which bind to the virus receptors in the cell membrane, may be regarded as a polyvalent ligand. After attachment of the virus to its appropriate receptor, the fluid bilayer flows around the virus particle. As contact is made with more receptors, the binding would be expected to become more secure until finally the virus is enveloped by the plasma membrane. Invagination of the membrane as a consequence of virus attachment may be considered to be the most thermodynamically stable configuration. For invagination to become an intracellular vacuole, fusion of the plasma membrane must occur by the mechanisms proposed by Wills *et al.* (1972).

Viropexis seems to represent the kind of "receptor-mediated" or "adsorptive" endocytosis recently reviewed by Goldstein *et al.* (1979). This process has been recognized as a general mechanism by which animal cells take up nutritional and regulatory proteins and other substances from extracellular fluid. Receptor-mediated endocytosis occurs at specialized regions of the surface membrane called coated pits that could invaginate rapidly into the cell during endocytosis to form coated vesicles. The coat of coated vesicles is composed predominantly of a single protein of molecular weight 180,000 called clathrin, which is noncovalently bound to the surface of the membrane (Pierce, 1975).

Receptor-mediated endocytosis is characterized by Goldstein *et al.* (1979) by the following properties: (1) The binding component on the cell surface is a receptor in the strict sense, that is, a molecule whose function is to bind an endogenous ligand. (2) Internalization of the ligand is effectively coupled to binding; the half-time for internalization is less than 10 minutes. (3) The receptor-bound substrates enter the cell through coated pits. (4) The internalized proteins are usually delivered to lysosomes where they are completely degraded into amino acids, or into cellular organelles other than lysosomes where they could accumulate in undegraded form. As an alternative to receptor-mediated endocytosis, extracellular substances could be trapped nonselectively in fluid droplets being internalized by cells through the invagination of noncoated regions of the membrane. Such a mode of entry occurs without binding to cell receptors ("bulk fluid endocytosis").

The features of virus engulfment are very similar to those of receptor-mediated endocytosis, with the exception that the viruses apparently may use cell surface receptors designed for physiological ligands.

The internalization of viruses by endocytosis has been described for a large number of enveloped and nonenveloped viruses, excluding perhaps only the picornaviruses.

Reovirus was the first virus used to show that endocytosis is involved in virus penetration (Silverstein and Dales, 1968). Virus particles were endocytosed into cytoplasmic vacuoles which fused with primary lysosomes. Lysosome enzymes removed some of the virus proteins but double-stranded viral RNA remained intact. Semliki Forest virus provided another example of endocytosis described in detail by Helenius *et al.* (1980) and Helenius and Simons (1980). The virus enters the cells through coated pits and coated vesicles and is delivered into cytoplasmic smooth-surfaced vesicles and then into lysosomes. The process is very effective, with more than 2000 virus particles being engulfed per cell per minute.

Adenoviruses (Dales, 1973), polyoma virus (Mackay and Consigli, 1976), and influenza virus (Dourmashkin and Tyrrell, 1974) have also been shown to be engulfed by endocytosis. Influenza virus particles have been observed to bind to coated pits which are transformed to coated vesicles (Patterson *et al.*, 1979).

As a route of infection, endocytosis obviously has more disadvantages than advantages when compared to the other mode of cell entry, fusion. The virions internalized by endocytosis are still outside the cytoplasm. To initiate infection, they need to pass the membrane barrier for uncoating and transportation to certain cell compartments. From this point of view, endocytosis differs from fusion which, in one step, solves the problem of penetration and uncoating. Nevertheless, as a route for virus entry, endocytosis may have some advantages for virus intracellular transport. Virus-containing vacuoles can move toward any cellular compartment and fuse with cellular membranes releasing the virus particles at the sites of their uncoating. By this means reoviruses and Semliki Forest virus are transported to lysosomes (Silverstein and Dales, 1968; Helenius et al., 1980; Talbot and Vance, 1980) and polyoma virus toward the nucleus (Mackay and Consigli, 1976).

In cases in which the strategy of the virus particle is to escape from the endocytic vacuole before its fusion with the lysosome, endocytosis perhaps may provide conditions for effective fusion with the vacuole membrane such as intimate contact between virus and cell membrane and a suitable "microclimate" within the vacuole.

It appears that the fate of virus particles internalized by endocytosis may be different. The following routes, hypothetical or experimentally shown, seem to be possible: (1) The part of the virus inoculum which has not undergone uncoating (when the virus is added in excess) is transported to lysosomes by fusion of the endocytic vacuole with the primary lysosome and is digested by lysosomal enzymes. (2) The virus particle escapes from the endocytic vacuole before its fusion with the lysosome. (3) The virus particle escapes from the secondary lysosome (Helenius et al., 1980a,b). (4) The virus particle is uncoated within the lysosome, and subviral components are released into the cytoplasm (Silverstein and Dales, 1968). (5) The virus particle is delivered within the endocytic vacuole to the sites of uncoating and then escapes from the vacuole (Bolen and Consigli, 1979). The mechanisms that determine the fate of the virus particle within the vacuole are obscure; they may be provided by the virus particle itself in combination with specific cellular factors involved in the virus–cell interaction.

The role of endocytosis as a pathway leading to infection is widely debated. Recently, Coombs et al. (1981) showed that cytochalasin B, an inhibitor of microfilament formation and phagocytosis (MacLean-Fletcher and Pollarg, 1980; Tanenbaum, 1978), though completely blocking endocytosis of Sindbis and vesicular stomatitis viruses by BHK cells had no effect on the ability of viruses to infect and replicate in these cells. In contrast, chloroquine, which did not inhibit endocytosis, greatly reduced the yields of virus produced. Similarly, vaccinia virus was shown to penetrate into the cell under conditions that restricted endocytosis, in the presence of sodium fluoride and cytochalasin B (Payne and Norrby, 1978). These observations are compatible with the suggestion that endocytosis is not essential for infection, and virus particles which are capable of initiating infection enter the cell by direct interaction with the plasma membrane (or can use this portal of infection when endocytosis cannot occur).

2. Entry by Fusion

In membrane fusion, an alternative model to endocytosis for virus entry into the host cell, enveloped viruses fuse with the cell surface through the interaction of the virus envelope and cell plasma membrane. Such a mechanism of cell entry has been demonstrated for orthomyxoviruses and paramyxoviruses electron microscopically (Morgan and Rose, 1968; Morgan and Howe, 1968) and biochemically. The integration of paramyxovirus glycoproteins with the plasma membrane during virus penetration was detected with ferritin-labeled antibodies. As a result of virus protein integration, the surface of the cell was antigenically modified so that the cells could be killed by antiviral antibodies and complement (Fan and Sefton, 1978). Using this phenomenon, Fan and Sefton compared the fusion activity of three enveloped viruses, Sendai, Sindbis, and vesicular stomatitis viruses, and found that while all the Sendai virions entered the cell by fusion, Sindbis virus entered by fusion 0.6% of the time and vesicular stomatitis virus 3% of the time.

Fusion occurs via lipid–lipid interactions between adjacent virus and cell membranes. The regions of the protein denuded lipid bilayer on the plasma membrane are involved in the process eliminating the role in fusion of host cell membrane proteins (Knutton, 1978). The involvement of virus proteins has been directly demonstrated by *in vitro* experiments using liposomes. For the first time, Haywood (1974b) observed fusion of Sendai virus envelope with the liposome membrane when the virus was mixed with the liposomes containing gangliosides as virus receptors. The requirement for fusion was the presence of sphingomyelin in the liposomes suggesting the role of specific plasma membrane lipids in fusion. Later, Huang *et al.* (1980a,b) demonstrated the fusion of liposomes which contained both influenza virus glycoproteins, hemagglutinin, and neuraminidase, with the plasma membrane of tissue culture cells, and White and Helenius (1980) described fusion between Semliki Forest virus envelope and liposome membrane. The fusion was demonstrated electron microscopically by the formation of large vesicles containing viral glycoproteins on the surface and nucleocapsids inside, and biochemically by trapping ribonuclease or trypsin within the liposomes. The digestion of viral RNA by trapped ribonuclease within liposomes was used as a quantitative assay for fusion.

Profound alterations in the virus and host cell membranes have been described during fusion, such as the change of fluidity of host cell membrane (Nicolau *et al.*, 1979) and lateral diffusion of cell membrane proteins and virus glycoproteins over long distances (Fan and Sefton, 1978), structural rearrangement of membrane lipids and exchange of lipids between viral and cell membranes, in particular, depletion of cholesterol from the membrane of vesicular stomatitis virus into the membrane of phosphatidylcholine vesicles (Moore *et al.*, 1978), reduction of the rotational mobility of proteins linked to glycophorin A in erythrocyte membranes, and aggregation of intramembrane particles in Sendai virus membrane (Nigg *et al.*, 1980).

The elucidation of the fusion mechanism greatly depends on the identification of virus proteins promoting fusion. Up to now, this protein has been identified only in paramyxoviruses. Fusion-promoting protein (F protein) is the second virus glycoprotein (the other is the HN glycoprotein) responsible for attachment to the host cell plasma membrane. An absolute requirement for fusion is proteolytic cleavage of inactive precursor, F_0, to form F protein consisting of F_1 and F_2 polypeptides linked with disulfide linkage (reviewed by Choppin and Scheid, 1980). Proteolytic cleavage of F_0 necessary for fusion occurs by specific trypsin-like proteinases and results in the generation of hydrophobic amino terminus of F_1 which is involved in the fusion reaction (Gething *et al.*, 1978).

There is increasing evidence that influenza virus penetrates the host cell membrane by the same fusion mechanism, and that this process is triggered by virus hemagglutinin. The activation of influenza virus infectivity also requires a highly specific cleavage of the HA precursor polypeptide (Klenk, 1980), and amino terminal residues of HA_2 glycoprotein, which are released after proteolytic cleavage, bear a striking resemblance to F_1 amino terminal residues in paramyxoviruses in their sequence and hydrophobic nature (Gething *et al.*, 1978; Waterfield *et al.*, 1979). By analogy with paramyxoviruses, this fragment of HA_2 apparently interacts with cellular membranes promoting fusion. One could speculate that in paramyxoviruses the role in infection which formerly resided in the HN protein has been superseded by the F protein, whereas in the orthomyxoviruses the HA protein has retained this function (Samson *et al.*, 1980).

By analogy with paramyxoviruses and orthomyxoviruses, it seems likely that hydrophobic terminal residues in one of the surface proteins of other enveloped and nonenveloped viruses could be responsible for the fusion reaction with cellular membranes.

One way to identify these regions is the synthesis of amino acid sequences that resemble the terminus of polypeptides involved in fusion. Richardson *et al.* (1980) showed that synthesized oligopeptides that resembled the N-terminus of paramyxovirus F_1 and orthomyxovirus HA_2 polypeptides were highly effective inhibitors of virus replication. It seems possible to use specific oligopeptides labeled with radioactive or fluorescent probes to elucidate the precise site and mechanism of the fusion.

In addition to F_1 and HA_2 proteins, the neuraminidase of paramyxoviruses (Yasuda *et al.*, 1980) and orthomyxoviruses (Huang *et al.*, 1980a) is also required for hemolysis and fusion. Its function in the fusion process has not yet been defined.

Details of the association of the virus proteins responsible for fusion with the lipid bilayer of the host cell are unknown. Data indicate that the F protein of paramyxoviruses induced destabilization of the lipid bilayer at the site of virus attachment, and, once envelope fusion has occurred, the implanted F protein diffuses laterally and enhances phospholipid exchange and envelope fusion with the other virus particles attached to the membrane sites (Kuroda *et al.*, 1980). As a consequence of fusion with the plasma membrane, the membrane permeability changes causing the cells to leak K^+ and to take up Na^+. Loss of cation symmetry leads to cell swelling. In the case of erythrocytes, the plasma membrane ruptures and hemolysis occurs. It is likely that some of the virus-mediated morphological changes are a consequence of cell swelling but not plasma membrane modifications induced by virus-cell fusion. Since the mobility and distribution of many integral membrane proteins are controlled by cytoplasmic cytoskeletal elements associated with the plasma membrane, disruption of this system of filaments during cell swelling provides the membrane perturbation which allows rapid diffusion of viral components to occur (Knutton, 1978).

To understand the mechanisms of fusion, it is important to know the factors influencing this process. Among the factors, pH value and lipid composition of host membranes seem to be the most important.

It was shown that hemolysis induced by alphaviruses (Väänänen and Kääriäinen, 1979, 1980; Väänänen *et al.*, 1981) and influenza viruses (Maeda and Ohnishi, 1980; Huang *et al.*, 1981; Lenard and Miller, 1981) occurred only at low pH. The strains of influenza virus had different optimal pH values, which varied from 5.0 to 5.75. Hemolysis sharply decreased with an increase of 0.5 pH units. Similarly, cell fusion induced by alphaviruses (White and Helenius, 1980; White *et al.*, 1980) and influenza virus (Huang *et al.*, 1981) also required low pH. By contrast, paramyxoviruses induce hemolysis and cell fusion at higher pH values varying from 6.25 to 7.0. This explains why under neutral pH conditions paramyxoviruses possess hemolytic and fusion activities whereas orthomyxoviruses and alphaviruses do not. Helenius *et al.* (1980a,b) suggest that this low pH makes possible the fusion of Semliki Forest virus particles with the lysosome membrane after the virus has been delivered to the lysosome.

The strict pH dependence of fusion is not understood. Probably, low pH changes the conformation of virus glycoproteins so that they can interact with host membrane lipids.

Neutral liposomes have been shown to be unable to fuse with the plasma membrane of cultured cells or with other liposomes (Poste and Papahadjopoulos, 1976). In contrast, charged liposomes are able to fuse with cellular membranes and are able to undergo rapid and extensive fusion with other vesicles. Whether phagocytosis or membrane fusion predominates on viral attack seems in this system to be determined by the presence of phosphatidylethanolamine and sphingomyelin in the liposomes (Haywood, 1974b). Conflicting observations on the mode of entry into the cell may reflect cell-specific or membrane-specific differences in membrane composition at the site of viral attack.

As for temperature, it was not shown to be an absolute requirement for fusion. White and Helenius (1980) reported that fusion between the membrane of Semliki Forest virus and liposomes, although it occurred rapidly at 37°C, was observed over a wide range of temperature; some fusion (25%) even occurred at 0°C. This suggests that fusion is a physicochemical process.

The same mechanisms which are responsible for fusion of virus and cell membranes are presumably involved in the virus-induced fusion of adjacent cell plasma membranes leading to the formation of polykaryocytes.

Viruses are known to induce two types of cell fusion (Bratt and Gallaher, 1969): (1) "fusion from without," which occurs in cells at a high multiplicity of infection within a few hours of infection. This type of fusion is due to the parental virus components and does not require intracellular synthesis of virus constituents (reviewed by Hosaka and Shimizu, 1977); (2) "fusion from within," which occurs at a low multiplicity of infection. The process extends from primarily infected cells to surrounding uninfected cells. This type of polykaryocytosis requires the penetration of the virus into the first cell and the synthesis of virus constituents, and consequently occurs relatively late in the replicative cycle of the virus. Fusion from within has been described for a number of viruses which had not been known to induce membrane fusion, such as some strains of herpes simplex virus (Haffey and Spear, 1980), Mason-Pfizer monkey retrovirus (Chatterjee and Hunter, 1979), respiratory syncytial virus (Dubov *et al.*, 1980), some members of the slow virus group, for example, visna virus, and others. The data obtained with paramyxoviruses (reviewed by Homma and Ohuchi, 1973) and herpes simplex virus (Para *et al.*, 1980; Haffey and Spear, 1980) show that fusion from within is promoted by the same viral glycoproteins that are involved in virus entry.

The mode of cell entry could be shown indirectly by the test of transition from photosensitivity to photoresistance for viruses grown in the presence of heterocyclic dyes. Herpes simplex virus treated with fluorescein isothiocyanate developed resistance to light and to antibody not longer than 5 minutes after infection. If entry involves the engulfment of the virus particles by endocytosis, there might be a longer period of time before the development of resistance to antibody while the virus remains sensitive to light. Thus, the data support fusion rather than endocytosis as the mode of entry for herpes simplex virus (DeLuca *et al.*, 1981).

One can see that the list of viruses which possess the mechanisms for membrane fusion is considerably larger than one would think, and includes representatives of different virus families. In addition, the data of Coombs *et al.* (1981) obtained with cytochalasin B and chloroquine support the view that fusion but not endocytosis is the route of successful infection. By this pathway a minority of viruses could penetrate which are not detectable by electron microscopic studies. The data described concern enveloped viruses. Surprisingly, nonenveloped viruses such as adenoviruses (Brown and Burlingham, 1973) and picornaviruses (Dunnebacke *et al.*, 1979) have been reported to follow a similar pathway of entry by direct interaction with a plasma membrane. So far, fusion seems to be a universal mechanism of cell entry.

Though two modes of virus entry into the host cell, by endocytosis and by fusion, are regarded generally as alternatives, it is probable that in a number of infections they both might subsequently be involved in penetration. Virus particles which are engulfed by endocytosis may escape from endocytic vacuoles or lysosomes by fusion with the membrane of the vacuole, or virus particles reaching the nuclear membrane within endocytic vacuoles may penetrate the nuclei by fusion with the nuclear membrane. In this case the enigma of how the viral genetic material escapes from cellular vacuoles has a simple solution. An example of the combination of both pathways is described by Helenius *et al.* (1980) for Semliki Forest virus. From this point of view, penetration through the cell plasma membrane ("penetration from without") and through the membrane of cytoplasmic vacuoles ("penetration from within") are the same kind of processes.

It could be concluded that membrane fusion seems to be a common mechanism which ultimately mediates the final penetration of the viral genetic material into the cytoplasm or the nucleus of the host cell. The concept of fusion envisages the viruses not as passive passengers during penetration but as active contributors to the process by interacting with specific cell receptors.

C. Uncoating

The stage following penetration is uncoating, i.e., the removal of surface protective components (the lipoprotein membrane in the case of enveloped viruses and surface proteins in the case of nonenveloped viruses) and the release of the inner component which is capable of initiating infection. Uncoating is manifested by the loss of virus infectivity, by sensitivity to nucleases, by resistance to photosensitizing effects of various dyes and to neutralizing antibodies, and by other features specific for certain groups of viruses.

The final products of uncoating are cores, nucleocapsids, or nucleic acids. Uncoating of viruses containing polymerase is aimed at the provison of an adequate viral template possessing active polymerase in order to initiate infection. Such components (usually cores and nucleocapsids) possess polymerase activity *in vitro*. Uncoating of viruses which do not contain polymerase results in the release of viral nucleic acids which function in translation. For these groups of viruses, the assumption that nucleic acid fully devoid of protein initiated infection has been recently reconsidered, since a series of findings suggests that the final products of uncoating of picornaviruses, adenoviruses, polyoma viruses, and some other viruses are nucleic acids associated with viral inner protein(s) but not free nucleic acids. The final product of uncoating of picornaviruses is RNA covalently linked at its 5' termini to VPg protein (Flanegan *et al.*, 1977; Lee *et al.*, 1977). For adenoviruses (Rekosh *et al.*, 1977; Challberg *et al.*, 1980) and polyoma virus (Bolen *et al.*, 1981) the final product of uncoating has been identified as a DNA–protein complex representing viral DNA covalently linked with internal viral protein. A similar complex has been described for simian virus 40 (SV40) retrovirus (Brady *et al.*, 1981). The function of the protein linked to nucleic acid is not yet clear. Recent experiments suggest that the protein is important for core conformation and transcription (Brady *et al.*, 1981).

Whereas viruses of certain groups are able to adsorb to and penetrate into both susceptible and resistant cells, uncoating occurs most often only in susceptible cells. It appears that the sensitivity of a cell to virus infection is determined by the ability of the cell to uncoat virus (reviewed by Smith, 1977). Thus, uncoating could be one of the events which determines host cell specificity. The role of uncoating as the factor limiting infection is also supported by the fact that there are substances which specifically inhibit this step (see below).

1. Sites of Uncoating and Intracellular Transport

Uncoating and modification of virus particles apparently start just after attachment of the virus particle to the cell surface receptors. In the case of picornaviruses, the main uncoating events occur at the plasma membrane sites whereas virus particles which penetrate through the plasma membrane have no chance to be uncoated and to be involved in the infectious process. The number of modifying sites per cell for Echovirus estimated by measurement of the OD_{260} and plaque-forming ability of the virus (assuming that 1 OD_{260} unit represents 9.4×10^{12} particles) was approximately 500–1000 (Rosenwirth and Eggers, 1979). The final uncoating of picornaviruses apparently occurs in the cytoplasm, although cardioviruses have been shown to be completely uncoated at plasma membrane sites (Hall and Rueckert, 1971).

With other viruses, the virus particles are directed to specific delivery sites within the cells (lysosomes, Golgi-associated structures, perinuclear space, nuclei). Data indicate that uncoating and intracellular transport could be interdependent events. Apparently, when intracellular transport is impaired and the virus particle misses the right direction to the uncoating sites, it could be taken up by the endocytic vacuole, delivered to the lysosome, and destroyed by lysosomal enzymes. As shown with polyoma virus, infected cells can distinguish between intact virus particle and capsids: virus particles are transported to the nucleus where their uncoating occurs whereas capsids are delivered to lysosomes and are there digested (Bolen and Consigli, 1979, 1980).

With certain viruses, lysosomal hydrolases were shown to be involved in uncoating. Reoviruses which enter the lysosome are converted to subviral particles which are resistant to further action of lysosomal enzymes.

In the case where the virus enters the cell by fusion, this event combines penetration and uncoating. During fusion, virus lipoprotein membrane or surface proteins which integrate with the host cell membrane are removed.

In the case where the virus is internalized by endocytosis, several routes of uncoating seem to be possible. Uncoating could occur during the fusion of virus membrane with the membrane of endocytic vacuole or lysosome as suggested for Semliki Forest virus (Helenius *et al.*, 1980; White and Helenius, 1980), or within lysosomes by lysosomal enzymes as shown for reoviruses (Silverstein and Dales, 1968). Another variant is that the virus particle escapes from the endocytic vacuole and is transported to the sites of uncoating by employing special mechanisms. For adenoviruses, such mechanisms of intracellular transport were shown to involve elements of the cytoskeleton. Dales and Chardonnet (1973) observed an association between virus particles and microtubules in adenovirus-infected cells and suggested that adenoviruses and possibly other viruses are transported from the cell surface to the nuclei along the microtubules. In support of this model, adenovirus particles were shown to associate *in vitro* with microtubules isolated from rat brain (Luftig and Weihing, 1973). The association was probably mediated by high-molecular-weight protein normally present as projections on the surface of microtubules (Weatherbee *et al.*, 1977). The physical basis for binding is unclear; it is not known whether this protein functions actively in virus transport or serves only as the site of attachment for particles that are then transported by another mechanism such as, for example, selective polymerization and depolymerization of microtubules.

Following the concept of Dales, adenoviruses escape from endocytic vacuoles as intact particles and attach to microtubules for transportation to nuclei. However, it remains unclear how virus particles conserve their full complement of proteins if they use the fusion mechanism to escape from the endocytic vacuole. One could suggest either that the virus particle released into the cytoplasm loses some surface proteins and is not as intact as electron microscopic observation allows us to suggest, or that mechanisms other than fusion are involved in escaping from the vacuole.

For the other viruses, for example, for influenza viruses, involvement of microtubules in intracellular transport has not been observed (Patterson *et al.*, 1979), and the mechanism of their transport remains obscure. Meanwhile, reovirus has been shown to associate with microtubules (Babiss *et al.*, 1979).

2. Uncoating Intermediates

Uncoating occurs through a series of virus- and cell-dependent reactions which sequentially remove surface protective structures from the virus particle. So far, uncoating intermediates should exist in infected cells. However, in most cases the intermediates are not isolated or are poorly investigated due to the high efficiency of the uncoating process and the absence of adequate approaches to isolate these components from the cell in a native state.

For enveloped viruses, the first step in uncoating, the removal of the lypoprotein membrane, presumably occurs during the fusion with the host cell plasma membrane or the membranes of cytoplasmic vacuoles. However, the final uncoating could hardly be achieved during this step and probably occurs at cell compartments other than the site of fusion. The end-products of uncoating of paramyxoviruses, orthomyxoviruses, and rhabdoviruses are nucleocapsids (Bukrinskaya, 1973), and after removing lipoprotein membrane by fusion they need to be stripped of M protein surrounding nucleocapsids. So far, uncoating of most enveloped viruses should proceed in at least two steps.

There is evidence that the uncoating of nonenveloped viruses is also not a single-step process. For example, picornaviruses undergo uncoating through a series of intermediate subviral particles with a progressive conversion of virus from 156 S to 12 S subunits (Cavanagh et al., 1978; Baxt and Bachrach, 1980). Echovirus uncoating has the following steps: virions (156 S) → A particles (130 S) → ribonucleoproteins (80 S) → empty capsids (80 S) + free RNA (Rosenwirth and Eggers, 1979). On the other hand, there are data that foot-and-mouth-disease virus uncoats in a single step without production of intermediate subviral particles (Baxt and Bachrach, 1980).

Uncoating of adenoviruses occurs at nuclear cores and results in rapid and efficient transfer of DNA into the nucleus. Three types of subviral structures presumably representing intermediates in uncoating are described: (1) subviral structures in the cytoplasm with a density greater than virions (1.46–1.48 instead of 1.34 gm/cm^3) which still contain all virus polypeptides; (2) corelike structures in the cytoplasm and nuclei composed of viral DNA and polypeptides VIa2, V, and PVII; (3) DNA–terminal protein complex in the nuclei apparently representing the end-product of uncoating (Mirza and Weber, 1979). A possible sequence of uncoating is suggested by these authors as follows. Virions penetrate the cytoplasm without the loss of any particular class of structural proteins (although the mechanism of such penetration is not clear). In the cytoplasm the virions lose some portion of the major shell proteins and increase in density (150 S particles). These particles are transported to the nucleus, associate with it, and the cores penetrate the nucleus leaving behind the empty shell. In the nucleus, the cores shed the proteins giving rise to the end-product of uncoating, the DNA–terminal protein complex.

Polyoma virus uncoats in the nucleus (Mackay and Consigli, 1976). The initial uncoating intermediates are 48 S particles which lack the coat proteins. Soon after their release, the particles associate with host components forming the 190 S complex. The host proteins in this complex apparently represent the protein of nuclear matrix and histones (Winston *et al.*, 1980).

3. *Inhibitors of Uncoating*

There are drugs which specifically block uncoating and provoke the artificial accumulation of intermediates in infected cells, such as amantadine and rimantadine for influenza virus, arildone and rhodanine for picornaviruses, and bis(5-amidino-2-benzimid-azolyl) methane (BABIM) for respiratory syncytial virus. The structure and antiviral spectrum of these inhibitors are different, but the mechanism of their antiviral effect is similar: these drugs do not interfere with adsorption and penetration of the viruses and with host cell function but specifically prevent uncoating. They produce a maximal inhibitory effect when added early in the replicative cycle, i.e., at the time of virus inoculation.

Amantadine hydrochloride and its structural analog rimantadine hydrochloride are effective inhibitors of influenza A virus replication. The generally agreed conclusion is that the drugs affect some initial step in the virus replication cycle, most probably input virus uncoating (Skehel *et al.*, 1977). In accordance with this idea, it has been shown that rimantadine interferes with virus uncoating, inhibiting the reduction of RNase-resistant material (Koff and Knight, 1979). In the presence of amantadine and rimantadine nucleocapsids did not penetrate into the nuclei, and uncoating intermediates were accumulated in the perinuclear area of infected cells (Fig. 1) and were identified as nucleocapsids associated with M protein ("cores") (Bukrinskaya *et al.*, 1980). Biochemical and electron microscopic studies suggested that the cores were localized in close proximity to the nuclear membrane, possibly in perinuclear space (E. Anisimova and A. Bukrinskaya, unpublished data). Further studies *in vitro* on the interaction of cores with membranes were made using the method of Gregoriadis (1980), who showed that M protein isolated from influenza virus possesses hydrophobic sequences which could interact with lipids. The data of Gregoriadis indicate that M protein readily integrated with liposome membranes. However, the integration did not occur when rimantadine was added to the mixture. These data suggest that in infected cells rimantadine prevents the interaction of M protein in virus cores with cellular membranes and "freezes" the cores. As a result of "freezing," M protein cannot dissociate from the core surface and nucleocapsids cannot be released and initiate infection. This concept is in accordance with the findings that the M protein gene is responsible for the sensitivity of virus strains to amantadine (Lubeck *et al.*, 1978; Hay *et al.*,

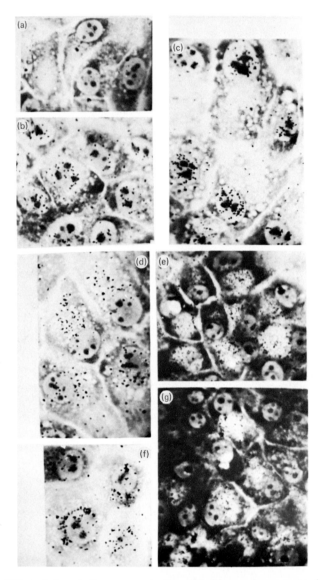

FIG. 1. Effect of rimantadine on penetration of influenza virus genome into nuclei. MDCK cells were treated with rimantadine (50 μg/ml) 30 minutes before infection, then infected with [³H]uridine-labeled influenza virus (strain WSN), and the same rimantadine-containing medium was added after infection. Thirty minutes after infection the cells were fixed and processed for autoradiography. (a) Cells treated with the mixture of WSN virus and WSN antiserum; (b, c) cells infected with WSN virus; (d–g) cells treated with rimantadine and infected with WSN virus.

1979). It seems surprising that lipophilic substances such as amantadine and its derivatives do not interfere with the events occurring at the plasma membrane but prevent uncoating in the perinuclear region. This situation has been explained by cell autoradiographic studies with tritium-labeled rimantadine, which revealed that large amounts of the drug were localized over the nuclei (presumably at the nuclear membrane).

Amantadine is also an inhibitor of Semliki Forest virus replication. Helenius *et al.* (1980a,b) interpret this inhibitory effect as follows: as a weak base, amantadine increases intralysosomal pH and prevents the fusion of virus membrane with the lysosome membrane. Therefore, the virus cannot escape from the lysosome and initiate infection.

It can be seen that both concepts concerning the mechanism of the amantadine effect on virus uncoating, as applied to influenza and Semliki Forest viruses, are very similar. They both explain antiviral effect by prevention of fusion with cellular membranes. The differences involve the site of drug application (perinuclear space or lysosome) and the factors which trigger the preventing mechanism. However, not to be excluded is the possibility that low pH could arise at the sites of fusion with cellular membranes other than the lysosome wall. As has recently been shown, amantadine rapidly concentrates in lysosomes; however, its antiviral effect is independent of intralysosomal concentration (Richman *et al.*, 1981).

Arildone [aryl-alkyl-diketone 4-6-(2-chloro-4-methoxyphenoxy)-hexyl-3,5-heptanedione] selectively inhibits the replication of some RNA and DNA viruses, such as picornaviridae and herpetoviridae, but not rhabdoviridae, myxoviridae, and poxviridae (Diana *et al.*, 1977). The drug does not interfere with the adsorption and penetration steps selectively blocking uncoating. The blockage of uncoating of poliovirus was demonstrated by a photosensitivity test: the virus grown in the presence of neutral red remained photosensitive in arildone-treated cells, thus indicating that the dye was preserved within the virus capsid. The similar effect of the drug on the early step in replication of herpes simplex virus (Kuhrt *et al.*, 1979) suggests a common mode of action of the drug for both viruses.

The mechanism of the drug effect is unknown. It is possible that arildone interacts with the icosahedral protein capsid, a structure common to both poliovirus and herpes simplex virus, and the changes in the capsid arrangement are sufficient to prevent the conformational alterations required to uncoat and release the viral genome into the cytoplasm. Such a model predicts that the other groups of viruses containing icosahedral protein capsids, such as togaviruses and adenoviruses, would also be sensitive to arildone unless the arildone – pro-

tein interactions are highly specific. In accordance with this prediction arildone was shown to act directly on poliovirus particles making
them more stable to alkaline pH (McSharry et al., 1979). On the other
hand, the drug was not found to be virucidal (Diana et al., 1977). It is
noteworthy that for antiviral activity, the lipophilic substituents on
the phenyl ring of arildone were shown to be essential. It was suggested
that the lipophilic nature of the drug accounted for its inhibition of
transport of thymidine and uridine across the plasma membrane
(McSharry et al., 1979). It could be responsible as well for interaction
with the plasma membrane and impairment of modifying sites for
poliovirus.

Rhodanine (2-thio-4-oxothiazolidine) is a specific inhibitor of echovirus 12 uncoating. As suggested, the effect of the drug is due to its
direct interaction with the virus capsid. Rhodanine-treated virions are
stabilized against heat inactivation and are not disrupted producing
130 S and 80 S particles and free RNA. In rhodanine-treated cells, no
elution of modified noninfectious A particles is observed (Rosenwirth
and Eggers, 1979), and a considerable portion of the virus preserves
infectious activity for hours after infection, accumulating in the cell
compartments other than the plasma membrane (Eggers, 1977; Eggers
et al., 1979).

Rosenwirth and Eggers (1979) represent the mechanism of the effect
of rhodanine on virus reproduction as follows. The virus particles are
stabilized by rhodanine in one single conformation so that their further
modification is impaired. These particles block the modifying sites at
the plasma membrane, and the remaining adsorbed particles have no
choice other than to penetrate into the cytoplasm. Rhodanine seems to
be irreversibly bound to the virus particles since even upon release of
the rhodanine block uncoating does not take place (Rosenwirth and
Eggers, 1979).

An aromatic diamidine, BABIM, was shown to inhibit infection induced by respiratory syncytial (RS) virus. When added to the cells
together with the virus, the compound blocked virus penetration. When
added after initiation of infection, the compound inhibited virus-
induced cell fusion (fusion from within). It did not interfere with virus
adsorption and replication and with cell fusion induced by other viruses
(paramyxovirus 3 and herpes simplex virus) (Dubov et al., 1980).
Dubov et al. suggest that the drug inhibits the fusion of RS virions with
cell membranes possibly by inhibiting arginine-specific esteropro-
teases which play a role in membrane fusion events.

Thus, all these drugs, in spite of differences in chemical nature and in
some manifestations, have similar effects on virus infection. They

"freeze" virus particles or uncoating intermediates preventing further uncoating and the release of a viral inner component which initiates infection. Most of them are lipophilic agents suggesting that their interaction with host cell lipids is essential for antiviral activity. It is possible that the universal mechanism of their action is the prevention of virus fusion with host cell membranes.

III. Individual Virus Groups

A. Enveloped Viruses

1. Orthomyxoviridae

a. *Virus Attachment Proteins.* The lipoprotein membrane has surface projections formed by two virus glycoproteins, hemagglutinin (HA) and neuraminidase (NA). Both glycoproteins are involved in interactions with the host cell plasma membrane. Nonglycosylated matrix (M) protein is believed to form a layer on the inside of the lipid bilayer (Lenard, 1978). Hemagglutinin accounts for approximately 30% of total virion protein, and is the major component of the viral envelope. It contains about 25% carbohydrate by weight (Waterfield *et al.*, 1979), the amino-terminal glycosylated region is external to the membrane, and a carboxy-terminal hydrophobic region is associated with the membrane. By the C-terminal end the spike is anchored in the lipid bilayer. Hemagglutinin subunits apparently do not penetrate through the lipid bilayer, unlike vesicular stomatitis and Sindbis virus glycoproteins (Oxford *et al.*, 1981). Functional HA is a trimer of HA polypeptides.

Hemagglutinin consists of either the precursor HA or the fragments HA_1 and HA_2 linked by disulfide bonds. Proteolytic cleavage of HA is not a precondition for hemagglutination, but is necessary for infectivity. This suggests that in addition to its role in adsorption, HA has another function. It has been recently shown that this function is penetration (see below).

The complete primary structure of hemagglutinin has been recently elucidated and turned out to be of great importance in understanding its biological functions. The conserved and variable regions were revealed by comparative analysis of different serotypes, and one of the conserved regions, consisting of about 20 hydrophobic amino acids, was located at the N-terminal end of HA_2 and was shown to be very similar to the N-terminal fragment of parainfluenza virus fusion (F) protein

(Gething *et al.*, 1978; Porter *et al.*, 1979; Waterfield *et al.*, 1979). This suggests a common function for these two proteins in virus penetration, a suggestion supported by other findings.

Since activation of infectivity requires a highly specific amino acid sequence at the N-terminus of HA_2, the cleavage of specific peptide bonds by specific enzymes is necessary for activation (although HA could be cleaved by proteinases of different specificity). The cleavage site of avian influenza virus (fowl plague virus) was shown to consist of six predominantly basic amino acids which are eliminated in the cleavage reaction (Porter *et al.*, 1979); with the other virus strains a single arginine residue is eliminated (Min-Jou, 1980; Sleigh *et al.*, 1980; Gething, 1980). The variation in the structure of the cleavage site may account for differences in susceptibility of HA to proteolytic enzymes in infected cells. Based on the correlation between the structure of HA and the pathogenicity of different influenza virus strains, Rott and his colleagues (Rott, 1979; Rott *et al.*, 1980; Klenk *et al.*, 1980; Klenk, 1980) suggested that the cleavage of HA determines host susceptibility and accounts for the spread of infection and, thus, for strain pathogenicity. Two enzymes are involved in the activation of the hemagglutinin, trypsin or a trypsin-like endoprotease furnished by the host, and an exopeptidase of the carboxypeptidase B type which appears to remove the arginine from the cleavage site, the second enzyme possibly being a constituent of the virus particle (Garten *et al.*, 1981).

The high specificity of recognition of the cleavage site by the host endoprotease is supported by the fact that HA could be cleaved only by an homologous enzyme. Thus, when influenza virus is grown in MDCK cells, HA in cell plasma membrane is cleaved only when the cells are incubated with canine sera, HA in porcine cultured cells is cleaved by the enzyme in porcine sera, and so on. Host-dependent variations in the structure of the cleavage site most probably involve carbohydrates (Zhirnov *et al.*, 1981).

The second glycoprotein on the surface of influenza virus is a neuraminidase. An important consequence of its presence is that no sialic acid can be detected in the virus particle. However, the structure and the functions of NA are less defined than that of HA. There are data suggesting that NA is necessary for virus entry into the host cell (Huang *et al.*, 1980b).

b. Cellular Receptors and Attachment. The virus attaches to the host cell surface by binding via the hemagglutinin spike to neuraminic acid-containing receptors. A large variety of animal cells possess sialic acid-containing molecules on the cell surface and hence cellular receptors for influenza viruses are not the primary determinants of host specificity. A rare example of an animal cell lacking receptors for influenza virus is a resistant line of MDBK cells that was initially persistently infected with the virus (Arens et al., 1981).

A question still being debated is whether glycolipids or glycoproteins or both are specific receptors for influenza viruses. The opinion that glycoproteins are universal influenza virus receptors, by analogy with erythrocytes, is based on the fact that the virus binds to soluble and erythrocyte-associated sialylated glycoproteins. On the other hand, influenza virus can initiate infection when sialylated glycoproteins are bound to hemagglutinin. This casts doubt on the assumption that receptors on host cells are like those on erythrocytes (Lakshmi and Schulze, 1978). The possibility that sialo glycolipids are specific receptors is supported by the data obtained for paramyxoviruses (Haywood, 1974a,b, 1975a,b; Holmgren et al., 1980). In accordance with these data, Schulze (1981) isolated the fraction from MDBK cell membranes which inhibited influenza virus binding and infectivity and identified the inhibitory component as glycolipid with terminal galactose residues.

The studies on virus adsorption were preferentially performed in model systems using liposomes or cell fractions, and the subsequent fate of the adsorbed virus particles was not precisely defined. To avoid these shortcomings in examining the receptor functions of gangliosides, we used the whole cells (Ehrlich tumor cells). Their native receptors were removed by Vibrio cholerae neuraminidase and the cells were loaded with certain gangliosides. Then the cells were infected with [^{3}H]uridine-labeled influenza virus and its adsorption and penetration were determined and compared to that of native cells and cells treated with neuraminidase. The adsorption of the virus on the cells loaded with three gangliosides, GM_1, GD_{1a}, and GT_{1b}, was fully restored up to the level of native cells while GM_2 was inactive (Fig. 2a). In accordance with the data of Schulze, three active gangliosides possessed terminal galactose while GM_2 possessed terminal galactosamine. One of three gangliosides, GT_{1b}, was highly active in virus penetration and accumulation of radioactive virus genomes in nuclei. The amount of ^{3}H radioactivity in the nuclei of the cells loaded with GT_{1b} was seven- to eightfold higher than in the nuclei of native cells (Fig. 2b and c). These results suggest, first, that certain gangliosides are specific receptors for influenza virus, and, second, that a ganglioside is involved in virus penetration and the delivery of virus genome to the nuclei (Bukrinskaya et al., 1982a; Bergelson et al., 1982).

The involvement of gangliosides in the interaction with the virus does not exclude the role of glycoproteins in adsorption. Virus hemagglutinin might initially react with sialic acid of cellular glycoproteins which could represent primary receptors. However, to bring the

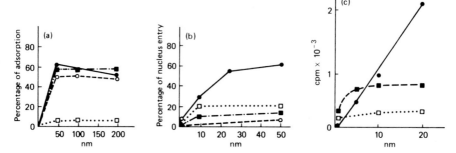

FIG. 2. Effect of gangliosides on influenza virus binding (a) and penetration into the nuclei (b, c) of the host cells. Ehrlich tumor cells (5×10^6 in the sample) were pretreated with neuraminidase (10 units per 5×10^6 cells), loaded with different concentrations of gangliosides isolated from beef brain (GM_1, GD_{1a}, and GT_{1b}) and mouse liver (GM_2), infected with a mixture of unlabeled and [^{3}H]uridine-labeled purified influenza virus (FPV) (6.0 and 0.6×10^2 hemagglutinin units, respectively, 5.5×10^3 cpm in the sample), and incubated for 45 minutes at 4°C. (a) The amount of cell-bound virus was determined by measuring unbound radioactivity in the supernatant after removing the cells by centrifugation. The radioactivity bound to neuraminidase-treated control cells was derived from the radioactivity bound to ganglioside-loaded cells, and the amount of bound virus was expressed as percentage of the virus added. (b, c) The cells were fractionated and the nuclei were purified by repeated 1% Triton X-100 washings; the radioactivity in the nuclei was measured and expressed in cpm (c) and as percentage of cell-bound radioactivity (b). ●, GT_{1b}; ○, GM_1; ■, GD_{1a}; □, GM_2.

viral envelope into closer contact with the host cell plasma membrane, hemagglutinin probably interacts with the other receptor which is involved in fusion of the virus envelope with the cell membrane, and this receptor is a ganglioside. Such a hypothetical sequence of events correlates well with the data of Huang *et al.* (1980b), that virus neuraminidase is involved in the adsorption and penetration of influenza virus, possibly removing neuraminic acid from the primary externally exposed receptor or unmasking a new receptor which reacts with the N-terminus of HA_2.

c. Penetration. In studies of influenza virus infection by electron microscopy, both mechanisms of cell entry, endocytosis (Dales, 1973; Dourmashkin and Tyrrell, 1974) and fusion (Morgan and Rose, 1968), were demonstrated. In support of fusion mechanism, Huang *et al.* (1980a,b, 1981) reported that liposomes containing both virus glycoproteins could fuse with tissue culture cells, and the virus is capable of inducing hemolysis and cell fusion under certain conditions (low pH). These data suggest that fusion is the mode of cell entry.

Virus particles containing cleaved hemagglutinin are able to integrate with cell membranes and to alter their fluidity (Nicolau *et al.,*

1979; Kurrle *et al.*, 1979). The role of HA proteolytic cleavage in fusion has been further demonstrated by experiments with liposomes containing cleaved and uncleaved HA. Liposomes with cleaved HA fused with cell membranes and the liposomal content was injected into the cytoplasm. By contrast, liposomes with uncleaved HA were adsorbed on the cell surface but did not fuse until treated with trypsin to cleave HA (Huang *et al.*, 1980). After proteolytic cleavage, the hydrophobic segment of HA_2 becomes exposed; this segment is probably involved in fusion. The situation looks very much like that existing for paramyxoviruses, where fusion activity is associated with proteolytic cleavage of F_0 into two F glycoproteins. The common biological role of two glycoproteins, HA_2 of influenza viruses and F_1 of paramyxoviruses, is further confirmed by the finding that both these proteins possess an N-terminal structure very similar in amino acid sequence and extreme hydrophobicity (Gething *et al.*, 1978).

In addition to a cleaved HA, neuraminidase is also required for fusion. Liposomes containing only the cleaved HA adsorb to cells without causing fusion. To induce fusion, soluble neuraminidases of bacterial or viral origin have to be added and are equally active, suggesting that the enzymatic but not structural role of this enzyme is important. Specific antineuraminidase antibody prevented fusion of liposomes and cells (Huang *et al.*, 1980b). In the light of these findings, the view that neuraminidase is not involved in the early stages of infection needs to be reexamined.

On the other hand, Oxford *et al.* (1981) did not observe the fusion of HA-containing liposomes with cultured cells. However, the presence of neuraminidase and the amount of neuraminidase attached to liposomes were not established in their experiments.

An absolute requirement for fusion and hemolysis was low pH; an optimum pH for different influenza virus strains varied from 5.0 to 5.75. Hemolysis decreased sharply with an increase of 0.5 pH units (in comparison, paramyxoviruses hemolyzed maximally at pH 6.25 or 7.00, and hemolysis remained at high levels within a wide pH range). Similarly, BHK-21 cells were extensively fused by influenza viruses at pH 5.75. Although the molecular basis of the pH dependence of hemolysis and cell fusion is not yet clear, the virus glycoproteins seem to be involved since each virus strain has its specific pH optimum for hemolysis and fusion (Huang *et al.*, 1981). As our unpublished data assume, low pH is required by virus neuraminidase.

These data are compatible with the suggestion that low pH could arise at the sites of cell membranes with attached virus particles to provide penetration by fusion. The delivery of the virus to lysosomes for

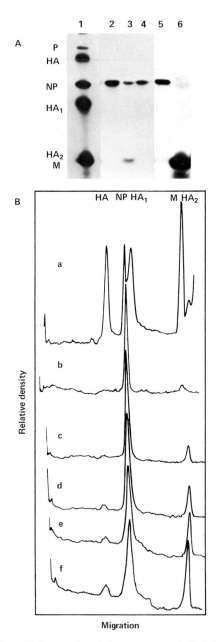

FIG. 3.   Polypeptides of influenza input virus components isolated from infected cells.
Chicken fibroblasts (A) and MDCK cells (B) were infected, respectively, with FPV and
WSN viruses labeled with ^{14}C-amino acids. Thirty minutes after incubation at 37°C the

fusion with lysosome membrane as suggested for Semliki Forest virus (Helenius *et al.*, 1980a,b) hardly occurs since lysosomes have been shown not to be involved in early virus–cell interaction (Dourmashkin and Tyrrell, 1974). One could suggest that a local drop in pH value along the cell membrane may occur to allow the virus particle to enter the cell. Again, the factors which trigger a drop in pH are obscure. One of the possibilities is the removal of arginine during hemagglutinin cleavage by specific carboxypeptidase B (Garten *et al.*, 1981), but how it could correlate with fusion is unclear.

Although it is generally agreed that animal viruses can attach to cells but do not penetrate the cells at subphysiological temperature such as 4°C, influenza virus appears to be an exception to this rule. Influenza virus (FPV) RNA (Stephenson and Dimmock, 1975) and internal virus proteins NP and M (Stephenson *et al.*, 1978; Hudson *et al.*, 1978) were found in the nuclei of the BHK-21 cells infected at 4°C. It is believed that micropinocytosis is the mechanism which accounts for virus penetration at 4°C.

The mechanism of the transport of influenza virus particles toward the nuclei is unknown. Entry of influenza virus into nuclei has been shown to occur in the presence of colchicine and cytochalasin B; that rules out a role for microfilaments and microtubules (Patterson *et al.*, 1979).

d. Uncoating. The final products of influenza virus uncoating are nucleocapsids which, after their release from the virus envelope, enter the nuclei and are involved as templates in primary transcription (Hudson *et al.*, 1978; Bukrinskaya *et al.*, 1979). However, the events which lead to virus uncoating are poorly defined. It is generally agreed that nucleocapsids are released from lipoprotein membrane and M protein in one step, and uncoating intermediates do not exist in infected cells. However, by treating the crude nuclei of influenza virus-infected cells with citric acid the input virus components were isolated, representing nucleocapsids in association with M protein ("cores") (Fig. 3).

---

cells were fractionated and input virus components were isolated from nucleus-associated cytoplasm and nuclear extracts by centrifugation in velocity glycerol gradients and CsCl density gradients. The polypeptides of subviral components were analyzed by SDS–PAGE in 10% polyacrylamide gels processed for autoradiography. (A) 1, FPV used as a marker; 2,4,5, nucleocapsids pelleted from the glycerol gradient fractions; 3, intermediate forms isolated from the nucleus-associated cytoplasm; 6, the proteins of perinuclear membranes from FPV-infected cells. (B) Polypeptides of WSN subvirus components of different densities isolated from nucleus-associated cytoplasm by centrifugation in CsCl gradients. (a) WSN virus used as a marker; (b) the 1.35 nucleocapsids; (c–f) intermediate forms of the density: 1.33 (c), 1.31 (d), 1.30 (e), and 1.28 (f) gm/cm^3.

Their number essentially increase when infected cells are pretreated with inhibitors of virus uncoating, amantadine and rimantadine (Bukrinskaya *et al.*, 1980), suggesting that the cores represent uncoating intermediates. Similar structures, free and associated with endocytic vacuoles containing engulfed virus particles, were observed early in infection by electron microscopy (Dourmashkin and Tyrrell, 1974). In electron micrographs of amantadine-treated cells, the cores were detected in perinuclear cytoplasm and in perinuclear space (Anisimova and Bukrinskaya, unpublished results). Biochemical studies showed that they were not free but were associated with cell membranes. This suggests that they might be transported toward nuclei along the endoplasmic reticulum. This agrees well with the highly hydrophobic properties of M protein.

The existence of intermediate forms could mean that uncoating occurs in two steps: first, the release of cores from the lipoprotein membrane, and second, the release of nucleocapsids from the M protein. This event apparently occurs in close proximity to the nuclei, since the cell membranes isolated from the crude nuclei fraction contained M protein only (Fig. 3A). On the basis of these data, the following concept of influenza virus penetration and uncoating could be proposed. Virus particles penetrate the cell by fusion of their lipoprotein membrane with the host cell plasma membrane or by endocytosis followed by fusion with the membrane of the endocytic vacuole. As a result of fusion, the cores are liberated into cytoplasm and transported to the nuclear membrane along the membranes of the endoplasmic reticulum. At the nuclear membrane, the second step of uncoating, the release of nucleocapsids from M protein, occurs, probably by fusion of M protein with perinuclear membranes. The fusion results in integration of M protein with the membranes while the naked nucleocapsids enter the nuclei (Fig. 4).

According to this model, both steps of uncoating occur by a fusion mechanism and, thus, do not require participation of cellular enzymes. Since fusion could occur even at 0°C (White and Helenius, 1980), such a mechanism correlates well with the findings that at low temperature influenza virus enters the cells (Stephenson and Dimmock, 1975) and nucleocapsids are released and accumulate in the nuclei (Hudson *et al.*, 1978). In accordance with this, we observed that both steps of uncoating occur at 4°C, although the second step (the release of nucleocapsids) proceeds much more slowly than at 37°C, and nucleocapsids mostly reside in the perinuclear region but not in the nuclei (Kornilayeva and Bukrinskaya, 1980). Apparently, the entry of nucleocapsids into the nuclei requires a higher temperature.

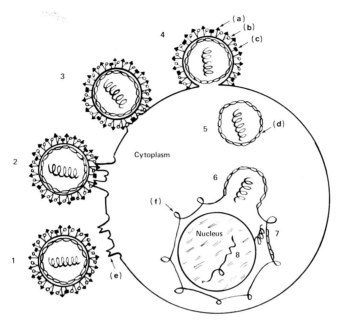

FIG. 4.   Hypothetical scheme of influenza virus uncoating in the host cell. (a) HA_1; (b) HA_2; (c) NA; (d) M protein; (e) cellular receptors; (f) perinuclear membranes; 1–8, steps of virus–cell interaction.

2. Paramyxoviridae

a. *Virus Attachment Proteins.* Two glycoproteins are inserted into the lipoprotein envelope forming spike-like projections from the surface of the virion. The larger glycoprotein, designated HN (hemagglutination–neuraminidase protein), possesses the hemagglutination and receptor-binding activity of the virus and also exhibits neuraminidase activity. In infected cells, inactive HN_0 is converted into biologically active HN by proteolytical cleavage, a glycopeptide of MW 8000 being removed during the cleavage process (Kohama *et al.*, 1981). The other glycoprotein, designated F (fusion protein), is involved in virus penetration through fusion with host cell membranes, as well as in virus-induced hemolysis and cell fusion. Among the viruses inducing fusion, this protein is a single protein whose involvement in the fusion process is well documented. Its activity is associated with proteolytic cleavage of a precursor glycoprotein F_0 and the production of glycoprotein F (Homma and Ohuchi, 1973; Scheid and Choppin, 1974; Nagai *et al.*, 1976) which consists of two disulfide-linked polypeptide chains F_1 and F_2 (Scheid and Choppin, 1977). The orientation of F glycoprotein in Sendai (Scheid and Choppin,

1977) and Newcastle disease viruses (Samson *et al.*, 1980) is NH_2–F_2–F_1–COOH. The hydrophobic carboxy-terminal region serves to anchor the protein in the lipid bilayer. Almost the entire polypeptide complement of F_0 is preserved in the cleaved glycoprotein $F_{1,2}$. This suggests that the activation is due to the rearrangement after cleavage in such a way that areas can interact that are far apart from each other in the uncleaved molecule. Therefore, the change in the three-dimensional structure of the F protein seems to be necessary to bring the amino terminus of F_1 into a functional position (Kohama *et al.*, 1981).

As in the case of influenza virus HA_1, the cleavage of paramyxovirus F and HN glycoproteins is necessary for infectivity, being of major significance for the spread of infection and thus for pathogenicity (Klenk, 1980; Garten *et al.*, 1980).

b. *Cellular Receptors and Attachment.* Sialic acid residues of membrane components are known to serve as the binding sites for paramyxoviruses (reviewed by Bächi *et al.*, 1977; Rott and Klenk, 1977). Receptors are uniformly distributed over a cell surface (Knutton *et al.*, 1977). Treatment with *Vibrio* cholerae neuraminidase reduces the susceptibility of cells to infection; endogenous regeneration of Sendai virus receptors is detectable within 30 minutes and is complete in 6 hours (Markwell and Paulson, 1980).

Since paramyxoviruses agglutinate erythrocytes of most mammalian and fowl species, the cellular receptors are intensively studied using this simple system. The main binding sites of Sendai virus to human erythrocytes are sialic acid residues of glycophorin, one of the major glycoproteins on the erythrocyte surface (Steck, 1974). As to animal cells, the possibility that the sialic acid residue is bound to glycolipids as substrates is also accepted at present, and a number of experimental findings indicate that gangliosides are specific receptors involved in virus binding and fusion with the cell membrane (Haywood, 1974a,b, 1975a,b; Holmgren *et al.*, 1980). On the other hand, Wu *et al.* (1980), using glycoconjugate-containing liposomes, found that virtually all the receptor activity for Sendai virus was in the glycoprotein rather than in the ganglioside fraction.

c. *Penetration.* Attachment of the virus particle to the cell surface is followed by the fusion of the viral envelope with the plasma membrane. For this virus group, it is generally agreed that fusion but not endocytosis is the normal mechanism whereby the viruses enter the cell, an opinion based on a series of morphological and biochemical findings (reviewed by Ishida and Homma, 1978; Choppin and Scheid, 1980; Knutton, 1978).

Fusion results in the integration of the viral envelope glycoproteins with the plasma membrane (Kohn, 1975) and antigen modification of the cell surface in such a way that the cell is killed by antiviral antibody and complement (Fan and Sefton, 1978). The process is followed by intracytoplasmic injection of the viral nucleocapsids, thus accomplishing in a single step penetration, uncoating, and transport of the virus genome to the cytoplasm of the host cell.

Fusion is mediated by the second viral envelope glycoprotein, F protein. For activation, the precursor polypeptide F_0 must be cleaved by specific host proteinases to form two disulfide-linked polypeptide chains, F_1 and F_2 (Scheid and Choppin, 1977). The virions which contained F_0 could not infect host cells, hemolyze erythrocytes, or mediate cell fusion. F_0-containing virions could be activated in vitro with trypsin: its effect is similar to that of host-specific proteinases. The proteinases contained in animal sera and in chick embryo allantoic fluid were identified as plasmin (Lazarowitch et al., 1973). For fusion activity, a new amino-terminus on F_1 arising after F_0 cleavage is required. This sequence is extremely hydrophobic and highly conserved among different paramyxoviruses (Scheid et al., 1978). There is evidence that this fragment of F_1 is involved in the fusion reaction (Gething et al., 1978; Richardson et al., 1980). There is also immunological evidence that F protein mediates penetration of the virus to the cell. Merz et al. (1981) showed that the monoclonal anti-F antibodies neutralize infectivity of SV5 and other paramyxoviruses when virus and antibody are mixed before infection and completely inhibit the spread of infection from cell to cell when added after infection of the first cell. In contrast, monoclonal anti-HN antibodies are incapable of preventing the spread of infection to adjacent cells. Similar immunological data with Newcastle disease virus were obtained by Avery and Niven (1979).

Recent data suggest that the presence of neuraminidase is also essential for fusion. Liposomes containing F glycoprotein of Sendai virus or F glycoprotein and HA of influenza virus did not possess fusion activity (Ozawa et al., 1979; Huang et al., 1980a). Liposomes containing F protein acquired hemolytic activity only after addition of lectin (Hsu et al., 1979). The replacement of Sendai virus HN by HA and NA of influenza virus but not HA alone preserved the fusion activity of liposomes containing F protein (Huang et al., 1980a). The exact mechanism of the cooperative role of neuraminidase in fusion is unknown. It could be suggested that the removal of neuraminic acid from cell membrane receptors is essential for closer contact between virus and cell membranes.

Infection with Sendai virus produces fusion between infected cells promoted by the same viral glycoprotein that promotes virus entry, that is, by F protein. This phenomenon has been reviewed by Hosaka and Shimizu (1977). Its morphology, as studied by the techniques of thin sections, negative staining, and freeze fracture etch, has been reviewed by Knutton (1978).* Cell fusion and hemolysis induced by paramyxoviruses could be inhibited by oligopeptides which possessed an amino acid sequence similar to that of the N-terminus of the F_1 polypeptide (Scheid *et al.*, 1978). A number of synthesized oligopeptides which resembled this region of the F_1 protein effectively inhibited membrane fusion activity of the viruses, possibly by competing with the fusion protein for specific sites in the cell plasma membrane or by altering the conformation of the F_1 N-terminus necessary for activity (Richardson *et al.*, 1980).

   d. Uncoating. Virus penetration by fusion coincides with partial uncoating of the virus particle. Both virus glycoproteins, HN and F, remain integrated with the plasma membrane, while the inner component is released into the cytoplasm. Logically, this component should represent virus cores (nucleocapsids surrounded by matrix protein), suggesting that the final uncoating, the release of nucleocapsids, occurs in the cytoplasm. However, the nucleocapsids were the single subviral component observed in paramyxovirus-infected cells. If uncoating intermediates do exist, they are unstable or the rate of uncoating is extremely high.

   Intact parental Sendai virus nucleocapsids are recovered from the cytoplasm at an early interval after infection, and are rapidly converted to structures of another kind. Such modified nucleocapsids are characterized by a lower sedimentation rate and a higher buoyant density and by sensitivity of their RNA to digestion by ribonuclease. Uncoiled threads of nucleocapsids are observed electron microscopically. These data are compatible with the suggestion that the final products of Sendai virus uncoating are modified (apparently uncoiled) nucleocapsids which have lost a part of their protein (Bukrinskaya, 1973).

3. Rhabdoviridae

   a. Virus Attachment Proteins. Among four to five major polypeptides, one is a glycoprotein which is inserted into the lipid envelope and forms spikes on the virus surface. The glycoprotein interacts with the cell plasma membrane and plays a critical role in virus infectivity. Removal of the G protein spikes with proteases abolishes viral infectivity of vesicular stomatitis virus (VSV) which can be restored by in-

---

   * Sendai virus is widely used as a tool for producing hybrid eukaryocytes employed in mammalian cytogenetic studies.

cubating the spikeless virions with detergent-extracted G protein. A possible explanation for the increase in infectivity is that the hydrophobic tail fragment of the glycoprotein is inserted into the spikeless virus membrane and enables it to attach to the cell surface.

Glycoprotein contains about 10% carbohydrate in the form of two to four oligosaccharide chains rich in sialic acid. The sialic acid is directly involved in the attachment of the virus to the cells and its agglutination of red blood cells. Removal of the sialic acid with neuraminidase resulted in a 100-fold drop in infectivity and a loss of the ability of the virus particles to agglutinate erythrocytes. These activities could be restored to their original level by resialylation of the neuraminidase-treated particles (Schloemer and Wagner, 1975a,b). On the other hand, Cartwright and Brown (1977) have reported that sialic acid of the virus is not essential for virus infectivity.

The reconstitution of virus membrane has been successfully carried out using the micellar form of the VSV glycoprotein and preformed phosphatidylcholine vesicles. G is spontaneously partitioned into sonicated vesicles so that the hydrophobic tail fragment becomes resistant to protease digestion (Petri and Wagner, 1980).

Glycoprotein of VSV has been shown to contain one or two molecules of covalently attached fatty acid bound to the hydrophobic tail which is apparently involved in the interaction of G protein with the virus membrane (Schmidt and Schlesinger, 1979). In reconstituted virus membrane, covalently attached fatty acid is found on this hydrophobic tail fragment (Petri and Wagner, 1980).

b. *Cellular Receptors.* Cellular receptors are not identified for this virus family. The major histocompatibility antigen is not involved as a specific receptor in virus attachment since cells lacking this antigen support reproduction of VSV (Haspel *et al.*, 1977; Oldstone *et al.*, 1980).

c. *Penetration and Uncoating.* Information concerning the mode of virus entry into the host cell is highly controversial. The belief that VSV particles enter the cell by endocytosis is based on the fact that unlike Sendai virus, the cell surface of VSV-infected cells is not antigenically modified, and the estimation of cell killing by antiviral antibodies and complement showed that fusion occurs only 3% of the time (Fan and Sefton, 1978). On the other hand, chloroquine, which does not inhibit endocytosis, reduces the virus yield, while cytochalasin B, which blocks endocytosis, has no effect on virus replication, suggesting that endocytosis is not essential for VSV infection (Coombs *et al.*, 1981). The penetration by membrane fusion has been demonstrated by electron microscopic and biochemical evidence showing that the envelope of infecting VSV fuses with the host cell plasma membrane and viral glycoprotein remains on the cell surface while nucleocapsids enter the cytoplasm (Heine and Schneitman, 1971). In some cells (mouse L cells, rat embryonic fibroblasts transformed by Rous sarcoma virus) VSV induces fusion from within (Nishiyama *et al.*, 1976; Chany-Furnier *et al.*, 1977). Fusion occurs at a low multiplicity of infection, and requires the uncoating of the virus in the cell, the synthesis of G and M proteins, and their incorporation into the cell membrane (Chany-Fournier *et al.*, 1977). The mechanism of this fusion process is not clear. Chany-Fournier *et al.* (1977) suggest that the insertion of proteins G and M into the cell membrane could not be directly responsible for cell fusion but could create in this area an imbalance of phospholipids which then diffuse through membrane junctions to the surrounding cells, thereby provoking cell fusion.

4. Togaviridae

This family contains two groups defined by their serological characteristics—alphaviruses and flaviviruses. Semliki Forest virus (SFV) and Sindbis virus are well-studied members of the alphavirus group.

a. *Virus Attachment Proteins.* The viruses contain only three or four structural proteins and are the simplest of the enveloped viruses. Among these proteins, two or three are glycoproteins which form spike-like projections on the external surface of the virus membrane and are involved in the interaction with the host cell plasma membrane. The structure of glycoproteins of SFV has been well studied by Garoff and colleagues (reviewed by Garoff and Simons, 1980; Garoff, 1981). In this virus, each spike glycoprotein is an oligomer of three glycopolypeptides, E1, E2, and E3, in a noncovalent association. About 90% of the spike protein is on the external side of the lipid bilayer; E3 is completely on the outside, whereas E2 and E1 have hydrophobic tails which attach them to the membrane. N-terminal regions of the proteins are external to the bilayer and carboxy-terminal regions penetrating into the bilayer.

The most outstanding feature of the SFV glycoprotein is that it spans the bilayer and a small carboxy-terminal segment of about 30 amino acids is located on the internal side of the membrane and could be involved in linkage with the nucleocapsid.

The spikes of Sindbis virus are similarly organized. They consist of two glycoproteins, E1 and E2. The proteins are similarly oriented in the lipid bilayer and span the virus membrane (Bonatti *et al.*, 1979). A covalent attachment of fatty acid to the hydrophobic fragment of glycoprotein as in the case of VSV has been demonstrated (Schmidt *et al.*, 1979; Schmidt and Schlesinger, 1979). All protein-bound carbohydrates are on the external side of the bilayer, perhaps mediating the close contact of spike proteins.

The role of the sialic acid of the virus glycoprotein in virus–cell interaction is not precisely established. Virus grown in cells which lack neuraminic acid residues on their surface, and thus fail to incorporate sialic acid molecules during maturation, is less infective than sialic acid-containing virus (Schloemer and Wagner, 1975b). A role for sialic acid is thereby suggested in the adsorption of these viruses to cells. On the other hand, there are data that indicate that the sialic acid of the virus glycoprotein does not play a major role in virus infectivity (Kennedy, 1976).

b. Cellular Receptors and Attachment. Helenius *et al.* (1978) reported that SFV and the glycoprotein isolated from the virus bound to MHC antigen on the surface of mouse and human cells. However, the fact that the cells lacking this antigen support virus reproduction (Oldstone *et al.*, 1980) indicates that this cellular glycoprotein is not a specific receptor. The chemical nature of specific receptors has not yet been identified.

Semliki Forest virus particles bind to multiple receptors, the preferential binding sites being microvilli (Helenius *et al.*, 1980). The binding capacity widely varies depending on cell type and virus strain. For SFV, it is from 1000 to 150,000 virus particles per cell. BHK-21 cells at 4°C bind about 5×10^4 virus particles with half-saturation at 10^{-10}–10^{-11} M virus concentrations (Fries and Helenius, 1979).

c. Penetration. The data on the mode of host cell entry for these viruses are controversial. The findings of Helenius *et al.* (1980a,b) and Fan and Sefton (1978) support the view that endocytosis is a general mechanism of cell entry while other observations are compatible with the model for direct penetration of the plasma membrane by membrane fusion.

Fan and Sefton found that penetration of Sindbis virus does not introduce viral antigens into the host cell plasma membrane and concluded that the virus enters the cell by another mechanism (presumably by endocytosis). This conclusion is supported by studies of Helenius *et al.* (1980a,b) who demonstrated that SFV entered the cells by a process morphologically similar to receptor-mediated endocytosis, i.e., through coated pits and coated vesicles.

On the other hand, Coombs *et al.* (1981), reported that Sindbis virus, although internalized by endocytosis, could not replicate in the presence of chloroquine, while cytochalasin B, which blocked endocytosis, did not interfere with virus replication. Coombs *et al.* (1981) conclude that the virus enters the cell by a mechanism other than endocytosis. It should be noted, however, that chloroquine inhibits receptor-mediated endocytosis (Fitzgerald *et al.*, 1980), and, in the presence of the drug, the virus might be internalized by nonspecific endocytosis.

d. Uncoating. In a series of papers, Helenius and his colleagues described the fate of SFV particles after their internalization by endocytosis and proposed a scheme of virus uncoating (Helenius *et al.*, 1980a,b). According to their scheme, engulfed virus particles are delivered to lysosomes where, because of a low pH value, the fusion between the virus envelope and lysosome membrane occurs, and virus nucleocapsids are released into the cytoplasm.

To prove the possibility of fusion, Helenius *et al.* used liposomes with a lipid composition similar to that of the lysosomal membrane. Fusion was demonstrated electron microscopically by the formation of large vesicles containing viral glycoproteins on the surface and nucleocapsids inside, and biochemically by trapping ribonuclease or trypsin within the liposomes. The digestion of viral RNA by trapped ribonuclease was proposed as a quantitative assay for fusion (White and Helenius, 1980). The fusion reaction was rapid and effective, and required a certain lipid composition of liposomes (the presence of cholesterol). An absolute requirement for fusion was low pH: fusion occurred only when the pH was 6.0 or lower, and the efficiency of fusion dropped dramatically at higher pH values. It was suggested that low pH might induce a change in the spike glycoprotein which is necessary to trigger the interaction with membrane lipids (although the binding of isolated SFV glycoproteins to the cell surface at low pH is nonspecific and nonsaturable) (Fries and Helenius, 1979). A low pH is critical as well for Sindbis virus attachment to protein-free liposomes (Mooney *et al.*, 1975) and for virus fusion with red blood cell membranes (Väänänen and Kääriäinen, 1979, 1980).

Helenius *et al.* (1980a,b) connect the inhibitory effect on virus reproduction of weak bases such as chloroquine, ammonium chloride, amantadine, methylamine, and tributylamine with their entering the lysosomes and raising the intralysosomal pH. As a result, fusion between viral and lysosomal membranes does not occur, and the viral genetic material cannot escape from the lysosome.

5. Retroviridae

a. Virus Attachment Proteins. The product of gene *env* is presumably involved in early virus–cell interaction. For mammalian types B and C viruses, it was shown that the *env* product is a glycoprotein with a molecular weight of 75,000 or 90,000–96,000 (type C viruses). It is a precursor of envelope proteins and is designated as gPr. The precursor is proteolytically cleaved to form a major glycosylated glycoprotein (MGP) of MW varying from 51,000 to 85,000, and nonglycosylated low-molecular-weight proteins (reviewed by Altstein and Zhdanov, 1979). Usually the second glycoprotein, in addition to MGP, is detected in the envelope, with the lower MW (gp20 to gp45).

The precursor gPr90 of mammalian type C oncoviruses is cleaved to produce glycosylated gp69/71 and a nonglycosylated p15(E) which is C-terminal to gp69/71. p12(E) is a product of proteolytic cleavage of p15(E) (Karshin *et al.*, 1977). Major glycosylated glycoprotein is bound by disulfide linkage to a highly hydrophobic glycosylated or nonglycosylated protein which apparently serves to fix viral glycoprotein complex (VGC) in the lipid membrane. Complexes from two or three VGCs comprise the projections of the envelope (reviewed by Altstein and Zhdanov, 1979). Major glycosylated glycoprotein is involved in virus binding to cells. Infection can be fully prevented by pretreating the virus with antibodies against this glycoprotein.

Using monoclonal antibody, several antigen sites have been identified in gp70 envelope protein of murine leukemia virus. gp70[f] antigenic determinant (epitope) has been in proximity to the site responsible for the ecotropic-specific binding of virus to the cell surface (Nowinsky *et al.*, 1981).

Polypeptide P12(E) is the most hydrophobic of the viral proteins (Brouwer *et al.*, 1979), the property characteristic for the proteins involved in virus–cell interaction. In this aspect and in their poor glycosylation, this protein is similar to HA_2 protein of influenza virus which is involved in fusion with cell membrane (Huang *et al.*, 1980b).

b. Cellular Receptors and Attachment. Major envelope glycoprotein specifically interacts with cellular receptors (Fowler *et al.*, 1977). Their presence on the cell surface determines host specificity for the viruses. For example, the host range of avian RNA tumor viruses was shown to be specified by the interaction of the envelope with the cell surface. This led to the division of this virus group into five "host range" subgroups lettered A to E (reviewed by Weiss, 1976). Host cell resistance might be overcome by direct intracellular inoculation of virions using, for example, Sendai virus-induced membrane fusion. Cellular receptors are under genetic control (reviewed by Weiss, 1975, 1976). They are not identified chemically.

The receptors on the surface of mouse and rat cells were shown to exhibit identical specificity for mouse mammary tumor virus binding and were able to distinguish different classes of the virus (Altrock *et al.*, 1981).

The number of cell receptors on murine cells for this virus was calculated. The different cell lines bound from 2.6×10^2 to 3.8×10^3 virus particles. The calculated numbers of receptors per murine cell were 140- to 2000-fold less than has been reported for the murine leukemia (MuLV) viruses (DeLarco and Todaro, 1976). The binding affinity constant reported for Rauscher MuLV gp70 $K_0 = 3.5 \times 10^8 \ M^{-1}$ (Kalyanaraman *et al.*, 1978). These data demonstrate that different receptors are recognized by mouse mammary tumor virus (MMTV) and MuLV.

c. Penetration and Uncoating. Virus particles most likely enter the cytoplasm by a fusion mechanism. Fusion from without and fusion from within have been described for a number of members of this family.

Early virus-induced syncytium formation has been demonstrated in the culture of indicator (F81 cells) infected with concentrated preparations of bovine leukemia virus. The process had all the features of fusion from without: syncytia appeared rapidly after infection, within 2 to 4 hours after virus inoculation, and their formation did not require *de novo* synthesis of DNA and protein and the functional virus genome (Graves and Jones, 1981). To enhance syncytium formation Graves and Jones (1981) pretreated the cells with DEAE-dextran or Polybrene. The same virus induced syncytia by a second mechanism, fusion from within, that requires infectious virus (Diglio and Ferrer, 1976; Benton *et al.*, 1978). Baboon endogenous virus induces syncytium formation in several transformed human cell lines (Tanaka *et al.*, 1981). Cell fusion is typical fusion from without resistant to cycloheximide treatment.

Fusion activity has been discovered as well in the virions of murine leukemia virus (Zarling and Keshet, 1979), simian sarcoma virus (Ocho *et al.*, 1980), and D-type viruses. Mason-Pfizer monkey virus is known to induce cell fusion in normal human and nonhuman primate cell lines (Chatterjee and Hunter, 1979), and this property is shared by the other two D-type viruses, Langur virus and Squirrel monkey retrovirus (Chatterjee and Hunter, 1980). The mechanism by which cell fusion is induced by these viruses seems to be different from that reported for paramyxoviruses and other retroviruses (visna and "foamy" viruses) in that *de novo* protein synthesis but not virus replication is required for this process. The possibility that cell fusion results from a translation product of the infecting viral genomic RNA is supported by the fact that it is inhibited by cycloheximide (Chatterjee and Hunter, 1979). The experiments with tunicamycin which inhibits glycosylation show that unlike Newcastle disease virus and herpes simplex virus glycosylation of putative fusion-inducing protein is not necessary for fusion activity (Chatterjee *et al.*, 1981).

The virus particles which penetrate the plasma membrane are rapidly transported to the nucleus where their final uncoating presumably occurs. Simian virus 40 has been found in the perinuclear region as early as 10 minutes after adsorption (Barbanti-Brodano *et al.*, 1970).

6. Herpetoviridae

a. Virus Attachment Proteins. Five glycoproteins are demonstrable in the viral envelope of herpes simplex virus (HSV), gA, gB, gC, gD, and gE. However, their functions are not precisely understood, and the glycoproteins responsible for attachment to the cell receptors are not determined. Studies with temperature-sensitive mutants of HSV-1 provide evidence that glycoprotein gB plays a role in the viral penetration of the host cell, being involved in the fusion process with membranes of infected cells (Sarmiento *et al.*, 1979), and also plays a role in virus-induced cell fusion (Manservigi *et al.*, 1977). Glycoprotein gC, carrying type specificity, also seems important in the process of virus-induced cell fusion suppressing this process. Glycoprotein gE has been identified as a virus-induced Fc receptor of infected cells which has an affinity for the Fc region of the immunoglobulin. It has been suggested that the most probable glycoprotein involved in attachment is gD (Vahlne *et al.*, 1979).

b. Cellular Receptors and Attachment. The adsorption rate of HSV varies with different virus strains and cells suggesting that various cell–virus systems exhibit differences in the density of cellular receptors and affinity between receptors and virus attachment proteins. The nature of cellular receptors has not been explored. It has been suggested that cell surface glycopeptides bear part of the cellular receptor for HSV type 1 (Lemaster *et al.*, 1978). Herpes simplex virus type 1 interferes with the attachment of subsequently added homotypic but not heterotypic HSV; thus, receptors are type-selective. The density of the plasma membrane of receptors and the predominance of receptors for type 1 or for type 2 has been shown to be cell line characteristics (Vahlne *et al.*, 1979). On the other hand, no competition for cellular receptors between HSV-1 and HSV-2 has been observed (Purifoy and Powell, 1977).

c. Penetration and Uncoating. The virus initiates infection from the nucleus; thus, it has to penetrate the host cell plasma membrane and the outer and inner nuclear membranes. Numerous findings support the view that the virus uses a membrane fusion mechanism for cell entry. The fusion between the viral envelope and the cell plasma membrane during virus penetration was demonstrated by electron microscopy of infected cells (Smith and de Harven, 1974). The occasional fusion of the envelope of progeny virus particles which moved from the nucleus to the cytoplasm with outer nuclear membrane was observed, suggesting that such a fusion mechanism could account for the presence of naked capsids in the cytoplasm (Smith, 1980). Similar images of physical continuity between the viral envelope and the outer nuclear membrane and the presence in the cytoplasm of naked capsids were described for the Lucke herpes virus of frogs (Stackpole, 1969). It is reasonable to conclude that the infecting virus uses the same mechanism to penetrate the nucleus.

One more piece of evidence supports the fusion mechanism for virus penetration. Herpes simplex virus treated with fluorescein isothiocyanate develops resistance to light and to antibody either simultaneously after addition to the cells, or within 5 minutes. If entry involved the engulfment of the virus particles by endocytosis, there should be a period of time after the development of resistance to antibody when the virus would remain sensitive to light (De Luca et al., 1981).

Herpes simplex virus is capable of inducing fusion from within. The fusion was reversibly blocked by adamantone and NH_4Cl (Konsoulas et al., 1978) which depends on the virus strain and on the cell type. As indicated, glycoprotein gB is implicated in this process (Haffey and Spear, 1980), the glycosylation of the protein being necessary for its fusion activity (Gallaher et al., 1973). Although only definite strains induce fusion from within, the products capable of inducing fusion are made by all the strains, suggesting that such products are required for entry into the host cell (Sarmiento et al., 1979; Para et al., 1980; Haffey and Spear, 1980).

The role of glycoprotein gC in fusion is more difficult to assess. The suppression of fusion-promoting activity could conceivably have selective advantage and facilitate the dissemination of infection because the appearance of gC in the membranes of infected cells could prevent progeny virus from fusing with and being eclipsed by an already infected cell.

The virus strains which induce fusion from within can produce the products responsible for the absence of cell fusion which affect primarily the capacity of infected cells to induce cell fusion rather than their capacity to be fused (Lee and Spear, 1980). Thus, fusion from within is a complicated process governed by activities of multiple viral and cellular products.

B. Nonenveloped Viruses

1. Picornaviridae

a. *Virus Attachment Proteins.* The virions contain 60 identical capsomers each of which can potentially bind to cellular receptors. The surface protein VP_4, the smallest capsid protein, is presumably involved in attachment to cells.

b. *Cellular Receptors and Attachment.* Cellular receptors for picornaviruses are not identified. The fractionation of plasma membrane isolated from susceptible cells showed that poliovirus receptors were present within a complex of glycoproteins and lipoproteins (McClaren et al., 1968). Cellular receptors were reported to be highly specific. However, different picornaviruses were shown to share receptors (Lonberg-Holm et al., 1976; Crowell, 1976; Baxt and Bachrach, 1980). The number of receptors on a susceptible cell was estimated as $1-2.5 \times 10^4$ (Lonberg-Holm and Korant, 1972; Baxt and Bachrach, 1980).

Picornavirus particles adsorbed to host cells were shown to be redistributed into patches and caps due apparently to the lateral rearrangement of cellular receptors on the cell surface (Gschwendler and Traub, 1979). As Gschwendler and Traub (1979) suggest, capping may be an early signal in picornavirus infection.

For this family of viruses, the presence of specific cellular receptors appears to be the primary factor in determining susceptibility of the host cell to the virus. For example, human and monkey cultured cells are highly sensitive to poliovirus while nonprimate cells do not adsorb virus and are insensitive to it. However, resistant cells could be infected if virus RNA is introduced directly into the cell or if infected cells are exposed to Sendai virus which presumably provokes the penetration of picornaviruses during the fusion process. The importance of virus interaction with cellular receptors in determining host and tissue specificity has been reviewed by Smith (1977).

The binding of the virus to the susceptible cell is essentially complete by 15 minutes when about 50% of the eventual adsorption occurs; the fastest binding occurs within the first 5 minutes. A rate constant of adsorption for foot-and-mouth disease virus (FMDV) and poliovirus is 20-fold less than that for human rhinovirus (Lonberg-Holm and Whiteley, 1976; Crowell, 1976; Baxt and Bachrach, 1980): for FMDV it is $4.35 \pm 1.04 \times 10^{-9} \text{ cm}^3 \text{ min}^{-1} \text{ cell}^{-1}$ (Baxt and Bachrach, 1980).

Foot-and-mouth disease virus, equine rhinovirus, and mengovirus attach to cells more efficiently at 4 than at 37°C, while human rhinovirus, poliovirus (Lonberg-Holm and Whiteley, 1976), and Coxsackie virus (Crowell and Siak, 1978) attach poorly at 4°C. There is no significant difference in the ability of the virus to bind to the cells between pH 6.4 and 7.8. At pH values less than 6.4 the virus does not adsorb (Baxt and Bachrach, 1980).

c. *Penetration and Uncoating.* After the virus particle has attached to the receptor in the plasma membrane, it either penetrates into the cytoplasm without being altered or is modified at the plasma membrane sites. To initiate infection the second pathway must be used. The virus particles which penetrate into the cytoplasm can be recovered as intact virions for a long period after adsorption, thus representing the virus portion which does not undergo uncoating. The reason why they are incapable of uncoating is poorly understood.

It is generally agreed that the initial virus cell interaction results in a reversible complex involving specific receptors on the cell surface and specific sites in the virus capsid. Experiments *in vitro* on the interaction of cellular membranes with poliovirus have shown that the virus undergoes modifications within the complex characterized by (1) loss of VP_4, (2) slight reduction in S value, (3) acquired sensitivity to proteases and detergents, and (4) loss of infectivity. These modifications are supposed to be initial stages of uncoating (De Sena and Mandel, 1976, 1977; Guttman and Baltimore, 1977; De Sena and Torian, 1980).

After formation of the complex, two alternative pathways are possible: elution from the cell surface or diffusion to the modifying sites of the plasma membrane. The spontaneous elution of cell-associated virus from the cell surface ("A particles") has been described for a number of picornaviruses (Lonberg-Holm and Whiteley, 1976; De Sena and Mandel, 1977; Rosenwirth and Eggers, 1979). In most cases, eluted picornaviruses were characterized by features similar to those of *in vitro*-modified particles: they altered their sedimentation characteristics, became noninfectious and could not reattach to susceptible cells, their RNA, although resistant to ribonuclease, was rendered sensitive by pretreatment of the particles with chymotrypsin, and, finally, they lost superficial protein VP_4 (Lonberg-Holm *et al.*, 1975; Lonberg-Holm and Whiteley, 1976; Rosenwirth and Eggers, 1979). However, FMDV if eluted was not altered in sedimentation and buoyant density characteristics and possessed a full complement of proteins (Cavanagh *et al.*, 1978; Baxt and Bachrach, 1980).

Plasma membrane from disrupted HeLa cells was shown to be able to modify infectious virus to noninfectious A particles (De Sena and

Mandel, 1976, 1977; Guttman and Baltimore, 1977), supporting the idea that the first irreversible alteration takes place at the plasma membrane, concomitant with penetration.

Following modification at the plasma membrane, the virus was shown to enter the cells and become resistant to antibody, either by virtue of inaccessibility or by change of the antigenic determinants on the virus surface. Subsequently, it is exposed to the cytoplasm and uncoated. There is evidence that cardioviruses completely uncoat at the plasma membrane, and free RNA is released to the cytoplasm (Hall and Rueckert, 1971).

Cell-associated picornaviruses undergo uncoating through a series of intermediate subviral particles from 156 S to 12 S. The penetration and uncoating seem to be coupled events but can be separated by their temperature dependence. While adsorption and penetration occurs at 20 and even at 4°C, uncoating of most viruses occurs at 37°C. For example, penetration of poliovirus into HeLa cells measured as a loss of neutralizability by specific antibody takes place at 20°C, but uncoating occurs only when the infected cells are warmed to 36°C, as measured by acquisition of light resistance of infective centers after infection with neutral red-sensitized virus (Mandel, 1967). Apparently, at 36°C when the membrane is semifluid, diffusion of the virus to the modifying sites and penetration into the cytoplasm occur.

The final product of uncoating is presumably viral RNA covalently linked at its 5' temini to a small basic viral protein designated VPg (Flanegan et al., 1977; Lee et al., 1977). It has been suggested that the protein functions as a primer for RNA synthesis. Covalently bound proteins have been isolated from the 5' ends of encephalomyocarditis virus and foot-and-mouth disease virus. VPg from FMDV and poliovirus has been resolved into two species, having different amino acid compositions (King et al., 1980) or different charges (Richards et al., 1981).

2. Reoviridae

a. *Virus Attachment Proteins.* The virus contains a double capsid shell. The outer capsid consists of four polypeptides, $\mu 1C$, $\sigma 1$, $\sigma 3$, and $\lambda 2$ (Hayes et al., 1981). The $\sigma 1$ polypeptide is a viral hemagglutinin (Weiner et al., 1978) which determines type specificity (Gaillard and Joklik, 1980). Viral proteins involved in virus–host interaction are $\sigma 1$ and $\mu 1C$ polypeptides. $\sigma 1$ polypeptide provides specific binding to cell surface receptors and thus is responsible for cell tropism (Finberg et al., 1979; Weiner et al., 1980; Lee et al., 1981).

This protein is the most type specific of all reovirus proteins. The S1 genes which code for σ1 of the three reovirus serotypes fail to hybridize with each other to any measurable extent, and NH_2- and COOH-terminal amino acid sequences of the three σ1s are also quite different, as are their antigenic determinants (Hayes *et al.*, 1981; Lee *et al.*, 1981). Therefore, σ1 interacts with the cell surface receptor via some domain that is conserved among the three reovirus serotypes.

Surprisingly, there is a very small amount of σ1 on the surface of reovirus particles. It constitutes no more than 1% of the virion protein components and only up to 24 molecules of it are present in the virion. The primary attachment of reovirus particles to host cell apparently involves no more than two molecules of σ1 (Lee *et al.*, 1981). σ1 is located close to λ2, the protein that is the principal component of the reovirus core projections penetrating through the outer reovirus capsid shell to the particle surface (Hayes *et al.*, 1981).

μ1C polypeptide plays a role in the entry of the virus into a host cell and its spread within the organism (Weiner *et al.*, 1977; Rubin and Fields, 1980). This polypeptide is generated by proteolytic cleavage of the μ1 polypeptide (Zweerink *et al.*, 1971; McCrae and Joklik, 1978; Gentsch and Fields, 1981). It is the most conserved of the reovirus outer capsid polypeptides (Gentsch and Fields, 1981). The role of the third outer capsid polypeptide, σ3, is not well defined, although it is the most abundant surface protein (about 900 molecules in the virion).

b. Cellular Receptors. Cellular receptors are not identified.

c. Penetration and Uncoating. According to Silverstein and Dales (1968), the virus enters the host cell by endocytosis and is then delivered to the lysosomes. Within the lysosomes, a part of the reovirus protein coat is digested by lysosomal hydrolases. Being double stranded, the viral RNA remains intact.

The products of lysosomal digestion are subviral particles of 1.38 to 1.39 gm/cm³ density in CsCl, which represents uncoating intermediates (Silverstein *et al.*, 1972; Chang and Zweerink, 1971; Galster and Longyel, 1976). It has been suggested that they escape from the lysosomes, since they are found free in the cytoplasm. The subviral particles generated in infected cells are very similar in density and protein composition to the subviral particles obtained *in vitro* by chymotrypsin (intermediate subviral particles, ISVP).

According to Borza *et al.* (1981), the uncoating process proceeds via two steps. The first step is mediated by proteolytic digestion of the intact virions; the virion is converted to an intermediate subviral particle which is completely refractory to further digestion with proteases. The transcriptase function in ISVP remains inoperative. In the second step, the ISVP are converted to particles with activated transcriptase. This step does not involve further proteolysis, and is mediated by a mechanism (probably enzymatic) which is probably triggered by intracellular concentrations of K^+ ions or by collisions between ISVP. Borza et al. demonstrated the existence of a second uncoating step in infected cells by infecting the cells with ISVP. Conversion of ISVP to active particles required incubation of infected cells at 37°C, suggesting that the second uncoating step requires some features of the intracellular environment.

ISVP gain entry into the cytoplasm by direct traversing the plasma membrane, inducing the leakage of radioactive chrome from the host cells (Borza et al., 1979). This suggests that a fusion mechanism is involved in virus penetration.

3. Adenoviridae

a. Virus Attachment Proteins. There are at least 10 polypeptides and 252 capsomers. Twelve vertex capsomers (or penton bases) carry filamentous projections (or fibers). The fibers are glycoproteins which are involved in interactions with the host cell plasma membrane (Philipson et al., 1968).

b. Cellular Receptors and Attachment. Plasma membrane components involved in virus–host interaction were studied by affinity chromatography using immobilized virions or highly purified fiber and penton structures. From the plasma membrane fraction of KB (Hennache and Boulanger, 1977) and HeLa cells (Svensson et al., 1981) the polypeptides recognizing Ad2 fiber were isolated and characterized. The proteins involved in reception were identified as glycoproteins (Philipson et al., 1968; Meager et al., 1976; Meager and Hughes, 1977) with molecular weights of 40,000 to 42,000 containing N-acetyl-D-glucosaminyl residues (Swensson et al., 1981). Removal of cellular sialic acid by neuraminidase does not prevent adenovirus attachment and even enhances it (Boulanger et al., 1972). Attachment to the cell surface has been shown to involve a salt bond between arginine in fiber protein and the side-chain carboxyl group of aspartic acid or glutamic acid in the cellular receptor (Neurath et al., 1970).

c. Penetration. There are two concepts regarding adenovirus penetration: (1) After contact with receptors the virions penetrate directly through the plasma membrane leaving the fiber penton structure behind at the surface (Lonberg-Holm and Philipson, 1974). (2) The inoculum is first internalized by endocytosis and subsequently, upon lysis of the vacuolar membrane, gains access to the cytoplasm (Dales, 1973).

In the act of lysing this membrane barrier, one or several of the intimately attached fibers or fiber + penton bases are broken away from the virion and remain on the vacuole membrane (Lyon *et al.*, 1978).

Virus particles penetrating into the cytoplasm appear to bind specifically to microtubules (Dales and Chardonet, 1973; Luftig and Weihig, 1975; Weatherbee *et al.*, 1977) and move vectorially along the microtubules toward the nuclear membrane, where they are associated with the nuclear core complex. From 2 to 31% of virus particles have been found to be associated with the microtubules of HeLa cells 1 hour postinfection (Miles *et al.*, 1980). The component of the virus involved in the interaction with the microtubules has been shown to be the hexon capsomere, not the pentons and fibers (Luftig and Weihing, 1975).

d. Uncoating. The first stage of adenovirus uncoating apparently occurs at the cellular membranes. Virus particles that cross membranes and enter the cytoplasm lack a portion of the coat protein and become susceptible to exogenous nucleases (Meager *et al.*, 1976). Further uncoating occurs at the nuclear pores and results in the penetration of virus cores into the nucleus where the final uncoating proceeds. The final product of uncoating is presumably viral DNA covalently linked to viral protein (Challberg *et al.*, 1980; Rekosh *et al.*, 1977).

Cores are 20–40 times more infectious than deproteinized DNA. Their high infectious activity is explained by Mirza and Weber (1980) as being due to their greater efficiency of uncoating as compared to the virus particles. Mirza and Weber (1980) described an interesting phenomenon concerning the temperature-sensitive mutant: it was blocked from uncoating at the core stage, but its cores, obtained by pyridine treatment, were infectious at the permissive temperature. The difference between the cores and the virus particle is in the reduced affinity of the hydrophobic region of the PVII virus protein for virus DNA; its binding to DNA in the cores was not as tight as in the virus particle.

4. Papovaviridae

The Papovaviridae consists of two genera, the papilloma virus and the polyoma virus. Information on early stages of infection is available primarily for the polyoma virus.

a. Virus Attachment Proteins. Polyoma virus contains three structural proteins, VP_1, VP_2, and VP_3, and host histones. The surface coat protein is VP_1 which presumably interacts with the host cell plasma membrane.

b. Cellular Receptors and Attachment. Polyoma virus agglutinate guinea pig and human erythrocytes as well as those of several other animal species, and hemagglutination has been used as a model reaction to study cell receptors. The agglutination was abolished by treating erythrocytes with *Vibrio* cholera neuraminidase and inhibited by glycoproteins containing sialic acid (Lomberg-Holm and Philipson, 1974). The specific receptors on the surface of erythrocytes were identified as certain sialyloligosaccharides attached to glycoproteins. In neuraminidase-treated erythrocytes the hemagglutination was fully restored by β-galactoside-α-2,3-sialyltransferase in the sequence AcNeu α2,3-Gal but not AcNeu α2,6-Gal or AcNeu α2,6-GalNac, suggesting that the receptors for polyoma virus are strongly specific (Cahan and Paulson, 1980). The neuraminidase prevented infection of cultured cells, and the same oligosaccharide sequence could restore their susceptibility to infection (Cahan and Paulson, 1980; Fried *et al.*, 1981).

VP_1 is composed of six species which differ in the degree of their posttranslational modifications (phosphorylation and acetylation). This is an example of a protein in which modifications allow a number of different functions. At least four functional roles could be served by these means. Two of them involve a structural function, and an association with the DNA–protein complex of the virus. The remaining two functions concern virus attachment to mammalian cells. Three of the VP_1 species, D, E, and F_1, are involved in such activities. Species D and F appear to be involved in adsorption and hemagglutination, while species E is apparently required for attachment to specific cellular receptors. This protein is found only on virions but not on capsids in accordance with the data that capsids do not compete with virions for cellular receptors (Bolen and Consigli, 1979). The attachment to cell receptors is inhibited by antibodies against the specific determinants presented on species E (Bolen *et al.*, 1981). Thus, VP_1 species E fulfils the role of the polyoma virus attachment protein. Bolen and Consigli (1979, 1980) reported that the receptors on the surface of mouse cultured cells are not identical to the receptors on the surface of erythrocytes. Polyoma capsids devoid of DNA and histones and containing only three structural proteins could adsorb to sialic acid residues, but they did not compete with virions for binding sites on the surface of mouse cells, although they competed for adsorption to erythrocytes. A mutant was obtained which lost the ability to agglutinate erythrocytes but specifically adsorbed to mouse cells. Accordingly, antigenic determinants, responsible for hemagglutination inhibitory activity and for neutralizing activity, were contained in two different polypeptides, both derived from VP_1. Bolen and Consigli conclude that the virus could bind to the surface of mouse cells in two different ways, by nonspecific attachment to the sialic acid residues and by specific attachment to cell receptors. The specific attachment represents about 10% of input virus. The maximal number of cell receptors on the surface of mouse kidney cells is approximately 10,000 per cell (Bolen *et al.*, 1981).

   c. Penetration. Interesting data on polyoma virus penetration and intracellular transport were reported by Mackay and Consigli (1976) and Bolen and Consigli (1979). Once the attachment to specialized receptors on the cell plasma membrane occurs, there is a rapid signal from the cell surface to initiate penetration. The cell membrane under the attached virus particle begins to invaginate, forming a monopinocytotic vesicle which is rapidly transported to the nucleus. The mechanism of rapid and specific transport of the monopinocytotic vesicle from the cell surface to the nucleus is unclear; however, it does not involve microfilaments and microtubules.

   Virion DNA and coat proteins arrive at the nucleus simultaneously, as early as 15 minutes postinfection. This suggests that virus uncoating occurs in the nuclei (Mackay and Consigli, 1976). Later Chlumecka *et al.* (1979) supported the conclusion that virus uncoating occurs exclusively in the nucleus. They proposed the following model of polyoma virus attachment, penetration, and nuclear entry. (1) The virus interacts on either its fivefold or twofold axis with the host cell receptor site. (2) The cell membrane undulates beneath the virus just enough to allow contact on its threefold axis and thereby signals penetration. (3) The virus penetrates the cell membrane which closes tightly around the virus particle to form a monopinocytotic vesicle. (4) The vesicle rapidly and selectively migrates to the outer nuclear membrane. (5) Upon contact with the outer nuclear membrane, the virus particle enters the nucleus.

   d. Uncoating. The virus particles were shown to enter the nucleus intact but they immediately became unrecognizable by thin section electron microscopy, presumably as a result of uncoating (Mackay and Consigli, 1976). An initial uncoating intermediate is a 48 S material recovered from the nuclei early after infection. The protein composition of this material is very similar to that of uncoating intermediates derived *in vitro* with EDTA and DTT; both components lack the coat proteins, primarily VP_1. The 48 S intermediates become associated with some host structures resulting in the formation of a 190 S complex. The nature and function of the host protein associated with the

190 S complex are unknown: the major protein species with a molecular weight of 53,000 could correspond to the protein associated with the nuclear matrix; the other proteins could correspond to host histones (Winston *et al.*, 1980).

IV. Concluding Remarks

Recent information concerning the early steps in virus–cell interaction has significantly expanded our appreciation of the role of these steps in virus infection and has demonstrated that these steps are crucial for the initiation of infection and are determinants of host susceptibility. The study of these early steps is important in understanding the events provoking infection.

Some success has been achieved in this field during the last decade; notably, in a more precise identification of virus attachment proteins and cellular receptors and their interaction, and in the mechanisms of virus penetration and intracellular modifications of virus particles.

A new approach in these studies is the elucidation of the fragments in virus protein molecules involved in attachment and penetration. The determination of the primary structure of virus proteins and the identification of conserved sequences within comparable proteins among closely related viruses that are involved in cell interaction are important events. An extended analysis of more distantly related viruses may lead in the future to the identification of sequences which functionally define a group of viruses with identical strategies of cell entry.

It has been shown that the cell surface is important in determining the attachment of virus particles as well as their further intracellular routes leading to initiation of infection. It is obvious that a more complete understanding of early virus–cell interaction will be achieved as more becomes known about the architecture of the cell surface.

The technical difficulties involved in defining molecular mechanisms of virus attachment, entry, and uncoating require new experimental approaches using more defined model systems. At present, one of the promising *in vitro* systems seems to be liposomes. Liposomes in which the lipid composition is manipulated to mimic that in specific cell membranes, artificial membranes with presumed cellular receptors incorporated into them, and viral membranes reconstituted from lipid and polypeptides possessing different determinants are now widely used. By these means the functions of individual viral components in attachment and penetration and their cooperative effects, and virus interaction with artificial membranes with different lipid compositions and cellular receptors, are successfully studied. However, although analysis *in vitro* will undoubtedly contribute to our overall knowledge, it still cannot fully replace studies *in vivo* using the whole cell.

In spite of the fact that the information gained from the study of early virus–cell interactions is remarkably diverse at present, several summarizing remarks can be made concerning the separate stages.

A. Attachment

Although different types of binding could occur between the cell surface and the virus particle, the attachment which leads to infection is a specific process involving certain determinants on the virus attachment protein(s) and specific receptors on the cell plasma membrane. In addition to specific adsorption, the viruses could bind to the cell surface nonspecifically, using nonspecific receptors, or they might even bind to the lipid bilayer devoid of proteins. However, such interactions are "noninfective," i.e., they are not followed by initiation of infection.

For hemagglutinating viruses, the erythrocyte membrane is a simple and convenient model system with which to study some aspects of virus–cell interaction. However, the data obtained on reception with erythrocytes should be generalized to other cell types with caution, since erythrocytes could differ from other animal cells in their surface structure.

Recent findings do not support the opinion that specific receptors for myxoviruses and paramyxoviruses are glycoproteins. The role of certain gangliosides as specific receptors has been demonstrated for these groups of viruses. Not to be excluded is the possibility that both glycoproteins and glycolipids are subsequently involved in virus attachment.

Data obtained with several virus groups indicate that specific receptors on the cell plasma membrane are involved in virus penetration and intracellular transport.

B. Penetration

There is much controversy concerning penetration. Both endocytosis and fusion are regarded as alternative modes of virus entry, and the question of which one is used by a given virus group is often debated. However, virus particles engulfed by endocytosis should still traverse the vacuole membrane to reach the cytoplasm and presumably use the fusion mechanism. The findings obtained with different virus groups support the idea that fusion is a universal mode of virus entry. What varies from one virus to another may be the site of fusion and the factors needed to trigger it. For certain viruses, low pH is an absolute requirement for fusion. As for endocytosis, perhaps its evolutionary role is to provide better conditions for membrane fusion than exist at the plasma membrane, such as intimate contact of the virus particle with the cell membrane and a suitable "microclimate" in the endocytic vacuole.

C. Uncoating

Uncoating is a complicated process provided by a series of virus–cell-dependent reactions. It most often consists of two or more subsequent steps, which might occur at different cell compartments. Special drugs which inhibit various steps in uncoating are widely used to elucidate the sequence of events during uncoating and to characterize intermediates. Since uncoating, being highly sensitive to factors within and without, is one of the steps which limits infection, the prevention of uncoating by specific drugs seems to be applicable in preventing virus infections through chemotherapy. Drugs such as amantadine and rimantadine are being successfully used in the prevention and treatment of influenza.

In general, many points concerning early virus–cell interaction remain obscure. The high efficiency of attachment, penetration, and uncoating, their asynchronous character, and the small proportion of uncoated particles and their rapid transport within the cell combined with the absence of adequate approaches for isolating viral components from the cell of input make this problem extremely difficult to study morphologically and biochemically.

For a large number of viruses, cellular receptors have still not been identified, and virus attachment proteins have been poorly studied. The high specificity of cellular receptors has not yet been adequately explained. The role of endocytosis in virus penetration is unclear. Little is known concerning the molecular basis of virus-induced fusion with the cell membrane and the role of low pH. The relationship between penetration and uncoating is not precisely understood. The least understood field is uncoating and intracellular transport of virus components. Uncoating intermediates and end-products are poorly characterized. It is not clear why virus particles move to certain cell compartments, how they recognize sites of their uncoating, how their inner components escape from surface protective structures and reach the sites where they could initiate infection, and, finally, why the cell cannot recognize its killer and efficiently help it.

The study of the molecular mechanisms of virus attachment, cell entry, and uncoating has just begun. These investigations could help not only in elucidating the problem of how the virus initiates infection and the cell helps it, but in elaborating a strategy for the prevention of the infectious process. Substances which can block or damage specific receptors on the cell surface, or imitate virus fusion mechanisms competing for virus penetration, or "freeze" uncoating intermediates could have wide applications in medicine. Undoubtedly, this area of research will attract a lot of attention in the coming years.

ACKNOWLEDGMENT

The author thanks Dr. N. V. Kaverin for his critical reading of the manuscript.

REFERENCES

Altrock, B. W., Arthur, L. O., Massey, R. J., and Schochetman, G. (1981). *Virology* **109**, 257.
Altstein, A. D., and Zhdanov, V. M. (1979). *Adv. Virus Res.* **25**, 451.
Arens, M., Lakshmi, M. V., Crecelius, D., Deom, C. M., and Schulze, J. T. (1981). *Symp. Genet. Variat. Influenza Viruses, Salt Lake City, Utah.*
Avery, R. J., and Niven, J. (1979). *Infect. Immun.* **26**, 795.
Babiss, L. E., Luftig, R. B., Weatherbee, J. A., Weihing, R. R., Ray, U. R., and Fields, B. N. (1979). *J. Virol.* **30**, 863.
Bächi, T., Deas, J. E., and Hower, C. (1977). *In* "Cell Surface Reviews: Virus Infection and the Cell Surface" (G. Poste and G. L. Nicolson, eds.), Vol. 2, p. 83. North-Holland Publ., Amsterdam.
Barbanti-Brodano, G., Swetly, P., and Koprowski, H. (1970). *J. Virol.* **6**, 78.
Baxt, B., and Bachrach, H. L. (1980). *Virology* **104**, 42.
Benton, C. V., Soria, A. E., and Gilden, R. V. (1978). *Infect. Immun.* **20**, 307.
Bergelson, L. D., Bukrinskaya, A. G., Procasova, N., *et al.* (1982). *Eur. J. Biochem.*, in press.
Bolen, J. B., and Consigli, R. A. (1979). *J. Virol.* **32**, 679.
Bolen, J. B., and Consigli, R. A. (1980). *J. Virol.* **34**, 119.
Bolen, J. B., Anders, D. G., Trempy, J., and Consigli, R. A. (1981). *J. Virol.* **37**, 80.
Bonatti, S., Caneedda, R., and Blobel, G. (1979). *J. Cell Biol.* **80**, 219.
Borza, J., Morash, B. D., Sargent, M. D., Copps, T. P., Lievaart, P. A., and Szekely, J. G. (1979). *J. Gen. Virol.* **45**, 161.
Borza, J., Sargent, M. D., Lievaart, P. A., and Copps, T. P. (1981). *Virology* **111**, 191.
Boulanger, P. A., Houdret, N., Scharfman, A., and Lemay, P. (1972). *J. Gen. Virol.* **16**, 429.
Brady, J. N., Lavialle, C. A., Radonovich, M. F., and Salzman, N. P. (1981). *J. Virol.* **39**, 432.
Bratt, M. A., and Gallaher, W. R. (1969). *Proc. Natl. Acad. Sci. U.S.A.* **64**, 536.
Brouwer, J., Pluijms, W., and Warnaar, S. (1979). *J. Gen. Virol.* **42**, 415.

Brown, D. T., and Burlingham, B. T. (1973). *J. Virol.* **12**, 386.

Bukrinskaya, A. G. (1973). *Adv. Virus Res.* **18**, 195.

Bukrinskaya, A. G., Vorkunova, G. K., and Vorkunova, N. K. (1979). *J. Gen. Virol.* **45**, 557.

Bukrinskaya, A. G., Vorkunova, N. K., and Narmanbetova, R. A. (1980). *Arch. Virol.* **66**, 275.

Bukrinskaya, A. G., Prokasova, N. V., Shaposhnicova, G. I., Kocharov, S. L., Shevchenko, V. P., and Bergelson, L. D. (1982a). *Dokl. Akad. Nauk SSSR*, **263**, 1481.

Bukrinskaya, A. G., Vorkunova, N. K., Kornilayeva, G. V., Narmanbetova, R. A., and Vorkunova, G. K. (1982b). *J. Gen. Virol.*, in press.

Cahan, L. D., and Paulson, J. C. (1980). *Virology* **103**, 505.

Cartwright, B., and Brown, F. (1977). *J. Gen. Virol.* **35**, 1197.

Cavanagh, D., Rowlands, D. J., and Brown, F. (1978). *J. Gen. Virol.* **41**, 255.

Challberg, M. D., Desiderio, S. V., and Kelley, T. J. (1980). *Proc. Natl. Acad. Sci. U.S.A.* **77**, 5105.

Chang, C.-T., and Zweerink, H. J. (1971). *Virology* **46**, 544.

Chany-Fournier, F., Chany, C., and Lafay, F. (1977). *J. Gen. Virol.* **34**, 305.

Chatterjee, S., and Hunter, E. (1979). *Virology* **95**, 421.

Chatterjee, S., and Hunter, E. (1980). *Virology* **107**, 100.

Chatterjee, S., Bradae, J., and Hunter, E. (1981). *J. Virol.* **38**, 770.

Chlumecka, V., D'Obrenan, P., and Colter, J. S. (1979). *Virology* **94**, 219.

Choppin, P. W., and Scheid, A. (1980). *Rev. Infect. Dis.* **2**, 40.

Coombs, K., Mann, E., Edwards, J., and Brown, D. T. (1981). *J. Virol.* **37**, 1060.

Crowell, R. L. (1976). *In* "Cell Membrane Receptors for Viruses, Antigens and Antibodies, Polypeptide Hormones, and Small Molecules" (R. F. Beers, Jr. and E. G. Basset, eds.), pp. 179–202. Raven, New York.

Crowell, R. L., and Siak, J. S. (1978). *In* "Perspectives in Virology" (M. Pollard, ed.), Vol. 10, pp. 39–55. Raven, New York.

Dales, S. (1973). *Bacteriol. Rev.* **37**, 103.

Dales, S., and Chardonnet, Y. (1973). *Virology* **56**, 465.

DeLarco, J., and Todaro, G. J. (1976). *Cell* **8**, 365.

DeLuca, N., Bzik, D., Person, S., and Snipes, W. (1981). *Proc. Natl. Acad. Sci. U.S.A.* **78**, 912.

DeSena, J., and Mandel, B. (1976). *Virology* **70**, 470.

DeSena, J., and Mandel, B. (1977). *Virology* **78**, 554.

DeSena, J., and Torian, B. (1980). *Virology* **104**, 149.

Diana, G. D., Salvador, U. J., Zalay, E. S., Johnson, R. E., *et al.* (1977). *J. Med. Chem.* **20**, 750.

Diglio, C. A., and Ferrer, J. F. (1976). *Cancer Res.* **36**, 1056.

Dourmashkin, R. R., and Tyrrell, D. A. J. (1974). *J. Gen. Virol.* **24**, 129.

Dubov, E. J., Geratz, J. D., and Tidwell, R. R. (1980). *Virology* **103**, 502.

Dunnebacke, T. H., Levinthal, J. D., and Williams, R. C. (1979). *J. Virol.* **4**, 505.

Eggers, H. J. (1977). *Virology* **78**, 241.

Eggers, H. J., Bode, B., and Brown, D. T. (1979). *Virology* **92**, 211.

Fan, D. P., and Sefton, B. M. (1978). *Cell* **15**, 985.

Fazekas de St. Groth, S. (1948). *Nature (London)* **162**, 294.

Fields, S., Winter, G., and Brownlee, G. G. (1981). *Nature (London)* **290**, 213.

Finberg, R., Weiner, H. L., Fields, B. N., Benacerraf, B., and Burakoff, S. J. (1979). *Proc. Natl. Acad. Sci. U.S.A.* **76**, 442.

Fitzgerald, D., Morris, R. E., and Saelinger, C. B. (1980). *Cell* **21**, 867.

Flanegan, J. B., Pettersson, R. F., Ambros, V., Hewlett, M. J., and Baltimore, D. (1977). *Proc. Natl. Acad. Sci. U.S.A.* **74**, 961.

Fowler, A. K., Twardzik, D. R., Reed, C. D., Weislow, O. S., and Hellman, A. (1977). *J. Virol.* **24,** 729.

Fried, H., Cahan, L., and Paulson, J. (1981). *Virology* **109,** 188.

Fries, E., and Helenius, A. (1979). *Eur. J. Biochem.* **97,** 213.

Gaillard, R. K., and Joklik, W. K. (1980). *Virology* **107,** 533.

Gallaher, W. R., Levitan, B., and Blough, H. A. (1973). *Virology* **55,** 193.

Galster, R. L., and Longyel, P. (1976). *Nucleic Acids Res.* **3,** 581.

Garoff, H. (1981). *In* "International Cell Biology 1980–1981" (H. G. Schweiger, ed.), p. 572. Springer-Verlag, Berlin and New York.

Garoff, H., and Simons, K. (1980). *In* "Biological Chemistry of Organelle Formation" (T. Bücher, W. Sebald, and H. Weiss, eds.). Springer-Verlag, Berlin and New York.

Garten, M., Berk, W., Nagai, Y., Rott, R., and Klenk, H.-D. (1980). *J. Gen. Virol.* **50,** 135.

Garten, M., Bosch, F. X., Linder, D., Rott, R., and Klenk, H.-D. (1981). *Virology,* in press.

Gentsch, J. R., and Fields, B. N. (1981). *J. Virol.* **38,** 208.

Gething, M. J. (1980). *In* "Structure and Variation in Influenza Virus" (W. G. Laver and G. Air, eds.). Elsevier, Amsterdam.

Gething, M. J., White, J. M., and Waterfield, M. D. (1978). *Proc. Natl. Acad. Sci. U.S.A.* **75,** 2737.

Goldstein, J. L., Anderson, R. G., and Brown, M. S. (1979). *Nature (London)* **279,** 679.

Graves, D. C., and Jones, L. V. (1981). *J. Virol.* **38,** 1055.

Gregoriades, A. (1980). *J. Virol.* **36,** 470.

Gschwendler, H. H., and Traub, P. (1979). *J. Gen. Virol.* **42,** 439.

Guttman, N., and Baltimore, D. (1977). *Virology* **82,** 25.

Haffey, M. L., and Spear, P. G. (1980). *J. Virol.* **35,** 114.

Hall, L., and Rueckert, R. R. (1971). *Virology* **43,** 152.

Haspel, M. V., Pellegrino, M. A., Lampert, P. W., and Oldstone, M. B. A. (1977). *J. Exp. Med.* **146,** 146.

Hay, A. J., Kennedy, N. C. T., Skehel, J. J., and Appleyard, G. (1979). *J. Gen. Virol.* **42,** 189.

Hayes, E. C., Lee, P. W. K., Miller, S. E., and Joklik, W. (1981). *Virology* **108,** 147.

Haywood, A. M. (1974a). *J. Mol. Biol.* **83,** 427.

Haywood, A. M. (1974b). *J. Mol. Biol.* **87,** 625.

Haywood, A. M. (1975a). *In* "Negative Strand Viruses" (B. W. J. Mahy and R. D. Barry, eds.), Vol. 2. Academic Press, New York.

Haywood, A. M. (1975b). *J. Gen. Virol.* **29,** 63.

Heine, J. W., and Schnaitman, C. A. (1971). *J. Virol.* **8,** 786.

Helenius, A., and Simons, K. (1980). *In* "The Molecular Basis of Microbial Pathogenicity" (H. Smith, J. J. Skehel, and M. J. Turner, eds.), pp. 41–54. Dahlem Conference, Berlin.

Helenius, A., Fries, E., and Kartenbeck, J. (1977). *J. Cell Biol.* **75,** 866.

Helenius, A., Morein, B., Fries, E., Simons, K., *et al.* (1978). *Proc. Natl. Acad. Sci. U.S.A.* **75,** 3846.

Helenius, A., Kartenbeck, J., Simons, K., and Fries, E. (1980a). *J. Cell Biol.* **84,** 404.

Helenius, A., Marsh, M., and White, J. (1980b). *Trends Biochem. Sci.* **5,** 104.

Hennache, B., and Boulanger, P. (1977). *Biochem. J.* **166,** 237.

Holmgren, J., Svennerholm, L., Elwing, H., Fredman, P., and Strannegard, O. (1980). *Proc. Natl. Acad. Sci. U.S.A.* **77,** 1947.

Homma, M., and Ohuchi, M. (1973). *J. Virol.* **12,** 1457.

Hosaka, Y. (1975). *In* "Negative Strand Viruses" (B. W. J. Mahy and R. D. Barry, eds.), pp. 885–903. Academic Press, New York.

Hosaka, Y., and Shimizu, K. (1977). *In* "Virus Infection and the Cell Surface" (G. Poste and G. L. Nicolson, eds.). Elsevier, Amsterdam.

Howe, C., and Lee, L. T. (1972). *Adv. Virus Res.* **17,** 1.
Hsu, M. C., Scheid, A., and Choppin, P. W. (1979). *Virology* **95,** 476.
Huang, R. T. C., Wahn, K., Klenk, H.-D., and Rott, R. (1979). *Virology* **97,** 212.
Huang, R. T. C., Rott, R., Wahn, K., Klenk, H.-D., and Kohama, T. (1980a). *Virology* **107,** 313.
Huang, R. T. C., Wahn, K., Klenk, H.-D., and Rott, R. (1980b). *Virology* **104,** 294.
Huang, R. T. C., Rott, R., and Klenk, H.-D. (1981). *Virology* **110,** 243.
Hudson, J. B., Flawith, J., and Dimmock, N. J. (1978). *Virology* **87,** 167.
Ishida, N., and Homma, M. (1978). *Adv. Virus Res.* **23,** 349.
Kalyanaraman, V. S., Sarngadharan, M. G., and Gallo, R. C. (1978). *J. Virol.* **28,** 686.
Karshin, W. L., Areement, L. J., Naso, R. B., and Arlinghaus, R. B. (1977). *J. Virol.* **23,** 787.
Kennedy, S. (1976). *J. Mol. Biol.* **108,** 491.
King, A. M. Q., Sangar, D. V., Harris, T. J. R., and Brown, F. (1980). *J. Virol.* **34,** 627.
Klenk, H.-D. (1980). *In* "The Molecular Basis of Microbial Pathongenicity" (H. Smith, J. J. Skehel, and M. J. Turner, eds.), pp. 55–66. Dahlem Conference, Berlin.
Klenk, H.-D., Garten, W., Keil, W., Niemann, H., Schwarz, R. T., and Rott, R. (1980). *In* "Biosynthesis, Modification and Processing of Cellular and Viral Polyproteins" (G. Koch and D. Richter, eds.), pp. 175–184. Academic Press, New York.
Knutton, S. (1978). *Micron* **9,** 133.
Knutton, S., Jackson, D., and Ford, M. (1977). *J. Cell Sci.* **28,** 179.
Koff, W. C., and Knight, V. (1979). *J. Virol.* **31,** 261.
Kohama, T., Garten, W., and Klenk, H.-D. (1981). *Virology* **111,** 364.
Kohn, A. (1975). *J. Gen. Virol.* **29,** 179.
Kornilayeva, G. V., and Bukrinskaya, A. G. (1980). *Vopr. Virusol.* **6,** 701.
Konsoulas, K. G., Person, S., and Holland, T. C. (1978). *J. Virol.* **27,** 505.
Kuhrt, M., Fancher, M. J., Jasty, V., Pancic, F., and Came, P. E. (1979). *Antimicrob. Agents Chemother.* **15,** 813.
Kuroda, K., Maeda, T., and Ohnishi, S. (1980). *Proc. Natl. Acad. Sci. U.S.A.* **77,** 804.
Kurrle, R., Wagner, H., Rollinghoff, M., and Rott, R. (1979). *Eur. J. Immunol.* **9,** 107.
Lakshmi, M. Y., and Schulze, J. T. (1978). *Virology* **88,** 314.
Lazarowitz, S. G., Compans, R. W., and Choppin, P. W. (1973). *Virology* **52,** 199.
Lee, G. T.-Y., and Spear, P. G. (1980). *Virology* **107,** 402.
Lee, P. W. K., Hayes, E. C., and Joklik, W. K. (1981). *Virology* **108,** 156.
Lee, Y. E., Nomoto, A., Detjen, B. M., and Wimmer, E. (1977). *Proc. Natl. Acad. Sci. U.S.A.* **74,** 59.
Lemaster, S. L., Baron, C., and Blough, H. A. (1978). *Int. Congr. Virol., The Hague,* (Abstr.), p. 522.
Lenard, J. (1978). *Annu. Rev. Biophys. Bioeng.* **7,** 139.
Lenard, J., and Miller, D. K. (1981). *Virology* **110,** 479.
Levanon, A., and Kohn, A. (1978). *FEBS Lett.* **85,** 245.
Lonberg-Holm, K., and Korant, B. D. (1972). *J. Virol.* **9,** 29.
Lonberg-Holm, K., and Philipson, L. (1974). *In* "Monographs in Virology" (J. L. Melnick, ed.), Vol. 9. Phiebig, White Plains, New York.
Lonberg-Holm, K., and Whiteley, N. M. (1976). *J. Virol.* **19,** 857.
Lonberg-Holm, K., Gosser, L. B., and Kauer, J. C. (1975). *J. Gen. Virol.* **27,** 329.
Lonberg-Holm, K., Crowell, R. L., and Philipson, L. (1976). *Nature (London)* **259,** 679.
Lubeck, M. D., Schulman, J. L., and Palese, P. (1978). *J. Virol.* **28,** 710.
Luftig, R. B., and Weihing, R. R. (1975). *J. Virol.* **16,** 696.
Lyon, M., Chardonnet, Y., and Dales, S. (1978). *Virology* **87,** 81.
McCrae, M. A., and Joklik, W. K. (1978). *Virology* **89,** 578.
McClaren, L. C., Sclaletti, J. V., and James, C. C. (1968). *In* "Biological Properties of the

Mammalian Surface Membrane" (L. Manson, ed.), pp. 123–135. Wistar Institute, Philadelphia, Pennsylvania.

Mackay, R. L., and Consigli, R. A. (1976). *J. Virol.* **19**, 620.

MacLean-Fletcher, S., and Pollard, T. D. (1980). *Cell* **20**, 329.

McSharry, J. J., Caliguiri, L. A., and Eggers, H. J. (1979). *Virology* **97**, 307.

Maeda, T., and Ohnishi, S. (1980). *FEBS Lett.* **122**, 283.

Mandel, B. (1967). *Virology* **31**, 702.

Manservigi, R., Spear, P. G., and Buchar, A. (1977). *Proc. Natl. Acad. Sci. U.S.A.* **74**, 3913.

Markwell, M. A., and Paulson, J. C. (1980). *Proc. Natl. Acad. Sci. U.S.A.* **77**, 5693.

Meager, A., and Hughes, R. C. (1977). *In* "Receptors and Recognition" (P. Cuatrecasas and M. F. Greaves, eds.), Vol. 4, Ser. A. Chapman & Hall, London.

Meager, A., Butters, T. D., Mauther, V., and Hughes, R. C. (1976). *Eur. J. Biochem.* **61**, 345.

Merz, D. C., Scheid, A., and Choppin, P. W. (1981). *Virology* **109**, 94.

Miles, B. D., Luftig, R. B., Weatherbee, J. A., Weihing, R. R., and Weber, J. (1980). *Virology* **105**, 265.

Miller, D. K., and Lenard, J. (1980). *J. Cell Biol.* **84**, 430.

Min-Jou, W. (1980). *In* "Structure and Variation in Influenza Virus" (W. G. Laver and G. Air, eds.). Elsevier, Amsterdam.

Mirza, M. A. A., and Weber, J. (1979). *J. Virol.* **30**, 462.

Mirza, A., and Weber, J. (1980). *Intervirology* **13**, 307.

Mooney, J., Dalrymple, J., Alving, C., and Russell, P. (1975). *J. Virol.* **15**, 225.

Moore, N., Patzer, E. J., Shaw, J. M., Thompson, T. S., and Wagner, R. R. (1978). *J. Virol.* **27**, 320.

Morgan, C., and Howe, C. (1968). *J. Virol.* **2**, 1122.

Morgan, C., and Rose, H. M. (1968). *J. Virol.* **2**, 925.

Nagai, Y., Klenk, H.-D., and Rott, R. (1976). *Virology* **72**, 494.

Neurath, A. R., Hartzell, R. W., and Rubin, B. A. (1970). *Virology* **42**, 789.

Nicolau, C., Klenk, H.-D., Reiman, A., Hildebrand, K., and Bauer, H. (1979). *Biochim. Biophys. Acta* **511**, 83.

Nigg, E. A., Cherry, R. J., and Bächi, T. (1980). *Virology* **107**, 552.

Nishiyama, Y., Ito, Y., Shimokato, K., Kimura, Y., and Nagato, I. (1976). *J. Gen. Virol.* **32**, 85.

Nowinski, R. C., Pickering, R., O'Donnell, P. V., Pinter, A., and Hammerling, U. (1981). *Virology* **111**, 84.

Ocho, M., Ogura, H., Tanaka, T., and Oda, T. (1980). *Exp. Cell Biol.* **48**, 421.

Oldstone, M. B. A., Tishon, A., Dutko, F. J., Kennedy, S. J. T., Holland, J. J., and Lampert, P. W. (1980). *J. Virol.* **34**, 256.

Oxford, J. S., Hockley, D. J., Heath, T. D., and Patterson, S. (1981). *J. Gen. Virol.* **52**, 329.

Ozawa, M., Asano, A., and Onada, Y. (1979). *Virology* **99**, 197.

Para, M. F., Baucke, R. B., and Speur, P. G. (1980). *J. Virol.* **34**, 512.

Patterson, S., Oxford, J., and Dourmashkin, R. (1979). *J. Gen. Virol.* **43**, 223.

Paulson, J. C., Sadler, F. J., and Hill, R. L. (1979). *J. Biol. Chem.* **254**, 2120.

Payne, L. Y., and Norby, E. (1978). *J. Virol.* **27**, 19.

Peterhaus, E. (1980). *Virology* **105**, 445.

Petri, W. A., and Wagner, R. R. (1980). *Virology* **107**, 543.

Philipson, L., Lonberg-Holm, K., and Pettersson, U. (1968). *J. Virol.* **2**, 1064.

Pierce, B. M. F. (1975). *Proc. Natl. Acad. Sci. U.S.A.* **73**, 1255.

Porter, A. G., Barber, C., Carey, N. H., Hallewell, R. A., Threlfall, G., and Emtage, J. S. (1979). *Nature (London)* **282**, 471.

Poste, G., and Papahadjopoulos, D. (1976). *Proc. Natl. Acad. Sci. U.S.A.* **73**, 1603.

Purifoy, D. J. M., and Powell, K. L. (1977). *Virology* **77**, 84.

Rekosh, D. M. K., Russell, W. C., Bellett, A. J. D., and Robinson, A. J. (1977). *Cell* **11**, 283.

Richards, O. C., Hey, T. D., and Ehrenfeld, E. (1981). *J. Virol.* **38**, 863.

Richardson, C. D., Scheid, A., and Choppin, P. (1980). *Virology* **105**, 205.

Richman, D. D., Yazaki, P., and Hostetler, K. (1981). *Virology* **112**, 81.

Rosenwirth, B., and Eggers, H. J. (1979). *Virology* **97**, 241.

Rott, R. (1979). *Arch. Virol.* **59**, 285.

Rott, R., and Klenk, H.-D. (1977). *In* "Cell Surface Reviews: Virus Infection and the Cell Surface (G. Poste and G. L. Nicolson, eds.), Vol. 2, p. 47. North-Holland Publ., Amsterdam.

Rott, R., Reinacher, M., Orlich, M., and Klenk, H.-D. (1980). *Arch. Virol.* **65**, 123.

Rubin, D. H., and Fields, B. N. (1980). *J. Exp. Med.* **152**, 853.

Samson, A. C. R., Chambers, P., and Dickinson, J. H. (1980). *J. Gen. Virol.* **47**, 19.

Sarmiento, M., Haffey, M., and Spear, P. G. (1979). *J. Virol.* **29**, 1149.

Scheid, A., and Choppin, P. W. (1974). *Virology* **57**, 475.

Scheid, A., and Choppin, P. W. (1977). *Virology* **80**, 54.

Scheid, A., Graves, M. C., Silver, S. M., and Choppin, P. W. (1978). *In* "Negative Strand Viruses and the Host Cells" (B. W. J. Mahy and R. D. Barry, eds.), pp. 181–193. Academic Press, New York.

Schloemer, R. H., and Wagner, R. R. (1975a). *J. Virol.* **15**, 882.

Schloemer, R. H., and Wagner, R. R. (1975b). *J. Virol.* **16**, 237.

Schmidt, M. F. G., and Schlesinger, M. J. (1979). *Cell* **17**, 813.

Schmidt, M., Bracha, M., and Schlesinger, M. (1979). *Proc. Natl. Acad. Sci. U.S.A.* **76**, 1687.

Schulze, J. T. (1981). *Symp. Genet. Variat. Influenza Viruses, Salt Lake City, Utah.*

Sharom, F. J., Barratt, D. G., Thede, A. E., and Grant, C. W. M. (1976). *Biochim. Biophys. Acta* **455**, 485.

Silverstein, S. C., and Dales, S. (1968). *J. Cell Biol.* **36**, 197.

Silverstein, S. C., Astell, C., Levin, D. H., Schonberg, M., and Acs, G. (1972). *Virology* **47**, 797.

Skehel, J. J., Hay, A. J., and Armstrong, J. A. (1977). *J. Gen. Virol.* **38**, 97.

Sleigh, M. J., Booth, G. W., Brownee, G. G., Bender, V. J., and Moss, B. A. (1980). *In* "Structure and Variation in Influenza Virus, (W. G. Lewer and G. Air, eds.). Elsevier, Amsterdam.

Smith, H. (1977). *In* "Virus Infection and the Cell Surface" (G. Poste and G. L. Nicolson, eds.), pp. 1–46. Elsevier, Amsterdam.

Smith, J. D. (1980). *Intervirology* **13**, 312.

Smith, J. D., and de Harven, E. (1974). *J. Virol.* **14**, 95.

Stackpole, C. W. (1969). *J. Virol.* **4**, 75.

Steck, T. L. (1974). *J. Cell Biol.* **62**, 1.

Stephenson, J. R., and Dimmock, N. J. (1975). *Virology* **65**, 77.

Stephenson, J. R., Hudson, J. B., and Dimmock, N. J. (1978). *Virology* **86**, 264.

Svensson, U., Persson, R., and Everitt, E. (1981). *J. Virol.* **38**, 70.

Talbot, P. J., and Vance, D. E. (1980). *Can. J. Biochem.* **58**, 1131.

Tanaka, T., Ogura, H., Ocho, M., Namba, M., Omura, S., and Oda, T. (1981). *Virology* **108**, 230.

Tanenbaum, S. F. (1978). "Cytochalasins: Biochemical and Biological Aspects." North-Holland Publ., Amsterdam.

Tiffany, J. M. (1977). *In* "Virus Infection and the Cell Surface" (G. Poste and G. L. Nicolson, eds.), pp. 157–194. Elsevier, Amsterdam.

Väänänen, P., and Kääriäinen, L. (1979). *J. Gen. Virol.* **43**, 593.

Väänänen, P., and Kääriäinen, L. (1980). *J. Gen. Virol.* **46**, 467.

Väänänen, P., Gahmberg, C. G., and Kaariainen, L. (1981). *Virology* **110**, 336.

Vahlne, A., Svennerholm, B., and Lycke, E. (1979). *J. Gen. Virol.* **44**, 217.

Vengris, V. E., Reynolds, F. M., Hollenberg, M. D., and Pitha, P. M. (1976). *Virology* **72**, 486.

Waterfield, M. D., Espelie, K., Elder, K., and Skehel, J. J. (1979). *Br. Med. Bull.* **35**, 57.

Weatherbee, J. A., Luftig, R. B., and Weihing, R. R. (1977). *J. Virol.* **21**, 732.

Weiner, H. L., Druyne, D., Averill, D., and Fields, B. N. (1977). *Proc. Natl. Acad. Sci. U.S.A.* **74**, 5744.

Weiner, H. L., Ramig, R. F., Mustoe, T. A., and Fields, B. N. (1978). *Virology* **86**, 581.

Weiner, H. L., Ault, K. A., and Fields, B. N. (1980). *J. Immunol.* **124**, 2143.

Weiss, R. A. (1975). *Perspect. Virol.* **9**, 165.

Weiss, R. A. (1976). *In* "Cell Membrane Receptors for Viruses, Antigens and Antibodies, Polypeptide Hormones and Small Molecules" (R. F. Beers, Jr. and E. G. Bassett, eds.), p. 237. Raven, New York.

White, J., and Helenius, A. (1980). *Proc. Natl. Acad. Sci. U.S.A.* **77**, 3273.

White, J., Kartenbeck, J., and Helenius, A. (1980). *J. Cell Biol.* **87**, 264.

Wills, E. J., Davies, P., Allison, A. C., and Haswell, A. D. (1972). *Nature (London) New Biol.* **240**, 58.

Winston, V. D., Bolen, J. B., and Consigli, R. A. (1980). *J. Virol.* **33**, 1173.

Wolf, D., Kahana, I., Nir, S., and Loyter, A. (1980). *Exp. Cell Res.* **130**, 361.

Wu, P., Ledeen, R. W., Udem, S., and Isaacson, Y. A. (1980). *J. Virol.* **33**, 304.

Yasuda, H., Shimizu, K., and Ishida, N. (1980). *Acta Virol.* **24**, 98.

Zarling, D. A., and Keshet, I. (1979). *Virology* **95**, 185.

Zhdanov, V. M., and Bukrinskaya, A. G. (1961). *Vopr. Virusol.* **4**, 416.

Zhirnov, O. P., Ovcharenko, A. I., and Bukrinskaya, A. G. (1981). *Vopr. Virusol.* **6**.

Zweerink, H. J., McDowell, M. J., and Joklik, W. K. (1971). *Virology* **45**, 716.

ADVANCES IN VIRUS RESEARCH, VOL. 27

COMPARATIVE BIOLOGY AND EVOLUTION OF BACTERIOPHAGES

Darryl C. Reanney

Department of Microbiology
La Trobe University
Bundoora, Victoria, Australia

Hans-W. Ackermann

Department of Microbiology
Faculty of Medicine
Laval University
Quebec, P.Q., Canada

I. Introduction

Viruses have left no fossil record yet evolutionary relationships among certain viruses can now be assessed with a high degree of confidence. This is because "memories" of past relationships are preserved in the nucleotide sequences of viral nucleic acids and in the amino acid sequences of viral proteins. Where sequences cannot be cross-checked directly, possible relationships can still be discerned by comparing, for example, the immunological reactions of viral proteins (serology), the particles built by the interaction of structural proteins (morphology), or the virus-specific enzymes which direct the pattern of viral infection (strategies of infection).

Bacterial viruses or bacteriophages seem particularly well suited to an evolutionary analysis. Phages have been favorite tools of microbial geneticists since the rise of molecular biology and viruses such as T2 and λ are among the best-studied genetic systems on earth. The genome of a small, isometric phage (MS2) was the first naturally occurring, self-replicating RNA to be sequenced in full (Fiers et al., 1976) while the primary structure of the DNA of a further small phage (φX174) was determined shortly afterward (Sanger et al., 1977). A voluminous phage literature has accumulated across the years and many of these data are relevant to evolutionary issues.

Several factors prompted us to undertake a review of phage evolution at this time. The first was the very explosion of knowledge in the field of microbial genetics that phages themselves helped to bring about. The wealth of DNA/DNA interactions uncovered by recent research permits a rapid, saltatory evolution among the various replicons resident in bacterial cells; this makes it possible, perhaps for the first time, to discuss phages in the light of those mechanisms most likely to have brought about their evolution. Also, the concept of "selfish DNA" (Doolittle and Sapienza, 1980; Orgel and Crick, 1980) has given us a useful new way of looking at viral genes (see Reanney, 1981).

A second reason why a review of phage evolution seemed timely is because our concepts of bacterial evolution have recently undergone a revolution. Until recently prokaryotes were regarded as little-changed descendents of the oldest cells on earth (Jawetz *et al.*, 1970). This view was reinforced by morphological similarities between microfossils reported to be $3.0-3.3 \times 10^9$ years old (Muir and Hall, 1974; Schopf, 1978) and modern bacteria. However, this assumption of conservatism has now been challenged on genetic grounds. Whereas bacterial genes are continuous, eukaryotic genes are often split into coding modules or "exons" and intervening silent regions or "introns." Doolittle (1978) and Darnell (1978) have argued persuasively that this "genes in pieces" organization antedates the divergence of prokaryotes and eukaryotes. If this hypothesis is correct then modern bacteria have differentiated greatly from their ancient ancestors. The phylogeny of bacteria has been further upset by evidence which suggests that methanogenic bacteria and some other taxa constitute a different kingdom of organisms (Balch *et al.*, 1979; Fox *et al.*, 1980; Kandler, 1981). These developments have made it necessary for virologists to reexamine some common assumptions about the evolution of bacterial viruses.

Our final reason for compiling this review arises from the involvement of both authors in attempts to classify phages. While we accept that current viral classifications should be based on operational "working" criteria, we believe that any taxonomic scheme that is not ultimately built on a sound phylogenetic base will cause more problems than it resolves. By highlighting gaps and ambiguities in phage taxonomy we hope to stimulate other virologists to take a more active interest in the comparative aspects of phage morphology, biochemistry, and genetics.

II. Some Cautionary Remarks

This article attempts to cover an enormous field in a limited number of pages. Inevitably, therefore, we have had to be selective in our treatment. Much information relevant to our theme has been omitted, not because it was unreliable or uninteresting, but because it did not deal with the particular phages used to illustrate particular points. Some of the references that are quoted deal with phages of unusual or arcane groups of bacteria. This is of concern where the description of a given phage in a host species, for example, depends upon a single reference. Little-studied bacteria are easily misclassified and phages that would be unique in one host taxon may be common in another. Likewise pilot reports of "new" phages sometimes concentrate on "exciting" features which later turn out to be spurious: the *Bacillus* phage AP50 was first described as having an RNA genome (Nagy *et al.*, 1976), but further work showed that AP50, like morphologically similar viruses of *Bacillus*, contains duplex DNA (Nagy, 1981). Ideally, any unexpected piece of information should be checked by several independent laboratories before it is accepted. Unfortunately this is seldom possible in practice.

III. The Systematics of Phages

Bacteriophages are a diverse group of entities; they span an enormous size range, their nucleic acids may consist of DNA or RNA in single-stranded (ss) or double-stranded (ds) form, and they include cubic, filamentous, and pleomorphic members.

Attempts to classify phages have been similar to attempts to classify viruses of plants and animals. A primary separation is usually made on the basis of nucleic acid character and gross particle morphology. Using these and other criteria phages are currently divided into 10 major groups (Matthews, 1979) labeled A through G, in continuation of an earlier system of nomenclature (Bradley, 1967). They can then be further subdivided into 18 groups or types (Ackermann and Eisenstark, 1974; Ackermann, 1978a) (Fig. 1 and Table I).

Tailed phages are by far the most common type of phage, occurring in over 90 genera of bacteria and cyanobacteria (Section IV,A). Almost all known tailed phages contain linear, duplex DNA. Morphological features apparently unique to members of this group of viruses include contractile and noncontractile tail sheaths, various head and tail appendages, collars, base plates, tail fibers, and spikes. The extraordinary diversity of tailed phages arises in part from the various ways these different elements are combined and in part from variations in dimensions and physicochemical properties (Table II); for example the molecular weights of tailed phage DNAs range from 12 to 490 $\times$ 10^6 (Table II).

Tailed phages are subdivided into three basic groups according to tail structure (Bradley, 1967): Group A, Myoviridae—contractile tails;

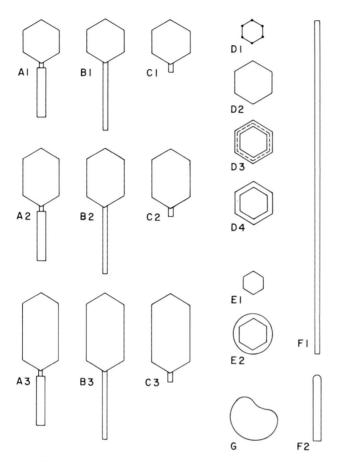

FIG. 1. Fundamental groups of bacteriophages. Modified from Ackermann and Eisenstark (1974). With permission of Intervirology and CRC Press.

Group B, Styloviridae (alternative name proposed, Syphoviridae)—long, noncontractile tails; and Group C, Podoviridae—short tails.

Groups D to G are made up of the cubic, filamentous, and pleomorphic phages. These taxa are distinctive and not related to each other in any obvious way. Unlike the tailed phages these groups of viruses are relatively rare and are confined to a fairly small group of host genera. They can be usefully arranged as follows:

1. Cubic DNA phages: (a) Group D1, Microviridae, ϕX type, ssDNA; (b) Group D2, provisional category for "naked," poorly known DNA

TABLE I

Basic Properties of Bacteriophages[a]

Group	Family	Example	Nucleic acid Nature	%	MW	Virion Symmetry	Envelope	Lipids
A–C	Tailed phages	T4, λ	D, 2, L	25–62	12–490	Binary	–	–
D1	Microviridae	φX174	D, 1, C	26	1.7	Cubic	–	–
D3	Corticoviridae	PM2	D, 2, C, S	13	5.8	Cubic	–	+
D4	Tectiviridae	PR4	D, 2, L	14	9.8	Cubic	–	+
E1	Leviviridae	MS2	R, 1, L	31	1.2	Cubic	–	–
E2	Cystoviridae	φ6	R, 2, L, seg	23	10.4	Cubic	+	+
F1, F2	Inoviridae	fd, MV-L1	D, 1, C	6–12[b]	1.5–2.7[b]	Helical	–	–
G	Plasmaviridae	MV-L2	D, 2, C, S	8.4		Pleomorphic	+	+

[a] Adapted from Ackermann (1982) and Matthews (1979). D or R, DNA or RNA; 1 or 2, single- or double-stranded; L or C, linear or circular; S, supercoiled; seg, segmented; MW, molecular weight × 10⁶; + or –, present or absent.

[b] *Inovirus* genus only (F1).

TABLE II

DIMENSIONS AND PHYSICOCHEMICAL PROPERTIES OF TAILED PHAGES[a]

Property	Average	Range
Virion		
Head diameter (nm)[b]	72	37–180
Tail length (nm)	—	3–539
Molecular weight ($\times 10^6$)	92	47–490
Sedimentation velocity ($S_{20,w}$)	551	262–1042
Buoyant density (gm/ml), CsCl	1.48	1.41–1.54
DNA		
Percentage	46	25–62
Molecular weight ($\times 10^6$)	59	12–490
Guanine–cytosine (%)	48	28–70

[a] Computed from 112 species descriptions including 662 phages of actinomycetes, *Agrobacterium, Bacillus, Clostridium,* enterobacteria, gram-positive cocci, pseudomonads, and *Rhizobium* (Ackermann, 1974, 1975, 1976, 1978c; Berthiaume and Ackermann, 1977; Liss *et al.,* 1981; Reanney and Ackermann, 1981).
[b] Isometric heads only.

phages of large size (60 nm); (c) Group D3, Corticoviridae, PM2 type, dsDNA, capsid includes lipid bilayer; and (d) Group D4, Tectiviridae, PR4 type, dsDNA, double capsid, inner coat contains lipids, tail equivalent appears upon DNA ejection.

2. Cubic RNA phages: (a) Group E1, Leviviridae, MS2 type, ssRNA; and (b) Group E2, Cystoviridae, ϕ6 type, dsRNA, enveloped phages containing three RNA segments and virion-associated RNA polymerase.

3. Filamentous phages—Inoviridae: (a) Group F1, genus *Inovirus,* fd type, ssDNA, long rods of 800–1950 × 8 nm; (b) Group F2, genus *Plectrovirus,* MV-L1 type, ssDNA, short rods of 84 × 14 nm.

4. Pleomorphic phages: Group G, Plasmaviridae, MV-L2 type, dsDNA, enveloped phages without detectable protein capsid.

IV. INFERENCES FROM DISTRIBUTION

The frequency with which the various phage morphotypes occur in different prokaryotic hosts is given in Tables III and IV. These tables are based on an earlier review (Ackermann and Eisenstark, 1974) and on the internal records of the International Committee on Taxonomy of Viruses (ICTV). Because many enterophages cross generic boundaries the enterobacteria are treated as a single host group. The large number of references, about 230 of which have been published since 1974, makes it virtually impossible to document this table in the usual way. Readers seeking further information on points not tabulated in the reference section of this article should contact the Phage Subcommittee of the ICTV.

A. Tailed Phages

1. General Inferences

Phages with isometric heads comprising the A1, B1, and C1 morphotypes constitute over 80% of known isolates. Of these the B1 morphotype seems especially prevalent. Table III shows that, of the 2025 phages listed, 856 belong to this category. These B1 phages are found in an extraordinary diversity of host genera, ranging from small wall-less bacteria such as *Mycoplasma* to stalked forms such as *Caulobacter*.

How does one account for this widespread distribution? It is statistically very unlikely that genes for structurally *similar* phage particles could arise independently in 51 *different* host genera. Convergent evolution as a general explanation is unlikely for the same reason. Part of the present day distribution could be due to horizontal spread across taxonomic borders. The temperate coliphage P1, for example, can be

TABLE III

DISTRIBUTION OF TAILED PHAGES

Bergey's Manual, part	Host genus	Morphotype									Total
		A1	A2	A3	B1	B2	B3	C1	C2	C3	
1	*Rhodopseudomonas*		1		3			2			6
2	*Myxococcus*	8						3			11
	Cytophaga	28			2						30
	Flexibacter	1									1
	Saprospira	1									1
3	*Sphaerotilus*					1					1
4	*Hyphomicrobium*				1			3			4
	Caulobacter	4			6	2	32	3			47
	Asticcacaulis				17	4	3				24
	Ancalomicrobium				3						3
5	*Treponema*	1			1						2
	Borrelia (?)	1									1
	Leptospira	1									1

(*continued*)

TABLE III (*continued*)

Bergey's Manual, part	Host genus	Morphotype									Total
		A1	A2	A3	B1	B2	B3	C1	C2	C3	
6	*Campylobacter*	3			9						12
	Bdellovibrio	3	1								4
7	*Pseudomonas*	48			27	6		40			121
	Xanthomonas	2			3	8					13
	Gluconobacter	1									1
	Azotobacter	12			2			6			20
	Rhizobium	24	1		31	17	1	17			91
	Agrobacterium	1			22	8		5			36
	Methylomonas	1									1
	Halobacterium	2									2
	Alcaligenes	3			7			1			11
	Brucella							45			45
	Bordetella				1						1
	Thermus	1									1
8	Enterobacteria	118	76	6	143	11		136		10	500
	Vibrio	12?	1		3			11?	1		28
	Beneckea	3	1		2			1			7
	Aeromonas	4	1					1			6
	Lucibacterium				1						1
	Flavobacterium				13						13
	Haemophilus	2			1						3
	Pasteurella				5						5
9	*Bacteroides*				10						10
	Fusobacterium	1						1			2
	Desulfovibrio	1									1
10	*Neisseria*	3?									3
	Acinetobacter	9	1			4		3			17
11	*Veillonella*				2			2			4
12	*Thiobacillus*	1									1
14	*Micrococcus*				12			1			13
	Staphylococcus	17			39	18		1			75
	Streptococcus	13			131	23	5	4	3	1	180
	Leuconostoc				1						1
15	*Bacillus*	75			105	8	2	1	16		207
	Clostridium	69			39	8		20			136
16	*Lactobacillus*	15			25	1					41
	Listeria				6						6
	Erysipelothrix				1						1
	Caryophanon	1			2	1					4

(*continued*)

TABLE III (*continued*)

Bergey's Manual, part	Host genus	A1	A2	A3	B1	B2	B3	C1	C2	C3	Total
17	*Corynebacterium*				61						61
	Arthrobacter				5			1			6
	Propionibacterium				10						10
	Actinomyces							1			1
	Bifidobacterium				1						1
	Mycobacterium	1			45	17					63
	Actinoplanes	1									1
	Dactylosporangium				1						1
	Nocardia				10						10
	Streptomyces	1			22	12		2			37
	Micromonospora				3						3
	Thermoactinomyces				1	2					3
	Thermomonospora		1		2			2			5
	Micropolyspora				1						1
18	*Porochlamydia*[b]							1			1
19	*Mycoplasma*	1			1						2
	Acholeplasma							1			1
	Spiroplasma				1			1			2
Host genera of uncertain status or position											
?	*Achromobacter*	13			9			2			24
	Brevibacterium				5						5
	Hydrogenomonas	1									1
	Methylosinus[b]							11			11
	Kappa-particles							1			1
Cyanobacteria and green algae (*Chlorella*)											
	Anabaena	2						2			4
	Anacystis	2									2
	Chlorella[c]							2			2
	Nostoc	1									1
	Plectonema							6			6
	Synechococcus				2			1			3
	Total	513	84	6	856	151	43	341	20	11	2025
			603			1050			372		

[a] Excluding shadowed phages, known defective phages and particulate bacteriocins. Nomenclature is of Bergey's Manual, 8th Ed. (1974). Completed January 31, 1981.

[b] Not in Bergey's Manual.

[c] In view of the difficulty in obtaining bacteria-free algal cultures, the existence of *Chlorella* phages (Moskovets *et al.,* 1970; Tikhonenko and Zavarzina, 1966) needs confirmation (Section XI,B).

accepted by species of *Erwinia, Proteus, Pseudomonas* and *Serratia* (Murooka and Harada, 1979) and can transfer genes to myxobacteria (Kaiser and Dworkin, 1975), as well as to strains of *Agrobacterium, Alcaligenes,* and *Flavobacterium* (Murooka and Harada, 1979). However, P1 combines the features of a phage and a plasmid and may not be typical of the majority of phages. Most temperate phages in nature integrate into host DNA. Since the integration of phage DNA into the bacterial chromosome depends on a specific region of homology between the interacting DNAs, for example the *att* region of λ, it seems unlikely that true temperate phages can lysogenize bacteria that differ markedly from their preferred hosts. Jones and Sneath (1970) found that phages crossed generic boundaries in only 0.3% of 26,681 attempted reactions between bacterial strains at various taxonomic levels.

Overall, the widespread distribution of B1 phages is best explained on the assumption that the development of genes for these particles *antedates the divergence of many of the host taxa listed.* Since there is no confirmed report of a phage of gram-negative bacteria which can productively infect gram-positive cells and vice versa (Jones and Sneath, 1970), it is possible that genes for typically phage functions, such as tail biosynthesis, arose before the division of the bacterial world into gram-positive and negative lines. Similarly, since no known phages are jointly active on bacteria and blue-green algae, we can infer that tailed phage genes existed before the blue-greens diverged from the true bacteria. Perhaps the most significant feature of Table III is the observation that phages with contractile tails occur in *Halobacterium* (Torsvik and Dundas, 1974; Wais *et al.,* 1975). *Halobacterium* is considered to be a representative of an extremely ancient phylogenetic category, the archaebacteria (Balch *et al.,* 1979; Fox *et al.,* 1980; Kandler, 1981). Table III suggests therefore that genes for tailed phages were already present in the common ancestor of the archaebacteria, eubacteria, and cyanobacteria or that transfer from one evolutionary line to another occurred before these three categories of organism became genetically isolated from each other.

These observations imply strongly that tailed phages are very old indeed. However it is impossible to fix an approximate origin for phages in time without invoking unwarranted assumptions or speculations. For example, the discovery of putative stromatolite remains in sedimentary rocks $3.4-3.5 \times 10^9$ years old (Lowe, 1980; Walter *et al.,* 1980) does not mark out a time of first appearance for cyanobacteria and hence for cyanophages for two reasons: (1) organisms other than blue-green algae can build stromatolite laminations (Walter, 1977), and (2) it is virtually certain that these early cells differed dramatically from their specialized descendents (Reanney, 1974, 1976a) and hence cannot be classified as "cyanobacteria." All we infer with any confidence from the distribution data is that genes for "phage" heads and tails were widely disseminated on a global scale at least 2000 million years ago.

2. Inferences from Specific Morphotypes and Individual Phages

The discussion to date has concentrated on the presence of major morphotypes (A1 and B1) in different host genera. However evolutionary information can also be gleaned from the *absence* of certain morphotypes. Inspection of Table III shows that the A3 type has been reported only six times among the 500 tailed phages of enterobacteria while no B3 phages occur in enterics. No C3 phages have been described among at least 207 known *Bacillus* phages or 121 *Pseudomonas* phages. Indeed the C3 type is found only in enterobacteria and *Streptococcus* (Grimont and Grimont, 1982; Moazamie *et al.*, 1979; Saxelin *et al.*, 1979).

It seems unlikely that C3 phages have some special survival advantages in these particular hosts in view of the evolutionary distance between streptococci and the enterobacteria. Overall, the type 3 morphology seems as rare as the type 1 is common. This rarity suggests (1) that phages with elongated heads have been less successful than competing phages with isometric capsids and/or (2) that the particular types or combinations of genes which make such particles have arisen only a few times during evolution, perhaps after the isometric types had become established in most host taxa. Type 3 phages, based on this thesis, arose from type 1 or type 2 phages (Section X,B).

The distribution of individual phage species among different host bacteria also has evolutionary implications for phages and hosts. Figure 2 shows a few examples of phages of characteristic morphology. (1) T-even type phages occur in both enterobacteria and *Pseudomonas*. (2) SP50 type phages, recognizable by their conspicuous capsomeres, have been reported in *Bacillus, Staphylococcus,* and *Streptococcus* (Ackermann, 1974, 1975). (3) Viruses of the 3A type, originally described in *Staphylococcus* and characterized by prolate heads and about 300-nm-long tails (Ackermann, 1975), occur in *Caryophanon, Clostridium,* and *Streptococcus* (Nauman and Wilkie, 1974; Dolman and Chang, 1972; Krämer and Lenz, 1975). The distribution of these phages correlates broadly with bacterial phylogenetic trees (Fox *et al.*, 1980; Kandler, 1981; Margulis, 1972). This suggests that these phages, or more precisely their distinctive morphological features, may act as surviving "indicators" of the past evolution of their bacterial hosts.

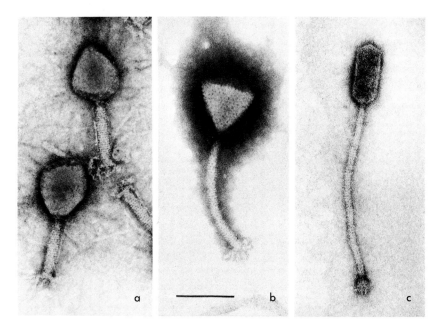

Fig. 2. Phages of characteristic morphology occurring in different bacterial groups.
(a) T-even-type phages of *Pseudomonas aeruginosa,* uranyl acetate; note the folded tail
fibers forming a "jacket." (b) SP50-type phages of *B. thuringiensis,* phosphotungstate;
note the capsomeres on the phage head. (c) 3A-type phage of *S. aureus,* uranyl acetate.
×157,410. The bar indicates 100 nm.

Some distinctive phage features seem almost "ornamental." *Bacillus*
phage $\phi 29$ for example produces viable mutants which are devoid of
head fibers (Reilly *et al.,* 1977). The exceptional and dispensable nature
of these structures suggests that they are of recent origin and/or that
the horizontal spread of such phages among widely different host taxa
must be subject to quite severe constraints. In assessing Table III it is
important to bear in mind the biased nature of the data. Much of the
information comes from intensely studied test bacteria such as *Escherichia coli* or *Bacillus subtilis.* This very concentration of information makes the *absence* of certain morphotypes in these taxa all the
more striking.

B. Cubic, Filamentous, and Pleomorphic Phages

The cubic, filamentous, and pleomorphic phages are relatively rare
and therefore of special evolutionary interest. The distribution of these
phages is shown in Table IV. Certain features are immediately appar-

TABLE IV

DISTRIBUTION OF CUBIC, FILAMENTOUS, AND PLEOMORPHIC PHAGES[a]

Bergey's Manual, part	Host genus	D1	D2	D3	D4	E1	E2	F1	F2	G	Total
4	Caulobacter					8					8
6	Spirillum		1								1
	Bdellovibrio		1			7?					8
7	Pseudomonas			2?		2	1	2			7
	Xanthomonas							3			3
8	Enterobacteria	26				16		13			55
	Vibrio							1			1
	Polyvalent phages[b]				5	1		1			7
15	Bacillus				3						3
18	Acholeplasma								12	3	15
	Spiroplasma								1		1
	Total	26	2	2?	8	34?	1	20	13	3	109

[a] Excluding shadowed phages and phage-like particles of uncertain nature or host, found in *Bacillus, Clostridium, Nitrobacter,* and mycoplasma-like plant pathogens. Nomenclature is of Bergey's Manual, 8th Ed. (1974). Completed January 31, 1981.

[b] Plasmid-dependent phages active on enterobacteria, *Pseudomonas, Vibrio,* and *Acinetobacter* (see Section XII,C).

ent. For example, 26 isolates of D1 type phages (Microviridae) are known in the enterobacteria, but none occurs elsewhere despite the large amount of literature on, for example, phages of *Bacillus* and *Pseudomonas.* This suggests that Microviridae arose after enteric-type organisms had diverged from other bacteria. By contrast, E1 type phages (Leviviridae) are found in four gram-negative taxa. In enterics and *Pseudomonas,* the multiplication of such ribophages is dependent on the presence of specific conjugative plasmids (Section XII,C). This dependence suggests that a search of *Caulobacter* and *Bdellovibrio* for plasmids capable of supporting phage growth would be rewarding. This suggestion is reinforced by the observation that *Caulobacter* can accept and maintain the plasmid RP1 (Alexander and Jollick, 1977) which provides the attachment potential for the E1 phage PRR1 (Olsen and Shipley, 1973). Similar considerations apply to the F1 group which includes fd and its relatives (genus *Inovirus*).

One of the most interesting groups in Table IV is the D4 type (Tectiviridae). Members of this unusual group possess a double coat, are sensitive to lipid solvents, and produce a tail from the inner coat (Section III). Some of these phages propagate in gram-negative taxa which contain plasmids of the incompatibility group P. Since these plasmids have a wide host range, the occurrence of D4 phages in enterobacteria and *Pseudomonas* is not surprising. What is remarkable is the simultaneous occurrence of morphologically and chemically similar phages in various *Bacillus* species, namely, *B. acidocaldarius, B. anthracis,* and *B. thuringiensis* (Sakaki and Oshima, 1976; Sakaki *et al.,* 1977; Nagy, 1974; Nagy *et al.,* 1976; Ackermann *et al.,* 1978a). It would be of interest to determine whether these *Bacillus* hosts harbor a plasmid and whether cells "cured" of this putative plasmid are still susceptible to the phages. Because of their distinctive morphology and composition these *Bacillus* phages appear to provide a striking example of divergent evolution: the tectiphage of the thermophile, *B. acidocaldarius,* propagates at 60°C, whereas the tectiphage of *B. thuringiensis* is immediately inactivated at this temperature (Ackermann *et al.,* 1978a; Sakaki *et al.,* 1977).

The remaining phage groups, D3, E2, F2, and G have very narrow host ranges and are very small, sometimes consisting of a single virus. (1) PM2, type phage and sole certain member of the D3 group or Corticoviridae, has been reported only in a marine pseudomonad (Espejo and Canelo, 1968). (2) ϕ6, the only representative of the E2 group or Cystoviridae, is specific for *Pseudomonas phaseolicola* (Vidaver *et al.,* 1973). (3) The rod-shaped phages of the F2 type (*Plectrovirus*) and pleomorphic phages (Plasmaviridae) have been found only in mycoplasmas. The limited occurrence of all these small phage groups suggests either that they were once more widespread and somehow died out or, as seems more likely, that they arose after the speciation of their host cells.

Cubic phages and filamentous viruses of the *Inovirus* or fd group are conspicuously associated with aerobic or facultatively aerobic hosts, hence it seems reasonable to assume that they evolved after the emergence of aerobic bacteria. The transition from an anoxygenic to an oxidizing atmosphere seems to have occurred about 2×10^9 years ago (Cloud, 1968; De Ley, 1968; Margulis, 1972; Schopf, 1972). Cubic phages and viruses of the fd type may date from this or later periods.

V. The Comparative Anatomy of Phage Nucleic Acids

We introduced the topic of phage evolution by way of comparative morphology because the architecture of a virus is usually the first aspect of a "new" isolate to be described and because morphological data are more numerous than molecular data. However morphological similarities do not necessarily imply evolutionary relatedness. The nature of viral nucleic acids is of prime importance in viral taxonomy and phylogeny.

A. General Properties

For cubic, filamentous, and pleomorphic phages the nucleic acid data reinforce what was obvious from morphology, namely, the lack of significant relationships between the groups (Table I). This suggests a polyphyletic origin. The cystovirus $\phi6$, with its genome of segmented double-stranded RNA, is unique among phages; indeed $\phi6$ resembles some animal viruses more than it does other phages. In a similar way Corticoviridae and Plasmaviridae are sharply set apart from other phages by the presence of supercoiled DNA. Closed circular DNA is a rare feature in general virology (Matthews, 1979) and occurs elsewhere in phages only in the tailed *Rhodopseudomonas* phage, R$\phi6$P (Pemberton and Tucker, 1977; Tucker and Pemberton, 1978, 1980; see Sections VIII,B and XII,C).

The presence of small single-stranded DNA genomes in Microviridae, long Inoviridae of the fd type, and short Inoviridae of the MV-L2 type superficially seems to be a linking feature (for a discussion of relationships between ϕX174 and fd groups see Denhardt, 1975). While a common ancestry cannot be ruled out it is possible that any genetic system under heavy pressure to reduce its genome size could negotiate the transition from double-stranded to single-stranded DNA. The presence of single-stranded DNAs in the Microviridae and *Inovirus* groups could thus be due to convergent evolution. Both groups are further set apart by the presence of overlapping genes in ϕX174 in which one gene can code for two proteins by using different reading frames (Section VI,A).

Since all tailed phages contain duplex DNA, nucleic acid data tell us less about possible large-scale evolutionary relationships among this group. Features which seem significant in a less dramatic way are (1) the presence or absence of terminally redundant sequences as in the T-even coliphages and (2) the presence or absence of specific "nicks" at constant positions in one strand of duplex DNA as in coliphage T5 (McCorquodale, 1975; Thomas and MacHattie, 1967), especially if these features can be related to some functional aspect of the viral life cycle.

B. Functional Genetic Maps

Well-studied tailed phages such as λ and P22 show a marked cluster-ing of genes for head and tail proteins (see Section VIII,C). A similar linkage occurs in coliphages P2 and 186 (Goldstein *et al.*, 1974; Young-husband and Inman, 1974) and in some phages of *Bacillus* (Hemphill and Whiteley, 1975), *Corynebacterium* (Barksdale and Arden, 1974), *Rhizobium* (Sik and Orosz, 1971), and *Staphylococcus* (Kretschmer and Egan, 1975). Such a grouping is less obvious in *Bacillus* phage φ29 (Murialdo and Becker, 1978) and in coliphages T3–T7 (Beier and Hausmann, 1973) and T4 (Champe, 1974). Indeed, the genome of T4 appears almost disorganized.

Concentrations of genes for specific functions in particular areas of the genome have marked advantages for transcriptional timing and sequential translation. In λ, for example, 10,000 base pairs separate the genes for the main head protein from those encoding the tail fibers. This arrangement of head and tail genes permits capsid proteins to accumulate—and to accommodate newly formed DNA—before tail as-sembly is far advanced. The clustering of genes for phage structural proteins could therefore represent some kind of "optimization" of the organization of the genetic map. This makes it difficult to ascribe such clustering to either a common ancestral pattern or convergent evolu-tion. In view of the diversity of tailed phages and the time scale in-volved it seems almost certain that both divergence and convergence have occurred many times in many ways, some pressures ordering structural genes into the clusters noted while other pressures en-couraging the disruption of such groupings and the rearrangement of functional sequences.

C. Guanine–Cytosine Contents and Base Doublet Frequency

Bacteria have an extraordinary range of G–C values which extends from 27 to 73. The range seen in tailed phages (28 to 75) is closely comparable (Table V). This coincidence suggests that phages in general have coevolved with bacteria for very long periods of time and provides independent evidence for their extreme antiquity. The parallelism is just as remarkable when the G–C contents of individual bacterial gen-era are checked against the G–C contents of their specific phages: for *Bacillus* the range is 32–56 (bacteria) as against 31–60 (phages); for *Streptococcus* 28–46 (bacteria) as against 27–48 (phages); for *Mycobac-terium* 62–73 (bacteria) as against 61–72 (phages). It is significant

TABLE V

GUANINE–CYTOSINE CONTENT RANGE IN SOME BACTERIA AND TAILED PHAGES

Bacterial group	G–C (%) Bacteria[a]	G–C (%) Phages[b]	Number of phages	References
Agrobacterium	56–68	50–60	12	Guay *et al.* (1981)
Azotobacter	50–67	52–63	10	Guay *et al.* (1981)
Bacillus	32–56	31–60	42	Guay *et al.* (1981)
Enterobacteria	37–63	36–62	39	Guay *et al.* (1981); Dhillon *et al.* (1980)
Micrococcus	68–75	68–73	10	Guay *et al.* (1981)
Mycobacterium	62–73	61–72	15	Guay *et al.* (1981)
Pseudomonas	53–69	36–68	25	Guay *et al.* (1981)
Rhizobium	58–65	49–62	15	Guay *et al.* (1981)
Staphylococcus	30–38	27–37	15	Guay *et al.* (1981)
Streptococcus	28–46	27–48	34	Guay *et al.* (1981); Lopez *et al.* (1977)
Streptomyces	67–80	58–70	12	Guay *et al.* (1981)

[a] From Normore (1981), excluding species *incertae sedis*.
[b] Excluding phages with unusual bases.

that these last two sets of values do not overlap. These "fits" suggest that phage genes can be viewed as normal genetic components of their host taxa, reflecting the same divergent and convergent pressures as the bacterial DNAs. A similar parallelism between the G–C ranges of phages and hosts is seen in the cubic and filamentous phages.

Relationships between phages and bacteria are also evident in the patterns of nearest neighbor base frequency. The DNAs of *Escherichia coli* and coliphages λ, λdg, T1, T5, and φX174 (RF) show close resemblances and the same is true for P22 and its host, *Salmonella typhimurium*. T-even phage, T5, and T7 DNAs however differ more or less substantially from that of *E. coli* (Bellett, 1967; Elton *et al.*, 1976; Josse *et al.*, 1961; Subak-Sharpe *et al.*, 1966, 1974). Although the observed similarities point to a "bacterial" origin of some phages, they do not prove phylogenetic relationships (Bellett, 1967) and may reflect the rapidity of mRNA translation in the host (Elton *et al.*, 1976).

D. Unusual Bases

Tailed phages contain a variety of unusual bases which partially or completely replace the normal nucleotides (Table VI). In evolutionary terms the genes which modify such bases may share an origin with

TABLE VI

Unusual Bases in Phage DNA[a]

Host	Phage or species[b]	Morphotype	Unusual bases	Base replaced	Extent of change (%)
Bacillus	PBS1	A1	Uracil	Thymine	100
	SP8	A1	5-Hydroxymethyluracil	Thymine	100
	SP-10	A1	α-Glutamylthymine	Thymine	15–20
	SP-15	A1	5-Dihydroxypentyluracil	Thymine	40
Enterobacteria	*T2*	A2	5-Hydroxymethylcytosine	Cytosine	100
	N-17[c]	A2	5-Hydroxycytosine	Cytosine	100
	16–19	A3	5-Methylcytosine	Cytosine	24
Pseudomonas	ϕW-14	A1	α-Putrescinylthymine	Thymine	50
Synechococcus[d]	S-2L	B1	2-Aminoadenine	Adenine	100
Xanthomonas	*XP-12*	B2	5-Methylcytosine	Cytosine	100

[a] Modified from Ackermann (1982) and Warren (1980).
[b] Species in italics contain several members.
[c] Kchromov *et al.* (1980).
[d] Cyanobacterium.

restriction/modification systems which are widely distributed among bacteria (Roberts, 1976) and which are located on plasmid, phage, and chromosomal DNAs. Modification in these cases often occurs by methylation (Arber and Linn, 1969). The novel modifications listed in Table VI provide a variety of distinctive biochemical markers. When the presence of such markers coincides with particular morphotypes and common hosts as in some type A1 *Bacillus* phages (Table VI), one can begin to be confident that one is dealing with a group of phages which speciated from a common ancestor. On the other hand, the occurrence of 5-methylcytosine in morphologically different phages of enterobacteria and *Xanthomonas* may be due to an accident of convergent evolution. Finally, the fact that these unusual bases have been detected to date only in strict or facultative aerobes suggests that modified nucleotides became a feature of phage nucleic acids rather late in evolution.

VI. Relationships Inferred from DNA/DNA Comparisons

Morphology and gross nucleic acid character then give preliminary information about possible relationships among phages and remain the prime criteria for phage taxonomy. However the development of a true phylogenetic base for phage taxonomy demands evidence of relationships gathered from a variety of sources. Molecular biology has now provided virologists with a battery of techniques by which quick determinations of molecular relatedness can be made. They include (1) heteroduplex mapping, (2) Southern blots, and (3) restriction nuclease fingerprinting. Perhaps the most definitive information comes from a direct comparison of nucleotide sequences or the amino acids encoded therein, using modern computer technology for detection and quantification of putative homologies (Dayhoff, 1972, 1976; Fitch and Margoliash, 1967; Gibbs, 1980; Hartl and Dykhuizen, 1979; Moore and Goodman, 1977). In the next section of this article we have assembled available information on comparisons between selected phage DNAs. This information mainly comes from certain groups of phages for reasons that have little to do with their potential evolutionary interest. The MS2 ribophage was sequenced first (Fiers *et al.*, 1976) because its small genome size made it a suitable candidate. Similar considerations applied to the small DNA phages of the Microviridae and Inoviridae families. Also, heteroduplex studies have largely been confined to the well-studied coliphages of the T-series and the lambdoid phages.

A. Sequence Comparisons

Since no DNA from a tailed phage has yet been sequenced in full the only nucleotide sequences from related DNA phages available for comparison are those of the microviruses ϕX174 and G4 (Godson *et al.*, 1978; Sanger *et al.*, 1977) and the inoviruses fd and M13 (Beck *et al.*, 1978; van Wezenbeek *et al.*, 1980).

Prior to the advent of full nucleotide sequences for phages ϕX174 and G4, their evolutionary cousinship was somewhat ambiguous. The two phages do not recombine or complement each other efficiently (Keegstra *et al.*, 1979). G4 initiates DNA synthesis using three proteins (Zechel *et al.*, 1975) and G4 complementary DNA synthesis begins at a unique site (Martin and Godson, 1977), whereas ϕX174 requires eight proteins for the initiation and synthesis of its complementary DNA and initiation occurs at random in many different positions (Clements and Sinsheimer, 1975; Schekman *et al.*, 1975). On the other hand, heteroduplex studies indicated considerable homologies between the two DNAs (Godson, 1974).

An alignment of the two complete nucleotide sequences allows definitive evolutionary conclusions to be drawn. The gene organization of both phages is identical. In the coding regions, the average base sequence difference is only 33.1% (Godson *et al.*, 1978). The most convincing evidence of homology comes from the genes which overlap. In φX174, gene E is entirely encoded in gene D but is read in a different phase; likewise gene B is encoded within A. Both these overlapping gene systems are present in G4 (Godson et al., 1978).

Sequence data have made it possible to speculate about the origins of these overlapping genes: the + strand of φX174 DNA is rich in thymine (T) nucleotides which tend to occur in the third codon position (Sanger et al., 1977). In the D/E overlap area, 39.1% of codons end in T in the D phase while only 14.3% end in the T in the E phase. By moving the reading frame one place to the left, third position Ts in phase D become second position Ts in phase E. Owing to the conservative structure of the genetic code, such a shift generates a high frequency of codons which specify hydrophobic amino acids in E phase. A strongly hydrophobic domain may help the lytic function of the E gene product. Fiddes and Godson (1979) suggest therefore that the E protein gene evolved from a preexisting D protein gene. Similar arguments were advanced for the K protein gene. The distribution of T nucleotides in the corresponding regions of G4 DNA closely resembles that of φX174 despite the fact that, overall, G4 lacks the high T content of φX174. Fiddes and Godson (1979) speculate "possibly a common ancestor to φX174 and G4 was rich in T nucleotides and developed second phase proteins and while G4 has lost the high third position T in the nonoverlapping regions it has been constrained to maintain it in the overlapping regions."

Overlapping genes are now known in both DNA and RNA phages (Beremand and Blumenthal, 1979; Atkins et al., 1979; see Reanney, 1981) and phase overlaps have been reported in one E. coli protein (Smith and Parkinson, 1980). In viruses under strong pressure to economize on genome sizes such multiple-choice systems may be a means of squeezing the maximum of information from the minimum of DNA or RNA. Overlapping genes are only one aspect of the genetic parsimony of φX174. A *single* φX gene, G, generates *four* different polypeptides. This has apparently been achieved by varying the efficiency with which START and STOP codons are recognized. Indeed the φX system is so organized that some internal codons initiate protein synthesis with a significant frequency while the opal UGA codons which normally halt peptide synthesis are neutralized by suppressor tRNAs with a significant frequency (Pollock et al., 1978; Geller and Rich, 1980). The result of these combined "strategies" is that a genome of 5375 nucleotides, which could normally encode only 200,000 daltons of protein, is able to specify a diverse set of peptides whose total molecular weight is estimated to be about 400,000 daltons. In this way, as

Pollock *et al.* (1978) point out, ϕX174 has enjoyed the evolutionary advantages of heterozygosity without increasing the size of its genome.

A molecular comparison of the ϕX and G4 systems then has yielded a rich harvest of evolutionary information. The two phages are clearly related and have diverged from a common ancestor by multiple changes. The larger size of G4 DNA is mainly due to an insertion of 141 bases in the G4 *A* gene and to the increased sizes of untranslated intergenic sequences which contain 282 nucleotides in G4 as opposed to 217 in ϕX174. A duplication has also occurred at the 3′ end of the G4 gene *D* and there are other small-scale insertions and deletions (Godson *et al.*, 1978). The divergent effect of these "quantal" rearrangements in the two genomes has been enhanced by the accumulation of single base changes at positions not rigorously policed by selection. These "major" and "minor" changes in the two phage DNAs parallel those noted in comparisons between recently diverged mammalian genes such as the β globins (Nishioka and Leder, 1979; Konkel *et al.*, 1979).

The base sequences of the filamentous phages, fd and M13, resemble each other closely. M13 DNA appears to be a single nucleotide shorter than fd DNA and the two genomes differ in about 3% of sequences only. In contrast to the DNA of ϕX174, sequences involved in transcription, translation, and replication as well as the sizes of their gene products appear to be highly conserved (Beck *et al.*, 1978; van Wezenbeek *et al.*, 1980).

B. Heteroduplex Comparisons

One of the first heteroduplex analyses of phage genomes was that of Simon *et al.* (1971). Its subject was the lambdoid phages (Styloviridae). One important conclusion was that the genome of each phage studied could be divided into regions of total homology and total nonhomology (see Section VIII,C).

Simon *et al.* (1971) ascribed the retention of homology in conserved areas of the phages genomes to the fact that these phages recombine readily. Lambdoid phages in fact possess a specialized system of generalized recombination—the *red* system (Dove, 1968) and it has been suggested that such legitimate recombination constitutes a further mode of molecular proofreading, maintaining sequence homogeneity by the selective removal of mismatched bases (Reanney, 1977, 1978a). A similar pattern emerged from heteroduplex studies of the T-even coliphages T2, T4, and T6 (Myoviridae; Kim and Davidson, 1974). As with lambdoid phages, it was found that regions of homology or nonhomology were of gene size or larger. As noted by the authors, this is expected of phages whose genomes recombine. Specifically, Kim and Davidson found that the *r11* region and the *D* region, the lysozyme and *ac* genes and gene 52 were homologous in all three phages. The late gene region showed greatest homology while the host range region was heterologous.

A very different situation was found for coliphages T3 and T7 (Podoviridae) (Davis and Hyman, 1971). While sequences encoding the major structural proteins of phages had diverged very little, the genes for the ligase and lysozyme enzymes had undergone extensive divergence. Both enzymes are dispensible in T7. The large amount of partial homology observed between T3 and T7 DNAs was ascribed to the fact that, unlike the lambdoid phages, T3 and T7 do not recombine. T3 and T7 also destroy the DNA of their hosts during infection. The genetic isolation thus imposed on the two phages was considered akin to speciation in cellular organisms (Davis and Hyman, 1971). Given the genetic barriers between the two phages, some, presumably time-dependent accumulation of mutations within their genes was to be expected (see Section VI,D).

This study of T3 and T7 was later extended to other female-specific coliphages, named ϕI, ϕII, W31, and H (Brunovskis *et al.*, 1973; Hyman *et al.*, 1973, 1974). The results of this work confirmed the earlier findings, with nonessential regions showing much more heterogeneity than essential regions such as some minor structural components of the mature phage particles. It was concluded that phages ϕI, ϕII, W31, and H are more closely related to each other and to T7 than to T3.

A different pattern was observed in coliphages P1 and Mu and in the *Bacillus* phages ϕ105 and SPO2. P1 and Mu DNAs were nonhomologous except for a 3-kb-long segment, which corresponds to the *C* region of P1 and the *G* region of Mu (Chow and Bukhari, 1977; Toussaint *et al.*, 1978). The genomes of ϕ105 and SPO2 were partially homologous in a central region. This region probably corresponds to the tail genes and may account for the serological relatedness of these phages (Chow *et al.*, 1972). P1 and Mu appear to lack common ancestors (Chow and Bukhari, 1977), whereas the homologous regions of ϕ105 and SPO2 may reflect divergence from a common ancestor or a recombination between unrelated phages (Chow *et al.*, 1972). In conclusion, phage DNA heteroduplexes present the following matching patterns: (1) total homology, (2) total nonhomology, (3) a combination of both known as the insertion–deletion type, (4) point or partial homology corresponding to random mutations as in T3–T7, and, finally, (5) homology regions limited to one segment or gene (Table VII). The frequency of the insertion–deletion type points to a modular organization of phage

TABLE VII

PERCENTAGE HOMOLOGY AND MATCHING PATTERNS OF PHAGE DNA HETERODUPLEXES

Host	Morpho-type[a]	Phages	Homology			References
			Percent	Pattern[b]		
Agrobacterium	B1	PB6Δ1, PB6Δ2[c]	~100	~H	⎱	Vervliet et al. (1975)
		PB6, PB2A, PS8, PV-1	~100	~H	⎰	
Bacillus	B1	g6, g12, g18	~84	ID		Bogush et al. (1978)
		φ105, ρ6, ρ10, ρ14	~80	ID-P		Rudinski and Dean (1979)
		φ105, SPO2	14	S		Chow et al. (1972)
Corynebacterium	B1	β, γ	91–99	~H		Buck et al. (1978)
Enterobacteria	A1	P1, Mu		S		Toussaint et al. (1978)
		P1, P7, P1Cm, P1Amp	~90[d]	ID		Yun and Vapnek (1977)
		P2, 186		ID-P		Younghusband and Inman (1974)
	A2	T2, T4, T6	~85	ID		Kim and Davidson (1974)
	B1	λ		ID		Simon et al. (1971)
		λ, 434	65	ID?		Simon and Davidson (1969)
		λ, φ80		ID		Fiandt et al. (1971)
	C1	T3, T7	80	P		Davis and Hyman (1971)
		T3, φII	72	P	⎱	
		T7, φII	88	ID	⎰	Hyman et al. (1973)
		H, φII	~99	ID		Brunovskis et al. (1973)
		T7, φI		ID	⎱	
		T7, W31		ID	⎰	Hyman et al. (1974)
	D1	φX174, S13		ID?		Godson (1973)
		φX174, G4, G14		ID?	⎱	
		φX174, G6, G13		ID?	⎰	Godson (1974)
	D4	PR4, PR772		H		Coetzee and Bekker (1979)
Pseudomonas	B1	B3, D3112		ID		Krylov (1980)
Streptomyces	?	φC31, φC43, φC62	93–96	ID		Sladkova et al. (1979)

[a] Not always known for all phages of a given group or pair.
[b] Patterns: H, complete homology; ID, insertion–deletion type; P, partial or point homology; S, homology is limited to one segment or gene.
[c] Deletion mutants of PB6.
[d] P1 and P7 only.

genomes and indicates that DNA transposition is a major avenue of phage evolution (Sections VIII,B and C).

C. Nucleic Acid Hybridization

Nucleic acid hybridization has to date yielded less data of evolutionary interest than heteroduplex studies and a detailed review of this subject is beyond the scope of this article. In a general way, hybridization studies have confirmed the close relationships among members of groups such as T-even, hydroxymethylcytosine-containing coliphages and there is a reassuring correlation between the hybridization data and morphological, serological, and other properties (Cowrie *et al.*, 1971; Rutberg *et al.*, 1972; Schildkraut *et al.*, 1962; Truffaut *et al.*, 1970). Of particular interest is that the genomes of coliphages P2 and P4 have less than 1% homology (Lindquist, 1974), but have identical cohesive ends (Wang *et al.*, 1973). As in $\phi105$ and SPO2, this may be due to recombination between unrelated phages.

A significant recent contribution has come from Grimont and Grimont (1982), who compared the DNAs of six C3 morphotype phages with those of nine enterophages belonging to five morphological groups. No C3-type phage DNA hybridized with those of other morphotypes. The C3 group itself was heterogeneous and, on the basis of hybridization patterns, molecular weights, G–C contents, and host range was subdivided into three apparently independent "species." In view of the observed rarity of C3 type phages, two explanations can be advanced: (1) the phages are unrelated and represent an exceptional quirk of convergent evolution or (2) they are related, but their genomes have become so disturbed through multiple rearrangements that the ancestral genetic relationships are no longer readily recognizable.

D. Restriction Endonuclease Patterns

This technique was used by Studier (1979) to compare about 40 strains of T7 phage and related viruses. DNA was cleaved by endonuclease *Hpa*I and fragments were separated by gel electrophoresis. Among 19 T7 strains investigated, three were mixtures containing normal and deletion DNAs, five were pure deletion strains, and one was a host range mutant. The 0.7 gene region was particularly affected. These modifications must have occurred since 1945–1946 when T7 was isolated. Genetic drift by mutation thus seems to proceed at a rather fast pace. Provided that no vital genes are affected, "new" phages are emerging continuously.

VII. Comparative Relationships from Other Data

A. Patterns of Macromolecular Synthesis

The survey of female-specific coliphages conducted by Hyman *et al.* (1974) included a comparative study of patterns of RNA and protein synthesis in phage-infected cells. Comparisons were made by separating radioactively labeled material on gels. These data reinforced relationships evident from the polynucleotide analyses. A more wide-ranging survey of selected podoviruses was undertaken by Korsten *et al.* (1979). They investigated the evolutionary relationships of coliphages T3 and T7, *Serratia* phage IV, *Citrobacter* phage ViIII, *Klebsiella* phage No. 11, and *Pseudomonas* phages gh-1 and PX3 by the following techniques:

1. Cross-reactivity with antisera against T3 and T7
2. An estimation of DNA homology by the C_0t method
3. Assay for the phage-coded enzymes, RNA polymerase, and SAMase
4. Gel analysis of phage coat subunits
5. Gel analysis of patterns of phage-directed protein synthesis.

Their conclusions can be summarized as follows: phages IV, ViIII, No. 11, and gh-1 are related to T3–T7 while PX3 is not related to T3–T7 by any of the criteria listed. The morphological similarity between these apparently "unrelated" phages was ascribed to convergent evolution. In our view, however, such a conclusion is premature. If T3–T7 and PX3 shared common ancestral genes in the very remote past then it is possible that later genetic changes have virtually blurred out any evidence of relationship which would be detectable by the techniques used. A comparison of the nucleotide sequences for the coat protein genes (or the amino acid sequences of the coat proteins) should reveal whether or not these genes (or their proteins) share a common origin.

B. Coat Proteins

To date most coat protein data of potential evolutionary relevance have come from the well-studied cubic and filamentous groups. Early attempts to compare the coat proteins of these phages were made using serology. These experiments and others showed that the ϕX174-like phages form a homogeneous group in most respects (Denhardt, 1975,

1977). However the ribophages of the E1 group are much more diverse. On the basis of serological and other criteria, these phages have been categorized into four groups, I to IV, and two supergroups, A and B (Furuse *et al.*, 1979). The relationships between these groupings and the small RNA phages of *Caulobacter* and *Pseudomonas* remain obscure. The *Inovirus* group includes several categories of filamentous phages and is morphologically heterogeneous (see Section X,A). Type Ff, named after the capacity of the phages to adsorb to the tips of F pili, includes phages fd, f1, M13, AE2, Ec9, HR, and δA. The serological and other interrelationships among members of this group have been reviewed by Marvin and Hohn (1969). Coliphages of the If1 type adsorb to I pili. There are serological relationships between If1-type phages, the Ff phages fd and Ec9 (Meynell and Lawn, 1968), and the newly discovered virus, PR64FS (Coetzee *et al.*, 1980). Coliphage IKe is antigenically distinct from the Pf and If types (Khatoon *et al.*, 1972). In addition, the filamentous phages of *Pseudomonas* and *Vibrio* have no known serological relationships with other established groups (see Marvin and Hohn, 1969). As a whole, the filamentous phages appear to be a fairly heterogeneous group.

Data on the coat proteins of the small cubic and filamentous phages have been collected in Table VIII. This table emphasizes the differences between ϕX- and fd-type phages. The former have large protein subunits with a complete set of amino acids whereas the subunit proteins of the filamentous phages are small and always lack histidine and cysteine.

The RNA phages of the E1 group differ from each other in the size of their subunits, the number of constituent amino acids, and in amino acid composition. Overall, however, Table VIII strongly suggests that these ribophages are phylogenetically related. A close look at the coliphages of group I, for example, shows that they often differ by only one or two amino acid substitutions (Lin *et al.*, 1967).

In filamentous phages, the main difference is in amino acid composition. Because of their apparent relatedness, they provide a good example of divergent evolution. One interesting feature is the striking similarity of the major coat proteins of phages fd and If1 (Wiseman *et al.*, 1972). Wiseman *et al.* point out that only eight mutagenic events are required to transform the coat gene of fd into that of If1. Since If1 is longer than fd and contains more DNA, Wiseman *et al.* (1972) speculate that If1 originated by the insertion of half a complementary strand into the viral strand of an fd-like ancestor.

Definitive evidence for relationships comes from amino acid sequences. Dhaese *et al.* (1979, 1980) compared coat proteins of coli-

TABLE VIII
Coat Protein Subunits of Small Cubic and Filamentous Phages

Morpho-type	Host	Phages[b]	Molecular weight (× 10³)	Amino acid residues	Amino acids lacking	References
D1	Enterobacteria	φX174, S13	48	210	None	c
E1	Enterobacteria	Group I: fr, f2, f4, MS2, M12, R17, ZR, $\mu_2{}^b$	13.4–14.5	129–138	His	d
		Group II: GA, SD[b]	12.5–13.4	121–123	His, Met, Cys	e
		Group III: Qβ, VK[b]	15.5–16.9	132	His, Met, Trp	f
		Group IV: F1, SP[b]	17.3		Not determined	g
		ZIK/1	12.1	102	His, Met, Cys	h
	Polyvalent[a]	PRR1	14.5	131	Cys, Try	i
	Pseudomonas	PP7		127	Met	j
	Caulobacter	φCb5	12.0	83	Met	k
F1	Enterobacteria	fd, f1, M13, ZJ/2	5.17–5.24	49	His, Cys, Arg	l
		If1	6.2?	49	His, Cys	m
	Pseudomonas	Pf1			His, Cys, Phe, Trp	n
	Xanthomonas	Xf	4.85	42	His, Cys, Phe	o

[a] Enterobacteria, *Pseudomonas*, and *Vibrio* containing plasmids of P incompatibility group (see Section XII,C).
[b] For other members, see Furuse *et al.* (1979).
[c] Burgess (1969), Edgell, cited by Sinsheimer (1968), Poljak and Suruda (1969), and Sanger *et al.* (1977).
[d] Enger and Kaesberg (1965), Furuse *et al.* (1979), Lin *et al.* (1967), Modak and Notani (1969), Nishihara *et al.* (1969, 1970), Notani *et al.* (1965), Overby *et al.* (1966), Piffaretti and Pitton (1976), Weber and Konigsberg (1967), and Wittman-Liebold and Wittman (1967).
[e] Furuse *et al.* (1979) and Nishihara *et al.* (1969).
[f] Furuse *et al.* (1979), Konigsberg *et al.* (1970), Nishihara *et al.* (1969), and Overby *et al.* (1966).
[g] Furuse *et al.* (1979).
[h] Robinson (1972).
[i] Dhaese *et al.* (1979).
[j] Dhaese *et al.* (1980).
[k] Bendis and Shapiro (1970).
[l] Asbeck *et al.* (1969), Berkowitz and Day (1976), and Snell and Offord (1972).
[m] Wiseman *et al.* (1972).
[n] Nakashima *et al.* (1975) and Yasunaka (1977).
[o] Lin *et al.* (1971).

phages MS2, and Qβ, the polyvalent phage PRR1, and the *Pseudomonas* phage PP7. All coat proteins showed a clustering of basic residues in the same region. It was concluded that the four phages very likely descended from a common ancestor (Dhaese *et al.*, 1980). In filamentous phages, amino acid sequences are so far available only for a few Ff-type phages (Asbeck *et al.*, 1969; Snell and Offord, 1972). They show that Ff coat proteins are closely related.

VIII. MECHANICS OF PHAGE EVOLUTION

A. *Point Mutations*

Genomes of bacterial viruses consist of either DNA or RNA. This division has immediate consequences for the adaptive genetics of the phages. DNA is normally copied much more faithfully than RNA. This is because the enzymatic machinery responsible for DNA synthesis contains a "proofreading" component (Hopfield, 1974). *Escherichia coli* polymerase I edits the DNA it copies by excising mismatched bases and repairing the gaps thus created (Brutlag and Kornberg, 1972). No comparable process has been reported in systems which make RNA on either DNA templates or RNA templates. RNA synthesis is thus an intrinsically "noisy" process.

RNA phage genomes are usually much smaller than the genomes of tailed phages (Table I). The "noisiness" of RNA replicases may be one factor setting an upper limit to the amount of viral information which can be encoded in RNA. It seems likely that a genome the size of T2 could not be transformed into RNA without imposing a crippling degree of infidelity on the process of information transfer. Such infidelity is not a handicap to small genomes. Indeed, ribophages of the E1 type have been able to turn this situation to their own advantage. T1 nuclease fingerprinting of RNA purified from clones of Qβ has revealed that the population of RNA molecules is heterogeneous with respect to base sequence: about 15% of the clones derived from a multiply passaged population showed fingerprint patterns which deviated from that of the total RNA population (Domingo *et al.*, 1978). However, growth competition experiments in which uncloned phage and selected variants were mixed resulted in a reemergence of the wild-type pattern. These data suggest that selection has calibrated the error rate of the Qβ replicase to maintain an optimal balance between the existing, tested genome sequence and the variants which arise at a high frequency. In this way, the identity of the phage is maintained in a situation which exposes multiple options to the test of selection.

Selection can influence the variability of DNA phages. Drake (1974) lists 13 genes that affect mutation rates in T4. Genes whose products increase mutation rates, are termed *mutator* genes while those with the opposite effect are called *antimutator* genes. Other genes affect not only the frequency of mutation but its character; for example, gene 42 of T4 promotes the GC → AT pathway (Drake, 1974). The existence of these multiple mechanisms allows selection to raise or lower the percentage of variant phages of a population in response to environmental pressures.

B. DNA/DNA Interactions

While point mutations or genetic noise remain the accepted agency for rewriting the base sequences of genes, recent discoveries in microbial genetics have suggested that DNA/DNA interactions are the prime accelerators of evolution in bacteria and their viruses.

The process of generalized or homologous recombination results in the reciprocal exchange of synapsed DNA molecules whose nucleotide sequences have been brought into register by extensive base pairing. Such a mechanism works most effectively when DNAs are largely homologous; hence, as has been pointed out (Reanney, 1977, 1978b, 1979), the ability of generalized recombination to generate large-scale diversity is limited.

A second type of recombination—quantal or illegitimate re-combination—now tends to dominate the literature on microbial ge-netics. Unlike systems of homologous recombination, quantal systems do not require extensive regions of preexisting homology for DNA/DNA interactions and in *E. coli* they are independent of the *recA* gene. Quan-tal recombination is a major source of genetic diversity in bacteria. A well-studied example is provided by the phenomenon of genetic trans-position.

1. Transposable Modules in Bacteria

The smallest class of transposable modules is represented by the *insertion sequences;* IS1, for example, contains only 768 base pairs (Ohtsubo and Ohtsubo, 1978). Such *IS* modules lack associated genetic markers. *Transposons* or *Tn* modules are larger elements which carry an identifiable phenotype. Most known transposons encode resistance determinants to antibiotics and heavy metals. These transposable modules usually contain nucleotide sequences which are repeated in inverse order at each end of the molecule, giving rise to characteristic "stem and loop" structures seen when the separated DNA strands are allowed to reanneal intramolecularly and are examined under the electron microscope.

A detailed treatment of transposable elements lies outside the scope and intention of this article. Readers are referred to the reviews by Kleckner (1977) and Calos and Miller (1980), to the text by Bukhari *et al.* (1977), and to the *Cold Spring Harbor Symposium of Quantitative Biology* (1980).

Transposable modules replicate *in situ* and copies of the sequence "transpose" from the original locus to other target sites. Their adaptive significance centers around this fact. The transposition of IS1 from plasmid *A* to an unrelated plasmid *B* for example creates a region of *common* homology in DNAs from *different* evolutionary backgrounds. These two replicons may now combine to form a cointegrate molecule. Likewise the presence of the same IS unit on a plasmid and the host chromosome may allow the plasmid to integrate into the chromosome (see Holloway, 1979).

Doolittle and Sapienza (1980) have proposed that these self-replicating modules constitute "selfish DNAs" which colonize multiple sites in bacterial replicons as part of their survival strategy. Be that as it may, transposable elements, as an accidental consequence of their replication and spread, lower the barriers that normally isolate separately evolved DNA molecules and make possible a diverse array of DNA/DNA interactions among nonhomologous genetic units. These aspects are summarized in Table XI.

2. Quantal Recombination as a Mechanism of Phage Evolution

Quantal recombination in general and genetic transposition in particular has undoubtedly played an important role in phage evolution. When temperate phages are propagated in strains of bacteria containing antibiotic resistance transposons, Tn^+ phage DNAs can sometimes be recovered. Wild-type λ phages can incorporate short transposons such as Tn9 (2683 bp) into their genomes without loss of plaque-forming activity (Scott, 1973). When deletion mutants are used the phage genome can absorb larger transposons such as Tn5 (5400 bp) (Berg *et al.*, 1975). Similar variants have been described for the temperate phage P1; the derived phages include P1*Cm* which has acquired chloramphenicol resistance (Kondo and Mitsuhashi, 1964), P1*Kan* which has acquired kanamycin resistance (Takano and Ikeda, 1976), and P1*CmTc* which has acquired joint resistance to chloramphenicol and tetracycline (Mise and Arber, 1976). The mechanism by which P1 incorporates the Tn DNA was studied by Yun and Vapnek (1977) who found that while P1*Cm* prophage DNA exceeded prophage P1 DNA in size by 1.6 million daltons, there was no significant difference between the sizes of the P1 and P1Cm phage DNAs. Since the difference in size between the phage and prophage DNAs is equivalent to the length of the terminally redundant (TR) sequences in P1, Yun and Vapnek suggested that the Cm transposon in P1Cm displaces part of the TR sequences of the wild-type phage DNA.

It is highly unlikely that these derived phages are mere laboratory artifacts. A wild-type coliphage, P7, contains an ampicillin resistance determinant (Chesney and Scott, 1975; Smith, 1972). P7 is a relative of P1 (Chesney and Scott, 1975). When P7 and P1 were compared by heteroduplex mapping it was found that a 5.5-kb segment corresponding to the ampicillin determinant occupied most of the TR sequences (Yun and Vapnek, 1977). So P7 is a natural example of the same kind of process observed in the laboratory construction of transpositional derivatives of P1.

The situation is not confined to coliphages. The tailed phage Rϕ6P encodes resistance to penicillin in the facultative prototroph, *Rhodopseudomonas sphaeroides* (Pemberton and Tucker, 1977). The DNA of Rϕ6P existed as a supercoiled "plasmid" in the host lysogen and could be eliminated by mitomycin C treatment. This phage appears to have evolved from an unknown ancestor by acquiring a β-lactamase transposon. This hypothesis was supported by isolating similar phages which did not encode penicillin resistance (Tucker and Pemberton, 1980). Other phages with resistance determinants are listed in Table IX.

TABLE IX

Translocatable Antibiotic Resistance Genes in Phage Genomes[a]

Phage	Host	Resistance gene[b]	References
Mu	*Escherichia*	Chloramphenicol	Chow and Bukhari (1978)
ϕ34	*Proteus*	Kanamycin	Coetzee (1974)
5006M		Ampicillin, kanamycin	Coetzee (1975)
P22	*Salmonella*	Tetracycline	Dubnau and Stocker (1964)
ϵ15		Chloramphenicol, streptomycin, sulfanilamide	Kameda *et al.* (1965)
P11*de*	*Staphylococcus*	Erythromycin	Novick (1967)
S1		Tetracycline	Inoue and Mitsuhashi (1975)

[a] Rϕ6P and Tn$^+$ variants of λ and P1 are discussed in the text.

[b] The transposable nature of the resistance determinants has not been established in all cases.

Genes encoding antibiotic resistance can be considered as "adventitious" DNA in the sense that they are not obviously needed for the survival of their carrier phages. However, transpositional events appear to have played a more formative role in the evolution of certain phages. Coliphage Mu, itself a transposable element, contains an invertable segment, G. The G segment controls the infectivity of Mu since particles with G in the $-$ (flop) orientation fail to infect host cells (Bukhari and Ambrosio, 1978; Kamp et al., 1978). P1 contains a 3-kb invertable unit which is homologous to G (Chow and Bukhari, 1977; see also Section VI,B). These data are most easily explained by postulating that Mu and P1 acquired a common transposon at some stage of their evolutionary history.

It is difficult to estimate how frequent and hence how "significant" specific transpositional events have been in phage evolution. We know only one study which gives any insight into this problem. Elliott and Stanisich (1981) attempted to isolate Tn^+ derivatives of temperate phages of Pseudomonas aeruginosa PAO. Of 18 phage isolates able to infect this bacterium, one yielded Tn^+ progeny when the phages were propagated on host strains carrying Tn^+ derivatives of the plasmid RP1. This brief survey tends to suggest that many phage genomes in nature are capable of accepting transpositional units from other DNA molecules but a much larger pool of phages needs to be sampled before meaningful inferences can be drawn.

All the above phages except Mu and P1 conform to the definition of "conversion phages" because lysogenization of host cells by these phages results in the acquisition of new phenotypes in addition to homoimmunity. For example infection of P. aeruginosa strain 1 by phage D3 results in changes in somatic antigens (Holloway and Cooper, 1962), infection of Bacillus megaterium with phage Ipt 3 results in changes in colony morphology (Ionesco, 1953), and infection of nonpathogenic strains of Corynebacterium diptheriae with phage β^{tox} results in the production of diphtheric toxin (Freeman, 1951). A detailed list of conversion phages to which readers are referred for further information is given in the review by Barksdale and Arden (1974).

3. DNA Transfers between Cells as a Mechanism of Phage Evolution

The diversity of phenotypes specified by conversion phages suggests that "conversion" genes have been recruited from a variety of sources during phage evolution. Phages can acquire genes from (1) other extrachromosomal genetic units such as plasmids and (2) the host chromosome, for example, by faulty excision of prophage DNA. Ultimately, however, the variability generated by DNA/DNA interactions within one bacterium is limited by the relatively small genetic content of a single prokaryotic cell. The adaptive capabilities of bacteria (and hence of their viruses) have been hugely expanded by coupling this *intermolecular* transferability with *intercellular* transferability (Table X) (Reanney, 1976b, 1977).

Three known systems result in the passage of DNA from one bacterial cell to others, transformation, transduction, and conjugation. While it is almost impossible to give any kind of relative "weighting" to these processes, certain comments seem justified. Transformation usually requires extensive base homology between entering and resident DNA molecules. Thus transformation chiefly exchanges DNA between closely related cells. With the exception of phages of enterobacteria, few phages cross generic barriers (Section IV,A). Also the ability of phages to pick up and transduce novel sequences is severely limited by the constraints of encapsidation. Hence it is to conjugation that we must turn for an efficient system of cell-to-cell communication. Conjugative plasmids in enterobacteria are divided into 17 incompatibility groups (Datta, 1979). Some of these plasmids have an extraordinary host range; for example, the P group plasmid RP1 can colonize cells belonging to enterobacteria, *Agrobacterium, Azotobacter, Caulobacter, Pseudomonas, Rhizobium, Rhodopseudomonas,* and others (Alexander and Jollick, 1977; Olsen and Shipley, 1973; see Reanney, 1977). Because they lack an extracellular phase, plasmids can gain or lose DNA with little selective handicap; indeed plasmids range in size from about 2 to over 200 million daltons (Bukhari *et al.,* 1977).

Conjugative plasmids can "mobilize" small nonconjugative plasmids by (1) combining with them physically to form a plasmid cointegrate or

TABLE X

MECHANISMS THAT CAN RESULT IN GENETIC CHANGE IN BACTERIA AND PHAGES

Intracellular	Intercellular
Point mutations	Recombination dependent
Rearrangements of genome; deletions, duplications, inversions	Transformation by chromosomal DNA
	Transduction of chromosomal DNA
Transposition of defined modules of DNA	Integration of temperate phages, such as λ
Plasmid–plasmid interactions	Conjugal transfer of chromosomal DNA
Plasmid–phage interactions	
Plasmid–chromosome interactions	Recombination independent
Phage–chromosome interactions	Transformation by plasmid DNA
Phage–phage interactions	Transduction of plasmid DNA
	Infection by temperate phages, such as P1
	Conjugal transfer of plasmid DNA

(2) providing a passive transfer bridge. Conjugative plasmids can also "mobilize" the host chromosome (Holloway, 1979). When conjugative plasmids pick up chromosomal sequences these "fixed" genes automatically acquire the cell-to-cell communicability and expanded host range of their carrier replicon. Conjugative plasmids and other vector systems then lower the barriers that normally isolate chromosomal genes. As a result a bacterial cell can, in a qualified sense, be regarded as a genetically "open" system (Reanney, 1976b, 1977) potentially able to draw on the data bank of genes contained in a whole microbial ecosystem (Sonea and Panisset, 1980).

This pooling of genetic information in bacteria means that genes such as the ampicillin resistance gene of coliphage P7 may have originated in the chromosome of a soil organism such as an actinomycete (Benveniste and Davies, 1973; Courvalin *et al.*, 1977). Similarly, conjugative plasmids may occasionally transmit genes for structural phage components to a host taxon that is not susceptible to the phage in question. This continuing diffusion of "outside" genes through bacterial populations has undoubtedly been a factor in phage evolution resulting perhaps in the appearance of new phage morphologies or variant strategies of infection.

C. Modular Evolution

The DNA/DNA interactions described in the preceding section do not occur at random. While IS and Tn units can insert at multiple points in various DNA molecules they usually have preferred integration sites (Bukhari *et al.*, 1977). Transposing sequences themselves act as recombination targets (Kleckner, 1977); once a common transposing sequence has inserted into two nonhomologous replicons, the site of future recombinational crossovers is precisely fixed by the position of insertion. With time one would expect that harmful insertions would be eliminated by selection leaving "homology targets" positioned at "joints" around which a programmed exchange of heterologous DNA could occur to best advantage (Bukhari *et al.*, 1977; Reanney, 1977; Richmond and Wiedeman, 1974). The effect of this would be to impart a *modular substructure* to many bacterial replicons (see Davey and Reanney, 1980).

Do phage DNAs show evidence of a modular organization? The best affirmative evidence comes from heteroduplex studies of the lambdoid phages and the coliphages of the T series. Further evidence can be seen in the partial relationships between such phage pairs as P1 and Mu, P2 and P4, and the *Bacillus* phages φ105 and SPO2 (Sections VI,B and C;

Table VII). The lambdoid phages are particularly instructive. Heteroduplexes constructed between the DNAs of phages like λ and φ80 showed that regions of near perfect homology were interspersed with regions of total nonhomology. The boundaries between such modular regions were sharp (Simon *et al.*, 1971). On the basis of these observations Fiandt *et al.* (1971) proposed that the evolution of the lambdoid phages had proceeded by inversions, deletions, and substitutions. A "modular" construction for phages like λ has also been proposed by Campbell (1972) and by Szybalski and Szybalski (1974).

This concept of modular evolution has been developed and extended by Botstein (1980). Botstein's point of departure came from a comparative study of the genomes of λ and P22. These phages belong to different morphotypes (λ to B1 and P22 to C1), have different hosts (*E. coli* and *S. typhimurium*) and differently structured DNAs (linear or permuted), and are heteroimmune. However, the organization of their genetic maps is very similar (Fig. 3). Attempts to obtain viable hybrid phages

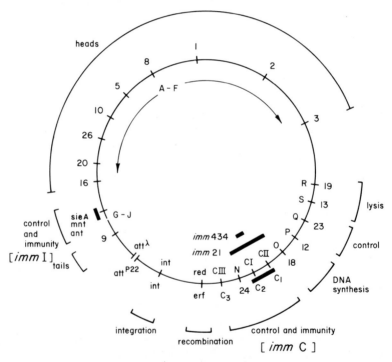

FIG. 3.   Comparison of the genetic maps of P22 and λ. P22 genes are indicated outside the circle, λ genes are shown inside. From Botstein and Herskowitz (1974), with permission of the authors and *Nature* (*London*).

showed that large segments of λ DNA could be exchanged for P22 DNA and vice versa (Botstein and Herskowitz, 1974; Susskind and Botstein, 1978). Significantly, the exchanged segments were often nonhomologous but they occupied identical positions in the hybrid genomes. The lytic functions of the phages are a case in point: P22 breaks down the walls of infected hosts with lysozyme, which cleaves glycosidic bonds, while λ uses an endopeptidase. The genes specifying these enzymes are completely nonhomologous (suggesting a different evolutionary origin for each) yet these genes are found at equivalent positions in the genetic maps of both phages and are interchangeable in interbreeding phage populations (Botstein, 1980).

Botstein equated these interchangeable units with a functional "module"; accordingly, it was inferred that the λ genome contains the following modules:

Module	λ Genes
DNA encapsidation	A–F
Tails	G–J
Antigen conversion	b2
Integration	*att, int, xis*
Recombination	*red, gam*
Early control	N
Immunity	*c*I
DNA replication	O, P, *ori*
Late control	Q
Lysis	S, R

According to Botstein's thesis, viruses found in nature represent favorable combinations of such modules. Evolution therefore "acts primarily not at the level of an intact virus but at the level of individual functional units (modules)" (Botstein, 1980). Diverse permutations and combinations among compatible modules will produce phages with new phenotypes without affecting viability.

We regard the modular concept as one of the most stimulating to emerge from recent discussions of viral evolution. Among others it explains the ubiquity and frequent resemblance of base plates, tail spikes, or tail fibers. An important part of phage evolution then has, in our view, taken place by shuffling and recombining modular elements whose products have been tested by long periods of selection. "Modules" in this context can be blocks of linked genes or individual genes. It is quite plausible that even smaller modules were shuffled during early phases of phage evolution. Recent studies by Matthews *et al.* (1981)

have revealed a remarkable correspondence between the backbone conformations and certain other features of phage T4 lysozyme and a lysozyme from the egg white of hens (see also Artymiuk *et al.*, 1981). The animal lysozyme gene is split into four coding modules (exons) by intervening sequences (introns). Each exon may correspond to a subdomain of the protein (see Blake, 1979). In bacteria and their viruses all intron sequences have been eliminated by pressures to economize genome sizes, but the exon–intron structure of the eukaryote gene may serve as a genetic "memory" of the primitive arrangement (see Section XII,A). We therefore suggest that early modules were of exon size and that the proteins of contemporary phage capsids, tails, and enzymes were built up by rearranging such simple exons into new combinations.

One of the intriguing aspects of the modular proposal is the suggestion that "viral" modules occur not only in phage genomes but in host chromosomal DNA. Botstein reports for example that a DNA segment indistinguishable from P22's DNA replication module exists in the genome of *E. coli* K12 as a cryptic genetic element (Botstein, 1980). This suggestion makes it possible to conceptualize how new phages could have been put together during evolution. If DNA/DNA interactions between different bacterial replicons result in the rearrangement of functional modules then some combinations will inevitably link origins for DNA replication (*rep* modules) for example with genes for capsid biosynthesis (head modules). The *rep* module could come from the chromosome or from various resident or introduced plasmids or from other phages. Modules for such functions as lysis and immunity could be recruited from a variety of sources. Most of these combinations would result in nonviable or abortive phages and the number of successful "experiments" would be further limited by the size of available phage capsids. However a few associations might yield novel phages which combine the properties of phages and plasmids (Section XII,C) or phages and bacterial DNA.

The modular concept causes some difficulties while resolving others. It is not hard to see how a system evolving toward autonomy could recruit origins for DNA replication. These are a feature of all replicons by definition. But the idea that bacteria contain cryptic "modules" for the biosynthesis of "phage" heads or tails separately seems implausible, until one examines the data. In fact induction of bacteria with UV or mitomycin C often results in the production of structurally complex nonviable entities which resemble phages or isolated phage tails. These entities are called defective phages or, in some cases, particulate bacteriocins. They are extremely widespread.

IX. EVOLUTIONARY POSITION OF DEFECTIVE PHAGES AND PARTICULATE BACTERIOCINS

A. Particulate Bacteriocins

Phage-like particles that adsorb to and kill sensitive cells without multiplying in them are called "particulate bacteriocins." They consist of heads and tails or tails without heads. The *Agrobacterium* "bacteriocins" and colicin 15, which look like conventional phages, have been shown to contain DNA (De Ley *et al.,* 1972; Vervliet *et al.,* 1975; Lee *et al.,* 1970). Particulate bacteriocins have been reviewed several times, most recently by Garro and Marmur (1970), Lotz (1976), and Ackermann and Brochu (1978). The morphology and dimensions of selected particles are listed in Table XI. Except for PBSX-type structures which will be discussed below, bacteriocins with heads and tails may be mutants of normal phages. Indeed, the rapid appearance of deletion mutants in coliphage T7 (Studier, 1979) suggests that morphologically complete defective phages are just that. The relationship of bacteriocidal phage tails to viable phages is less clear. Some are serologically related to ordinary temperate phages. This was demonstrated for a tail-like bacteriocin and a phage of *Yersinia enterocolitica* (Vieu *et al.,* 1967) and for the so-called R-type pyocins and several phages of *P. aeruginosa* (Homma and Shionoya, 1967; Kageyama *et al.,* 1979). Neutralization was reciprocal. In addition, the tail of *P. aeruginosa* phage PS17 was shown to be bacteriocidal for strains sensitive to the phage and mutants of this virus were found that produced bacteriocidal tails (Shinomiya and Shiga, 1979). It was concluded that genes for the phage tails and for R-type pyocins had common ancestors (Kageyama *et al.,* 1979).

B. Killer Particles of the PBSX Type

These phage-like entities are a special case of particulate bacteriocins and occur in several *Bacillus* species (Table XII). They are immediately recognizable by their idiosyncratic shape: namely, a small head, a long, thick, contractile tail with conspicuous striations, and a base plate. Upon contraction, the sheath becomes loose and slides along the core. The heads contain randomly packaged fragments of bacterial DNA (Hirokawa and Kadlubar, 1969; Okamoto *et al.,* 1968). The particles are not confined to *Bacillus* and also occur in *Actinomyces, Clostridium, Streptococcus,* and *Streptomyces* (Table XII). They may have relatives among conventional phages with similar tails and much

TABLE XI

Selected Particulate Bacteriocins

Morpho-type	Host	Name	Head diameter (nm)	Tail length (nm)	References
A1	Agrobacterium	PO362	70–76	130–150	De Ley et al. (1972); Vervliet et al. (1975)
	Bacillus	"Killer"	40	190–275	See Table XII
	Escherichia	Colicin H?	65 or 95	180 or 280	Bradley and Dewar (1967)[a]
		Colicin 15	60	100	Sandoval et al. (1965); Lee et al. (1970)[a]
	Listeria	Monocin	110	240	Bradley and Dewar (1967)
B1	Agrobacterium	PO396	67	204	Vervliet et al. (1975)
		P8149	76	320	De Ley et al. (1972); Vervliet et al. (1975)
	Escherichia	Colicin D?	30		Bradley (1967)
	Lactobacillus	056, others	52–60	204–243	Tohyama et al. (1972)
	Mycobacterium	—[b]			Imaeda and Rieber (1968)
A, tail only	Chromobacterium	A, B		219, 114	Ackermann and Gauvreau (1972)
	Clostridium	Boticin P		100	Lau et al. (1974)
		Clostocins O, M		125	Ogata et al. (1972)[a]
	Proteus	Bacteriocin 45			Smit et al. (1969)
		Mirabilicin D52		130	Tikhonenko et al. (1971)
	Pseudomonas	R-type pyocins		108–130	Homma et al. (1967); Kuroda and Kageyama (1979)[a]
	Rhizobium	INCO		138	Lotz and Mayer (1972)[a]
	Vibrio	Vibriocin		110	Jayawardene and Farkas-Himsley (1969)[a]
	Yersinia	—		75–80	Vieu et al. (1967)
B, tail only	Pseudomonas	Pyocin F1		106	Kuroda and Kageyama (1979); Kuroda et al. (1979)
		Pyocin 28		100–120	Takeya et al. (1969)[a]
	Rhizobium	Bacteriocin 16-2		200	Gissman, cited in Lotz (1976)

[a] Further references in Ackermann and Brochu (1978).
[b] No name.

TABLE XII

KILLER PARTICLES AND RELATED STRUCTURES[a]

Host	Name	References
Bacillus amyloliquefaciens	—[b]	Steensma *et al.* (1978)
B. cereus	—	Stiube and Dimitriu (1969)
B. funicularius	—	Steensma *et al.* (1978)
B. licheniformis	—	Bradley (1967)
	PBLB	Huang and Marmur (1970)
B. mycoides	—	Tikhonenko and Bespalova (1961)
B. pumilus	—	Steensma *et al.* (1978)
B. subtilis	SPα	Eiserling (1963)
	PBSX	Seaman *et al.* (1964); Okamoto *et al.* (1968)
	PBSH	Haas and Yoshikawa (1969)
B. stearothermophilus	—	Kelly and Reanney (unpublished)
Actinomyces israelii, others	—	Klein and Frank (1973)
Clostridium botulinum	—	Dolman and Chang (1972)
Streptococcus mutans, others	—	Greer *et al.* (1971);
	—	Klein and Frank (1973)
Streptomyces griseus	—	Rautenstein *et al.* (1971)

[a] Further references in Ackermann and Brochu (1978) and Steensma *et al.* (1978).
[b] No name.

larger heads. One potential "relative" is a defective phage of *Lactobacillus* with a 69-nm head and killing ability (Yokokura *et al.,* 1974). Others are the *Bacillus cereus* phage Cb and the *Bacillus licheniformis* phage 21 whose head diameters are 78–80 and 50 nm, respectively (Stiube and Dimitriu, 1969; Klipikova *et al.,* 1978). The PBSX group is itself heterogeneous and includes five types of different tail lengths, suggesting that some divergent evolution has taken place (Steensma *et al.,* 1978).

PBSX-type particles do not contain genes for their morphogenesis and replication, hence they cannot spread their own genotypes in the horizontal manner typical of phages. Since conjugal systems have not yet been reported in *Bacillus,* only the vertical transmission of these genes seems possible. Therefore, the widespread distribution of these particles suggests that they descended from a common ancestor and arose prior to the divergence of their present hosts, thus indicating bacterial phylogenetic relationships.

C. Evolution or Devolution?

The phage affinities of tail-like pyocins and killer particles raise the question of their origin. Are they relics of past phage evolution, degenerate fragments of once viable viruses? Or are they "stepping stones" on the way to new phages. Or both?

It has been suggested that killers are poorly evolved phages (Seaman et al., 1964) or ancestors of bacterial viruses in a general way (Maramorosch, 1972). Garro and Marmur (1970) speculated that defective phages arose either through the progressive addition of specific genetic components or through the loss of structural and replication genes. For headless phage tails, Bradley (1967, 1971) considered a devolutionary origin most probable, because a functional episome seems unlikely to code for an injection system *before* it becomes surrounded by a capsid. Bradley's hypothesis is strengthened by the observation that a mutant of PSBX is defective for head formation and produces tail structures only (Thurm and Garro, 1975).

The modular thesis and the concept of coupled evolution (Reanney, 1976b, 1977) are compatible with both progressive and regressive modes of evolution. Let us assume that genes for head and/or tail biosynthesis, like modules for DNA replication or lytic functions, exist as linked gene blocks in the pool of interbreeding bacteria in nature. On this basis, new conjunctions of viral modules might be expected to arise from time to time. A viable phage might enter into a lysogenic relationship with a cell which already contains the tail genes of a bacteriocin. Genetic events of the type described in Section VIII,C may rearrange the modules so that genes for capsid construction and DNA replication become linked to the genes for the bacteriocin. This could result in the appearance of a viable variant or a defective phage. Genes for particulate bacteriocins may thus constitute a repository of "spare parts" for phages, allowing occasional new phage morphologies to appear. There is no contradiction with the observed conservatism of PBSX-type particles. To focus our argument more sharply, genes may be conservative but their combinations are not.

Do these cryptic genes confer any selective advantage on the cells that contain them? Certainly their widespread distribution hints at a role in cellular functions (Garro and Marmur, 1970). λ, P1, P2, and Mu lysogens of *E. coli* reproduce more rapidly in certain conditions than nonlysogens (Edlin et al., 1975, 1977; Lin et al., 1977). Furthermore, defective phages, as well as their viable counterparts, may protect the bacterial cell against virus infections, either by modification of the cell membranes or by encoding restriction enzymes or by manufacturing repressors which inhibit infecting phage DNAs (Lotz, 1976). Bacteriocidal particles thus have the function of eliminating competitors. Although the cell which synthesizes these particles is destroyed in the process, the fate of the individual cell must be weighed against the survival fitness of the population.

X. Morphological Evolution

The various groups of cubic and pleomorphic phages are morphologically homogeneous, but this is not so in tailed and filamentous phages. Together with poxviruses, tailed phages are the most diversified of all virus groups and the existence of so many phenotypes is a major riddle. In this section we are concerned less with underlying genetic processes than with their phenotypic results.

Much work has been done on phage morphogenesis and the assembly pathways of phages T2–T4, T7, λ, P22, ϕ29, and ϕX174 are more or less known (see reviews by Casjens and King, 1975; Denhardt, 1977; Earnshaw and Casjens, 1980; Eiserling and Dickson, 1972; Murialdo and Becker, 1978; Steven and Wurtz, 1977; Wood and King, 1979). Phages are built of a limited number of protein species, present in a definite number of copies and assembled in an orderly, sequential fashion, largely through self-assembly once the relevant proteins become available. In tailed phages, assembly proceeds through several pathways and involves prohead formation and maturation by protein cleavage. Primers for the initiation of head assembly and factors governing particle size have been identified in a few cases. DNA is cut to size from concatemers and enters preformed capsids. An exception is coliphage P2 where the packaging substrates are circular DNAs of unit size (Pruss *et al.*, 1974). In a general way phage assembly includes several genome-saving mechanisms that may lower the barriers between morphotypes.

A. Genome Economy in Phage Morphogenesis

1. Self-Assembly

Self-assembly of equivalent subunits is probably the most important genome-saving mechanism in phage synthesis (review by Kushner, 1969). For example, assembly or disassembly through addition or deletion of rows of subunits may conceivably lead to enlargement or reduction of capsid or tail size. It probably happens in the production of aberrant phages (below) and of polyheads and polysheaths. This mechanism may also result in the combination of large building blocks such as heads and tails (Wood and King, 1979). Primers and sequential assembly are additional saving measures, because the only regulation mechanism needed is the availability of material.

2. Determination of Genome Size by the Capsid

In tailed phages and apparently in ϕX174 as well (Murialdo and Becker, 1978), the genome is cut to the size of the capsid, thus alleviating the need for a special DNA sizing mechanism. This "headful hypothesis," first proposed by Streisinger et al. (1967) for phage T4, is strongly supported by observations in the P2–P4 system and in variants of T4. Coliphage P4 is defective and needs the presence of P2 for lytic growth. It has a much smaller head than P2 and contains about one-third of its DNA mass (Goldstein et al., 1974). In vitro packaging experiments show that two or three copies of P4 DNA can be packaged into P2 heads (Pruss et al., 1974). In phage T4, Mosig et al. (1972) found correlations between DNA length and the size of normal, isometric, and intermediate heads.

3. Coordinate Assembly of Tail Sheath and Tail Core

Some phages or phage-like particles produce contractile tails of variable length, showing a correlation between the length of the core and that of the extended or contracted sheath. In these cases which may well reflect a general pattern of phage morphogenesis, the core obviously regulates the length of the sheath: (1) P2 mutants producing headless tails (Lengyel et al., 1974); (2) tail-like bacteriocins of Chromobacterium (Ackermann and Gauvreau, 1972); (3) T4 tails assembled in vitro (Tschopp and Smith, 1977); (4) B. subtilis phage SPO1 (Parker and Eiserling, cited by Tschopp and Smith); and (5) Several Shigella phages (Krzywy, 1972).

B. Morphological Variation

1. Tailed Phages

Some tailed phages produce aberrant structures that could indicate mechanisms of transition from one morphotype to another. There is a great wealth of aberrant particles and only those that could give rise to new phages will be discussed, not polyheads or polysheaths. We stress that all abnormal phage heads listed below appeared to be "full" when viewed under the electron microscope and presumably were filled with DNA.

  a. Enlargement of Head Size. Upon induction by canavanine followed by arginine chase, T-even phages produce particles with abnormally long heads dubbed "giants" or "lollipops." Similar structures are produced by certain mutants. These particles have heads up to 44 times the normal length and up to three tails, are infectious, and contain large quantities of DNA (Cummings and Bolin, 1975; Cummings *et al.*, 1973; Doermann *et al.*, 1973). Catenates of up to 20 times the normal genome length have been extracted (Uhlenhopp *et al.*, 1974). Similar particles occur in other phages with elongated heads, e.g., phages of *Staphylococcus* and *Streptomyces* (type B2), *Streptococcus* (type B3), and *Salmonella* (type C3; see Table XIII).

b. *Reduction of Head Size.* Isometric variants of elongated heads were first reported in T-even phages (Anderson and Stephens, 1964), and have been observed many times since. They are probably identical with T-even phage "light particles" and it is interesting that seven to eight of the latter can recombine and achieve a productive infection (Amati and Favre, 1968). Other isometric variants have been found in phages of *Acinetobacter* (type B2) and *Salmonella* (type C3). Heads of intermediate size exist too. T-even phages produce two size classes (Doermann *et al.*, 1973; Mosig *et al.*, 1972) and the *Salmonella* phage 7–11 produces eight of them (Moazamie *et al.*, 1979). Each size class seems to correspond to the loss of a row of capsomers. For phages with isometric heads, the only certain observation is from coliphage P1, which produces three types of capsids with diameters ranging from 86 to 47 nm (Walker and Anderson, 1970).

c. *Tail Length.* Abnormally long contractile tails with full heads were observed by Krzywy (1972) in *Shigella* phages of the A1 type. Others have been described several times, but the particles could be contaminants or always had empty heads, and are therefore beyond the scope of this article. Abnormally long noncontractile tails, first observed by Mount *et al.* (1968), have frequently been reported in phages with full or empty heads and seem to be rather common. Abnormally short tails have so far been seen only in B2 type phages of *Staphylococcus* (Ackermann *et al.*, 1976).

2. Cubic Phages

To our knowledge, the only aberrant structure reported to date is a single, elongated particle in the levivirus ZIK/1, which measures 37 × 23 nm (Bradley, 1964).

3. Filamentous Phages

Phages of the *Inovirus* genus offer many examples of size variations among normal phages and produce abnormal particles (Table XIV).

TABLE XIII

Morphological Aberrations Leading to Size Changes

Normal particle	Modification	Morphotype	Phage	Host	References
Elongated head	Further elongated	A2	T-even	Escherichia	Cummings et al. (1973)
		B2	6	Staphylococcus	Ackermann et al. (1976)
			MSP2	Streptomyces	Jones et al. (1974)
		B3	VD13	Streptococcus	Ackermann et al. (1975)
		C3	7-11	Salmonella	Moazamie et al. (1979)
	Laterally enlarged	A2	T-even	Escherichia	Boy de la Tour and Kellenberger (1965)
	Shortened	A2	T-even	Escherichia	Mosig et al. (1972)
		C3	7-11	Salmonella	Moazamie et al. (1979)
	Becomes isometric	B2	531	Acinetobacter	Ackermann et al. (1973)
			MSP2	Streptomyces	Jones et al. (1974)
		C3	7-11	Salmonella	Moazamie et al. (1979)
Isometric head	Becomes smaller	A1	P1	Escherichia	Walker and Anderson (1970)
Contractile tail	Elongated	A1	Several	Shigella	Krzywy (1972)
			a	Clostridium	Ackermann et al. (1978b)
Noncontractile tail	Elongated	B1	λ	Escherichia	Mount et al. (1968)
			Several	Lactobacillus	Accolas and Spillmann (1979)
			X	Micrococcus	Compton et al. (1979)
		B2	Several	Acinetobacter	Ackermann et al. (1973)
			3C	Staphylococcus	Ortel (1965)
			D29	Mycobacterium	Schäfer et al. (1977)
	Shortened	B2	Several	Staphylococcus	Ackermann et al. (1976)

[a] No name, not cultivated.

TABLE XIV

Length Variation in the *Inovirus* Genus

Phage	Host	Length (nm)	References
f1 miniphage	*Escherichia*	~270[a]	Enea and Zinder (1975)
M13 miniphage	*Escherichia*	~160–400[a]	Griffith and Kornberg (1974)
Pf3	*Pseudomonas*[b]	760	Bradley (1974)
fd[c]	*Escherichia*	809	Hoffmann-Berling et al. (1963)[d]
PR64FS	*Escherichia*	917	Coetzee et al. (1980)
Xf	*Xanthomonas*	971	Kuo et al. (1969); Wakimoto et al. (1979)
Cf	*Xanthomonas*	1008	Dai et al. (1980)
If1, If2, IKe?	*Escherichia*	1300	Meynell and Lawn (1968); Wiseman et al. (1972)[d]
Xf2	*Xanthomonas*	1573	Kamiunten and Wakimoto (1979)
f1 diploid	*Escherichia*	~1600[a]	Scott and Zinder (1967)
M13 diploid	*Escherichia*	~1600[a]	Beaudoin et al. (1974); Salivar et al. (1967)
Pf1, Pf2	*Pseudomonas*	1915	Takeya and Amako (1966); Bradley (1973a)[d]
VO-1 hybrid	*Escherichia*	~2000[a]	Ohsumi et al. (1978)
M13 triploid	*Escherichia*	~2600[a]	Beaudoin et al. (1974)
Pf1 diploid?	*Pseudomonas*	3600	Bradley (1973a)
Pf1 tetraploid?	*Pseudomonas*	~8000[a]	Yasunaka (1977)

[a] Recalculated or inferred.

[b] Phage replicates in *P. aeruginosa* and *E. coli*.

[c] Includes phages AE2, E9, f1, HR, M13, ZG/2, ZJ/2, and δA.

[d] Further references in Ackermann (1978b).

The normal phages show a wide range of particle lengths. It is noteworthy that phages fd, IF1, and PR64FS, which are serologically related and have similar coat proteins (Section VII-B), are of different lengths. "Miniphages" are deletion mutants which contain only a fraction of the normal genome and are shorter than the normal virion. Particle length and DNA content are closely correlated (Enea and Zinder, 1975; Griffith and Kornberg, 1974). In addition, inoviruses produce diploid and triploid particles of two and three times normal length which contain at least two DNA molecules (Beaudoin et al., 1974; Salivar et al., 1967; Scott and Zinder, 1967). Even particles of four times normal length seem to occur in Pf1, a *Pseudomonas* phage (Yasunaka, 1977). Finally, a hybrid of phage f1 and a plasmid has been constructed which contains both parent genomes and measures about 2.5 times the unit length of f1 (Ohsumi et al., 1978). Inoviruses therefore are a good example of morphological evolution. Furthermore, these observations, especially that of the hybrid phage, shed light on inovirus morphogenesis. They indicate (1) that particle length depends on DNA size and (2) that the DNA is synthesized first and coated later, perhaps as the virus is extruded from the cell (Marvin and Hohn, 1969).

The *Plectrovirus* genus presents a similar picture. The usual type multiplies in *Acholeplasma* and is about 84 nm long (Section III). However, similar rods, so far not cultivated and 230–280 nm in length, exist in the genus *Spiroplasma* (Cole *et al.*, 1974). In addition, preparations of both types contain rigid rods up to 1.5 μm long which may or may not contain DNA (Bruce *et al.*, 1972; Cole *et al.*, 1974; Liss and Maniloff, 1973). It is possible that at least some of them are polyploid in nature.

C. Evolution of Tailed-Phage Morphotypes

Based on the frequency of morphotypes and the nature of some abnormal particles, one may attempt to construct a set of schematic pathways which relate the basic morphotypes of tailed phages. This scheme (Fig. 3) is highly speculative and is offered in the hope that it will stimulate others to reappraise this topic. We begin by noting that noncontractile tails, whether long or short, are structurally simpler and occur more frequently than contractile ones. Hence it seems reasonable to consider them as ancestral features. The same argument applies to isometric heads which are more common than elongated capsids. This gives the B1 and C1 morphotypes as the preferred candidates for "ancestral" phages. Of these, we tend to prefer B1 because of its much greater frequency, but differences between these types are rather slight. The next step would be the elongation of the head through the addition of successive rows of subunits. This could have happened several times independently and very long heads might be derivative of shorter ones. A tentative sequence of events would then be as follows: tailed phages originated as structures like the B1 (or C1) morphotype which gave rise to types A1 and C1; all types then differentiated by head elongation (Fig. 4).

Transitions within one morphotype (A1 to A3 for example) do not seem to pose major difficulties. For example, phages with long heads produce long and short head variants, and thus appear to be relatively instable and may revert to the isometric "norm." Transitions between the basic types, A, B, or C, are more difficult but might be greatly aided by modular gene exchange. An A–B transition would require the loss or acquisition of a tail sheath. This may actually happen in several

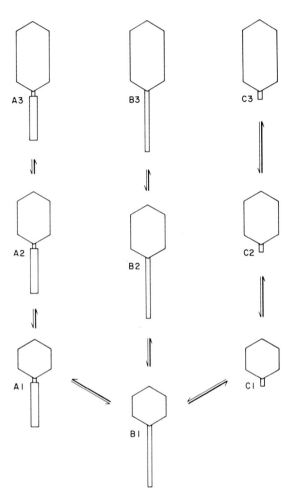

FIG. 4. Hypothetical evolutionary pathways of tailed phages. Bold arrows indicate main direction of evolution.

phages for *Clostridium botulinum* types B and C. These phages, widely known for converting nontoxigenic to toxigenic strains, have isometric heads and long tails provided with what seems to be a rudimentary, permanently contracted sheath of slightly variable length (Dolman and Chang, 1972; Eklund and Poysky, 1974; Eklund *et al.*, 1971). Defective phages of identical morphology occur in *Bacillus* (Steensma and Robertson, 1978). A transition between types B and C apparently requires a small step only, but it is curious that no intermediate forms seem to exist and that C-type tails are always very short. Phage T3 was claimed to produce a mutant with a long tail (Eisenstark *et al.*, 1961) but recombination with other phages cannot be excluded. Finally, "long distance" transitions between morphotypes, for example, between an A3 phage with elongated head and long tail and a C1 phage with isometric head and short tail, seem unlikely on statistical grounds.

D. Conclusions

The apparent lack of structural variation in cubic phages suggests that they are of relatively recent origin and/or that their capsids have severe, built-in constraints that do not permit misassembly. Tailed and filamentous phages on the other hand are or have been in a state of frequent morphological evolution. In inoviruses, DNA mass regulates particle size. The study of aberrant structures suggests that there is no impassable boundary between tailed phage morphotypes. We particularly stress the fact that heads can evidently be elongated (and otherwise enlarged?), because this makes room for additional DNA. Modular evolution, by gene transfer and acquisition, appears as a prime mechanism for the modification of morphology. Mutation has probably been important in the generation of new components, and perhaps in the modification of still hypothetical size determining factors.

XI. POSSIBLE RELATIONSHIPS BETWEEN PHAGES AND OTHER VIRUSES

A. Phages as Ancestors of Eukaryote Viruses?

There are interesting parallelisms in the organization and expression of the genomes of phages and their prokaryotic hosts. The mRNAs of both phages and cell genes are colinear with their DNA templates. Also the mRNAs of phages and bacteria are polycistronic, each messenger molecule containing information from more than one gene. This means that bacterial cells translate all mRNAs, viral and cellular, by means of a system of internal punctuation. The polycistronic RNAs of leviviruses conform to these general rules.

1. Split Genes in Eukaryotes and Their Viruses

The situation is very different in eukaryotes. Eukaryotic mRNAs in all known cases are monocistronic and bear a 5' "cap." Many eukaryote mRNAs have a poly(A) "tail." The most startling difference between

TABLE XV

SOME FEATURES WHICH DISTINGUISH PROKARYOTES AND EUKARYOTES

Feature	Prokaryotes	Eukaryotes
Genetic		
Chromosome(s)	Single, circular	Several, linear
Split genes	No	Common
RNA:RNA splicing	No	Mandatory
mRNA		
Cistrons	Polycistronic	Monocistronic
5' end	Unmodified	Cap
poly(A) at 3' end	No?	Frequent
Histones	No	Yes
Structural		
Nuclear membrane	No	Yes
Nucleolus	No	Yes
Mitochondria	No	Yes
Peptidoglycan wall	Yes	No
Lifestyle		
Divisions	Unlimited	Limited[a]

[a] Somatic cells appear to have a programmed limit of number of divisions (Hayflick limit).

prokaryotes and eukaryotes lies in the relationship between their DNAs and their transcribed RNA copies. Many eukaryotic genes are "split," the coding regions (exons) being separated by intervening regions of noncoding DNA (introns). Consequently the immediate transcript of an interrupted eukaryote gene must undergo a complex set of processing reactions which selectively remove the introns and religate the exons to give an mRNA which can be read in the correct phase (for reviews see Crick, 1979; Sharp, 1981). This process of RNA:RNA "splicing" has no known parallel in bacteria, even though the selective processing of transcribed RNAs occurs in both bacteria and some phages.

Since most eukaryote genes appear to be split, one might infer that the genes of eukaryote viruses would also be split. But here a clear-cut and interesting distinction emerges (Table XVI, and Reanney, 1981). Eukaryote viruses with an obligatory nuclear phase all contain split genes; this includes the orthomyxoviruses (influenza group), often considered as "cytoplasmic" viruses but recently shown to contain an in-

TABLE XVI

GENETIC ORGANIZATION OF SELECTED VIRUSES AND THEIR HOSTS[a]

Nucleic acid	Viral group or taxon	Replication site	Host		Virus	
			Split genes	mRNA	Split genes	mRNA
	Bacteriophages					
DNA	Tailed phages		No	Polycistronic	No	Polycistronic
	Microviridae[b]		No	Polycistronic	No	Polycistronic
	Inovirus[c]		No	Polycistronic	No	Polycistronic
RNA	Leviviridae		No	Polycistronic	No	Polycistronic
	Cystoviridae[d]		No	Polycistronic	No?	Polycistronic?
	Eukaryote viruses					
DNA	Adenoviridae	Nucleus	Yes	Monocistronic	Yes	Monocistronic
	Herpesviridae	Nucleus	Yes	Monocistronic	Yes	Monocistronic
	Papovaviridae	Nucleus	Yes	Monocistronic	Yes	Monocistronic
	Poxviridae	Cytoplasm	Yes	Monocistronic	No	Monocistronic
	Iridoviridae	Cytoplasm	Yes	Monocistronic	No	Monocistronic
RNA	Retroviridae	Nucleus	Yes	Monocistronic	Yes	Monocistronic
	Orthomyxoviridae[d]	Nucleus	Yes	Monocistronic	Yes	Monocistronic
	Picornaviridae	Cytoplasm	Yes	Monocistronic	No	Polycistronic
	Tobamovirus	Cytoplasm	Yes	Monocistronic	No	Polycistronic
	Tymovirus	Cytoplasm	Yes	Monocistronic	No	Polycistronic
	Reoviridae[d]	Cytoplasm	Yes	Monocistronic	No	Monocistronic

[a] See Reanney (1981).
[b] See Denhardt (1977).
[c] See Ray (1977).
[d] Genome is segmented.

tron in at least one mRNA (Lamb and Lai, 1980). However, viruses lacking a known nuclear involvement *without exception* lack split genes.

This observation has far-reaching implications. These are evident in the nature of the RNAs which are transcribed from eukaryotic viruses or which comprise such viruses. DNA viruses which propagate in eukaryote cytoplasm have monocistronic, capped RNAs like their host cells, despite the prokaryotic ancestry sometimes ascribed to large virions like the poxvirus group (Matthews, 1978). The genomes of cytoplasmic RNA viruses can, in some cases, be viewed as attempts to adapt a prokaryotic-type mRNA to a cellular environment tailored to process only monocistronic mRNAs (Jaspars, 1974). Consider the strategies of these RNA viral genomes. Negative strand viruses such as the rhabdoviruses are copied into unit length monocistronic mRNAs; here the negative RNA strand serves the same function as the negative strand of DNA. In reoviruses, for example, the division into monocistronic messengers has already occurred at the level of the genome itself and monocistronic mRNAs are transcribed from the negative strands of segmental duplex reovirus RNAs. This segmental character is typical of many RNA viruses (Matthews, 1979).

The most suggestive group of eukaryotic viruses are those whose genomes consist of polycistronic mRNAs. Eukaryote viruses with this bacterial organization seem very much out of place in nucleated cells and invite the speculation that they represent phages adapted to survival in higher cells.

Several other observations on these polycistronic eukaryotic RNA viruses (Table XVI) are consistent with the thesis of a phage ancestry. Poliomyelitis virus (see Wimmer, 1979) lacks the 5' 7^{me}GpppN cap that is a universal feature of eukaryote cellular mRNAs. Poliovirus RNA contains a low-molecular-weight protein covalently attached to the 5' end of its genome (Wimmer, 1979). The RNA of poliovirus is translated into one continuous polyprotein which is subsequently cleaved to give functional viral peptides. This suggests that the host cell treats the entire viral genome as one "monocistronic" messenger. It is possible to envisage how a polycistronic (phage?) genome accidently introduced into a eukaryote cell could acquire the ability to process the long polypeptide inevitably produced by the uninterrupted translation of its genome. It is almost impossible to see how the process could work in the other direction, i.e., how a series of monocistronic messengers could become covalently linked into one molecule, the uninterrupted translation of which was *subsequently* made possible by the development of special processing enzymes.

A phage ancestry for viruses like the *Picornaviridae* then is at least sensible. In our view the notion that some of the viruses listed in Table XVI could represent modified phage genomes is strengthened by ecological considerations. One prerequisite to any transfer of genes across wide evolutionary gaps is *opportunity*. The 10^{13} cells of an average human body carry a microflora estimated at 10^{14} microorganisms (Savage, 1977). This microflora sustains an astronomical population of phages. This is true not only of humans but of multicellular plants and animals, most of which sustain a complex, associated bacterial population. In view of the intimate physical association between bacteria and hosts in these situations and the enormous numbers of available phages involved it would be surprising if no genetic exchanges in either direction had occurred during evolution. Indeed the well-publicized case of crown gall has shown that extrachromosomal bacterial DNA can successfully invade the genomes of higher organisms (Chilton *et al.*, 1977; Schell and van Montagu, 1977; Tomashow *et al.*, 1980). Furthermore, the coat genes of ribophages R17 and Qβ have been translated in a mammalian cell-free system (Schreier *et al.*, 1973).

Such a process of gene transfer between higher and lower cells has a one-way character. A split eukaryotic gene could not colonize a bacterial cell because prokaryotes lack the splicing machinery required to express the gene. On the other hand a prokaryotic gene could be translated *in the cytoplasm* of eukaryotic cells since the splicing process occurs in the nucleus.

The above argument presupposes that the ancestor of present-day picornaviruses was already organized in a fashion typical of modern prokaryotes and that the cell it invaded was organized along the lines of modern eukaryotes; in other words both *virus* and *host* had already assumed their contemporary genetic structures *before* the postulated transfer from prokaryote to eukaryote took place. The time of origin of the first recognizable eukaryote cell is disputed, but estimates range from 850 million years (Bitter Spring fossils) to perhaps, 1500 million years (Schopf, 1978). If the phage ancestry theory is correct, then some eukaryote viruses must be *less* than 1500 million years old.

2. Antiquity of Splicing Machinery

In Section IV,A we suggested that typical phage genes such as those for isometric heads and helical tails were widely disseminated over 2×10^9 years ago. We cannot be certain that ancient "phage" genes possessed the genetic features which characterize their modern descendants. We can develop this argument by examining the antiquity of the splicing machinery. The canonical sequence (Catterall *et al.*, 1978), which flags out the splice sites in vertebrate and invertebrate transcript RNAs, has recently been reported in a plant pre-mRNA (Sun *et al.*, 1981). Thus the splicing machinery must antedate the common ancestor of plants and animals. Most significantly, certain genes such as those for apocytochrome b are spliced in some mitochondria (Mahler and Perlman, 1979), while rRNAs in some chloroplasts are spliced (Allet and Rochaix, 1979). These observations must be set in the context of theories which postulate that mitochondria and chloroplasts are endosymbiotically simplified bacteria and blue-green algae (see Margulis, 1972).

Overall these data support the view (Darnell, 1978; Doolittle, 1978; Reanney, 1979) that the very early genetic systems contained split genes. If correct, this assumption would cause us to revise some of our intuitive ideas concerning phage evolution. According to this revised scenario the earliest "viruses" contained interrupted genes, like their host cells. The correspondence between exons in the hen egg white lysozyme and domains in the lysozyme of phage T4 suggests that the ancestral phage gene was split (Section VIII,C). The pressure to compact information, which has been a conspicuous feature of viral evolution in general, caused one evolving line to streamline its genomes. This line was the ancestor of modern prokaryotes. A second line retained split genes both in nuclear DNA and perhaps in the DNA of certain endosymbionts. This line was the ancestor of the eukaryotes. Viruses propagating in the *nucleus* of such cells would have to retain a mode of gene expression compatible with the presence of a splicing machinery, while those propagating only in the cytoplasm could perhaps become "prokaryote-like" in some aspects of the organization and expression of their genetic material. In this new view the traditional "tailed phage," although of great antiquity, must be considered as specialized and "derived" as the bacterial cells it infects.

B. Phages in Eukaryotes?

Papers which describe phages able to propagate in eukaryotic organisms have appeared from time to time. They have far-reaching implications in view of the gap between prokaryotes and eukaryotes.

Viruses resembling phage T7 were reported in cultures of the green alga, *Chlorella pyrenoidosa* (Moskovets *et al.*, 1970); Tikhonenko and Zavarzina, 1966). Their existence has not been independently confirmed. In view of the known difficulties of obtaining axenic cultures of algae, a reinvestigation of these reports is urgently needed. The phage nature of these particles should be very carefully evaluated, as several investigators (below) have found algal viruses with some phage properties.

True bacteriophages able to multiply in *E. coli* were detected in *Penicillium* cultures (Tikhonenko, 1978; Tikhonenko *et al.*, 1974; further references in Lemke, 1976). They contained DNA and resembled λ and T7. It was claimed that their physicochemical properties were distinctive enough to rule out contamination. However, in our view, contamination is still the most likely explanation and we regard the "unique" properties of these phages as perfectly commonplace. Any report of phages in eukaryotes should prove the absence of contaminants, endosymbionts, and invading bacteria (Lemke, 1976).

The situation is very different with some phage-like viruses. A polyhedral particle with double-stranded DNA and a tail appendage was found in the green alga, *Uronema gigas* (Cole *et al.*, 1980; Dodds and Cole, 1980). However, the particle measured 390 nm in diameter and was thus far larger than any known bacteriophage; its tail, which was about 1 μm long, was present in 10% of particles only. Another phage-like entity was found in another green alga, *Chlorococcum* (Gromov and Mamkaeva, 1979). It had a hexagonal head of 220 × 180 nm and a tail 165 nm long. However, the tail was seen outside *or* inside the head, unlike any bacteriophage or indeed any other virus. The *Uronema* and *Chlorococcum* particles probably represent new virus categories, but their resemblance to phages is superficial and their tail seems to be a response to the unicellular nature of the host. It is remarkable that these particles, the *Tectiviridae,* and tailed phages independently seem to have evolved a common strategy for infection.

XII. Origins for Phages

A. General

We now arrive at the final and most speculative part of this article, the origin of phages. Three basic theories as to the origin of viruses have been put forward: (1) that viruses are specialized descendants of the earliest forms of life, (2) that they represent "degenerate" (prokaryote?) organisms, or (3) that they represent cell genes that became infectious. Although the first hypothesis has been discounted by most virologists, the concept that certain RNA viruses may be lineal descendants of primitive organisms that contained RNA as their genetic material instead of DNA has been suggested by Baltimore (1980) and others. It is almost impossible to obtain a concensus from virologists as to the relative merits of hypotheses (2) and (3). Some virologists (e.g., Matthews, 1978) view certain viruses as degenerate bacteria, while many others agree that most viruses, with the possible exception of the poxvirus group, evolved from cellular nucleic acids (Agol, 1976; Fenner, 1973; Joklik, 1974; Krieg, 1973; Luria and Darnell, 1968; Lwoff, 1957; Maramorosch, 1972; Nahmias and Reanney, 1977). For phages in particular, an origin from the bacterial chromosome (Adams, 1959; Bradley, 1967; Luria and Darnell, 1968; Lwoff, 1953, 1957; Maramorosch, 1972) or from extrachromosomal elements (Agol, 1976; Bradley, 1967, 1971; Maramorosch 1972) has been proposed. A monophyletic origin was considered as unlikely by some (Adams, 1959; Bradley, 1971; Maramorosch, 1972; Zinder, 1980), but was advocated by Joklik (1974) for tailed phages. Adams also suggested that phages originated as a means of genetic transfer from one cell to another. It is difficult to evaluate teleological arguments such as this.

Bradley (1967) suggested that filamentous phages of the fd type originated from R factors and proposed a hypothetical sequence according to which a portion of the bacterial chromosome became detached. This "episome" then acquired the ability to code for the synthesis of F pili. The pili were used for DNA transfer and later became autonomous. Pili could be considered as primitive forms of a capsid and filamentous phages as the result of a single evolutionary chain (Bradley, 1971). Possible analogies between filamentous phages and extrachromosomal elements were also stressed by Agol (1976). RNA phages of the *Leviviridae* family may have evolved from cellular mRNA (Bradley, 1971; Luria and Darnell, 1968; Zinder, 1980); Zinder also considered the possibility that RNA phages were primitive RNAs that became parasites or early "attempts" at sexuality.

Rather than evaluate the merits and demerits of these theories we have chosen to concentrate on three specific issues which bear on the question of phage origins. The first is an attempt to set the origin and development of phages in a wider context by asking the question, does an extracellular phase confer survival advantages on genes? The second investigates the possibility that the origins of phages may be intermeshed with those of plasmids. The third tackles the question of phage origins directly, in the only way possible—by looking for "molecular fossils" of very ancient relationships in nucleotide sequence data.

B. Why Phages?

Evolutionary theory has recently reentered one of its periodic bouts of self-criticism and reappraisal. One source of this renewed debate was Dawkins' book "The Selfish Gene" (1976) which emphasizes the central role of the gene as the basic unit of selection. This notion of genetic selfishness was developed by Doolittle and Sapienza (1980) and by Orgel and Crick (1980) and was specifically applied to viruses by Reanney (1981). The selfish gene concept provides a fresh insight into the kinds of selective forces that may have favored the origin and maintenance of viral genes.

Compare the survival chances of a set of "typical" chromosomal genes with those of the "selfish" genes of a broad host range plasmid such as RP1. The only cells which can acquire the chromosomal gene are those of its clonal descendents—unless the gene happens to be incorporated into the genome of a conjugative plasmid or a phage. By contrast the continued survival of the RP1 genes is *not* dependent on the survival of one single host species. An extended host range therefore represents an important element in the competitive survival strategy of selfish genes. It follows that *selection places a very high premium on the ability of a block of linked genes to move between cells (horizontal transmission) as well as from parent to progeny (vertical transmission)*. Within limits this argument gets stronger as the "mobile DNA" gets smaller. It is worth noting that there is a substantial overlap between the sizes of small phages and plasmids and known naturally occurring transposons (size range about 2500 to about 20,500 base pairs).

Conjugative plasmids move between cells as a result of specific cell-to-cell contacts (pair formation or mating aggregate formation) mediated by pili, followed by DNA exchange through a rolling-circle mechanism (Falkow, 1975; Clark and Warren, 1979; Willetts and Skurray, 1980). However, there are many situations in the natural world in which intimate contact between cells under conditions appropriate to DNA transfer is unlikely or impossible. In these situations, any mechanism which allows selfish genes to move from cell to cell *without physical contact between the cells* would enjoy a considerable selective advantage. Genes which encapsidate selfish DNA in stable extracellular particles fulfill this need. Thus the phage mode of DNA dissemination can be viewed as a corollary of the transfer function of many plasmids, having its maintenance, if not its origin, in the same selective pressures. The viral "strategy" has a further advantage which may help account for the ubiquity of viruses. In the extracellular particle state genes can survive stresses which would kill cells, e.g., low temperatures, long periods without nutrients, or high concentrations of cytocidal chemicals (Reanney, 1977, 1981).

C. Evolutionary Relationships between Phages and Plasmids?

The survival advantages of a horizontal mode of gene transfer apply equally to phage and plasmid genes. This raises the issue of possible evolutionary relationships between these two categories of extrachromosomal genetic element. Many bacteriologists agree that plasmids and phages have common evolutionary features. In a review of transposable elements, Cohen (1976) states, "thus, in this evolutionary continuum, plasmids can be viewed as primitive bacteriophages that have not yet acquired those specialized functions necessary for a complex replicative cycle or the production of infectious particles capable of existing outside the host cell. Conversely, bacteriophages can be viewed as plasmids in which a basic replication region has become linked to gene segments that specify the additional biological functions needed for the production of phage particles." Similarly Frédéricq (1963) comments, "A phage is a particle that can establish effective contact with a cell and inject into it genetic material. What else is a F^+ cell?" A detailed discussion of possible similarities between conjugative plasmids and phages is given by Brinton (1971).

The overlap between phages and plasmids is emphasized by reference to phage λ. Mutations in phage λ which suppress the lytic functions without affecting autonomous replication and autoreplicative control can convert the λ genome into self-replicating closed circular (cc) DNA species in infected cells (Szybalski and Szybalski, 1974). In the absence of prior knowledge, these ccDNA species would automatically be classified as nonconjugative plasmids. Nondefective phages such as P1 combine the properties of a plasmid and a phage. P1 DNA exists cytoplasmically as a nonintegrating ring but is transferred between cells in a typical phage head. Because the DNAs of P1 and related phages such as P7 cannot coexist stably in a single cell, P1 has been assigned a plasmid incompatibility group, Y (Hedges et al., 1975; Wandersman and Yarmolinsky, 1977). Perhaps the most apt example of a phage with plasmid features is the already mentioned virus RΦ6P of Rhodopseudomonas sphaeroides (Sections V,A and VIII,B). Unlike all other tailed phages reported to date, RΦ6P contains supercoiled circular DNA in the phage capsid; it is thus not difficult to speculate that RΦ6P was a plasmid at some stage of its evolution or that "plasmid" genes have contributed to the DNA of the present-day "phage."

Further evidence of relationships between phages and plasmids comes from the study of phage host ranges. Table XVII shows that the host range of certain phages is determined by plasmids of specific incompatibility groupings. The phages concerned are a highly heterogeneous group. Perhaps the most significant, in the context of this article, are those phages that adsorb selectively to the pili specified by conjugative plasmids. Pili are a phenotypic expression of the *transfer* genes that allow plasmid DNA to pass between mating cells (see Brinton, 1971). They consist of proteins arranged in a symmetrical way, not unlike the arrangement of subunits in a filamentous phage. It is quite possible that the pilus-specific leviviruses, inoviruses, and tectiviruses (Table XVII) *happen* to adsorb to a structure encoded in an extrachromosomal genome. In this case, the association is adventitious and without any evolutionary significance. On the other hand, it is also possible that the adsorption of these viruses to pili reflects a common distant ancestry for the genes encoding pili and phage proteins. Bradley (1971) has suggested that the RNA of leviviruses is derived from transcripts of the DNA plasmid which renders the host cell susceptible to infection by the phage. Furthermore, it is interesting to note that

TABLE XVII

PLASMID-DEPENDENT PHAGES[a]

Viral group or taxon	Phages	Plasmid (Inc) group needed for propagation	Additional references
Podoviridae	J	C, D, J	Coetzee *et al.* (1981)
Tectiviridae	PRD1, PR3, PR4, PR772	N, P, W	Coetzee *et al.* (1979)
Leviviridae	MS2, M12, R17, Qβ	F	Hoffmann-Berling *et al.* (1966)
	C-1	C	Sirgel *et al.* (1981)
	$F_0 lac$, UA6	$F_0 lac$	Coetzee *et al.* (1981)
	PRR1	P	
	t	T	Coetzee *et al.* (1981)
Inovirus genus	fd	C, F, $F_0 lac$	
	C-2	C	Coetzee *et al.* (1981)
	If1, If2	I	
	IKe	N, P	
	Pf3	P	
	PR64FS	I	Coetzee *et al.* (1980)
	X	M, N, P, U, W, X, R775	Coetzee *et al.* (1981)

[a] Modified from Bradley (1981).

pilus-dependent tailed phages are exceptional and are known for *Astic-cacaulis, Caulobacter, Pseudomonas,* and *Rhizobium* only (Auling, 1978; Bradley, 1973b; Bradley and Pitt, 1974; Johnson *et al.*, 1977; Lotz and Pfister, 1975; Pate *et al.*, 1979; Scholl and Jollick, 1980).

The plasmid relationships which may be most important for phage evolution are those processes of gene transfer that appear to *couple* plasmid and phage features. In mixed cultures of *Staphylococcus,* DNA transfer depends not only on the presence of temperate phages in the donor but also on cell-to-cell contact. Virtually no transducing particles were detected in the supernatant (Lacey, 1975). In "phage-associated gene transfer" in *Streptococcus pneumoniae,* host genes are packaged into particles that resemble normal phages: however, the transfer of these genes into sensitive recipient cells requires the development of competence for transformation (Porter *et al.*, 1979). While we do not wish to elevate any of these examples to the status of a "stepping stone" or "missing link" in phage development, their existence suggests that the three primary modes of DNA transfer in bacteria should not be considered as entirely separate phenomena.

The expansion of gene cloning has seen the development of techniques which marry aspects of plasmid biology with those of phage biology. Collins and Hohn (1978) have constructed a hybrid plasmid–phage entity which they dubb a "cosmid." Cosmids are plasmids in which identifiable markers such as antibiotic resistance have been joined to the *cos* region of phage λ (Collins and Bruening, 1978). The *cos* genes allow the cosmids to be packaged in preformed phage heads in a manner similar to the encapsidation of newly formed phage DNA.

The features which make cosmids attractive tools for genetic engineers also make them interesting models for evolutionary interactions between plasmid and phage genes. Cosmids have a considerable capacity for absorbing foreign (i.e., non-λ) DNA since large insertions result in molecules which may be *preferentially* packaged in order to fill the phage head with DNA. This fact in itself suggests that plasmids may become phages by a few recombinational steps which link structural phage genes with various plasmid genes.

D. Molecular Fossils in Sequence Data?

The only way to put factual constraints on evolutionary speculation is to look for evidence of relationships in the one place where they are likely to survive—nucleotide sequences from different evolutionary sources. This final section assembles the meager information on this topic available to us at this time.

Consider a key function such as the origin of DNA replication in some phages. The nucleotide sequence of the origin of DNA replication in *E. coli* K12 (the *oriC* locus) has been determined (Meijer *et al.*, 1979). Elements of this sequence are very similar to the origin of replication for λ (Styloviridae) and for phage G4 (Microviridae) (Fiddes *et al.*, 1978).

The similarities between the cellular and phage sequences seem sufficiently strong to justify use of the term "homologous." Convergent evolution toward a common sequence seems unlikely for three such different microorganisms. The implications of this putative homology are (1) that λ and G4 originated by a process that incorporated the cell's replicative origin into their own DNA and (2) that sequence homology has been retained, possibly over a very long evolutionary time span, by the requirement that λ and G4 replicate in an *E. coli* host using the same primase (the *dna* G product) employed by the cell.

A very interesting example of possible phage origin from cellular sequences is the case of the ribophages MS2 and Qβ (Leviviridae). The replicase of Qβ is a composite molecule containing four subunits, only one of which is virally coded (Blumenthal *et al.*, 1972; Wahba *et al.*, 1974): subunit I = ribosomal protein SI; subunit II = viral peptide; subunit III = elongation factor EF Tu; and subunit IV = elongation factor EF Ts.

Three subunits of the enzyme, all of them host proteins, are thus involved in the recognition of two cellular RNAs basic to the translation mechanism, namely, ribosomal RNA and transfer RNA. Do these data give us any clues as to the actual origins of ribophages?

The topologically complex 3'OH end of MS2 RNA bears some resemblance to the structures proposed for the 3' termini of aminoacylated plant and animal tRNAs (see Hall, 1979). Perhaps more significantly, there are analogies in the base doublet patterns of *E. coli* mRNA and the MS2 coat gene (Elton *et al.*, 1976) and extensive nucleotide homology between the 3' end of 16 S ribosomal RNA and segments of MS2 and Qβ RNA (Engelberg and Schoulaker, 1976):

$$\text{Q}\beta \ 3'—151\text{-CGUCUUCUGGCACCCUACGGGG}$$
$$\text{rRNA—20 -CACGGUCCAGACUCCUACGGGG}$$

These segments contain a region of direct homology preceded by a region of inverse homology. As Engelberg and Schoulaker note, the inverted homology can be most simply explained on the basis of an inversion in *DNA* since there are no known RNA recombination mechanisms in bacteria. *If* the homology is real and *if* the DNA assumption is valid, this would imply that ribophages such as Qβ evolved distantly from some sequence of DNA, perhaps through a recombinational event which linked genes for an isometric DNA capsid to sequences encoding ribosomal RNA. In view of the very limited amount of data however we feel it is important to resist the temptation to indulge in undue speculation.

A further intriguing suggestion on origins of phage DNA was made by Landy and Ross (1977) who sequenced the *att* region of coliphage λ DNA. They pointed out that the sequences in the OP region of the λ *att* site have a striking homology with elements of the terminal inverted repeat sequences of IS1. Their speculation that the λ *int* protein is related to the "integrase" protein(s) of IS1 is most attractive to us. In this context it is interesting to note that MacHattie and Shapiro (1978) have been able to integrate the DNA of phage λ into host DNA by means of a DNA insertion element. Campbell *et al.* (1977) have also speculated that the p_l promoter of λ "might (then) very well be the direct descendant of the promoter that controlled the insertion and excision functions of that (IS-like) element, which has secondarily modified its control properties to become well-integrated into the λ life cycle."

XIII. Conclusion

In conclusion, what can we say about phage evolution? In one sense this article has raised more questions than it has answered. However we are now able to bring certain questions into sharper focus and to put certain inferences on a sounder footing. For example, it has always been intuitively obvious that tailed phages are very old. We can now be more precise. Tailed phages and the various families of cubic, filamentous, and pleomorphic phages have little in common, are most probably polyphyletic in origin, and arose at different times. Cubic phages and the *Inovirus* genus may have appeared about 2×10^9 years ago or later (Section IV,B). On the other hand, the presence of tailed phages in archaebacteria, eubacteria, and cyanobacteria (Table III) and the remarkable coincidence of G–C values between the DNAs of tailed phages and those of their hosts (Table V) suggest that tailed phage genes predate the common ancestor of the three major divisions of the prokaryotic world. From this conclusion it follows that the genomes of temperate phages are part of the vertically inherited "survival kit" of bacterial DNA, subject to the same selective pressures as other "chromosomal" genes (Table V). Thus the traditional view of phages as intracellular parasites or "bacteria-*eaters*" (the literal translation of bacterio*phage*) is misleading. Genes for the construction of extracellular particles have probably been a factor in the evolution of host bacterial genomes and the speciation of phage, plasmid, and chromosomal DNAs must be viewed as a reticulate web of *coupled* processes (Anderson, 1966; Cohen, 1976; Reanney, 1976b, 1978b).

It will never be possible to reconstruct the actual historical path followed by a particular phage in its evolution. But the topics covered in this article have given us a better understanding of the mechanisms likely to have been involved: transposition, chromosome mobilization, the natural use of restriction nucleases or natural "genetic engineering" (Chang and Cohen, 1977; Reanney, 1976b), intercell transfers by conjugation, transduction, and transformation—the potential for genetic variation in these supposedly "simple" organisms is staggering. Set against this background we do not find it hard to see how a common pool of genes for morphogenetically complex entities like tailed phages could be directed into a variety of channels, generating the diversity of phage morphotypes set out in Fig. 1.

It is also possible to see *why* the phage strategy has been successful to the point where tailed phage genes are found in almost every known prokaryotic taxon. Consider the temperate phage P1. P1 genes have a vastly higher "survival rating" than genes for any single species of host organisms. There are two reasons for this. First, P1 genes in the extracellular particle phase can survive conditions inimical to cells. Second, P1 genes can colonize over 8 different genera of host bacteria (see Section IV,A). By spreading their survival options across so many taxonomic baskets P1 genes have taken an almost failsafe insurance policy on their future. Whole lineages of bacterial DNAs may become extinct without obliterating the viral genes. The selfish DNA concept then provides a rationale for the development and maintenance of the extracellular state.

The diversity of tailed phage morphotypes, in our view, was achieved by shuffling a limited group of ancestral genes into new combinations. This process of rearrangement is similar to the concept of "exon shuffling" advocated by Gilbert (1978, 1979) to explain the diversification of eukaryotic proteins. However, if the contemporary architecture of tailed phage morphotypes is a result of this "modular" mode of evolution, the present-day *abundance* of these phages reflects the filtering action of natural selection. The gene combination responsible for the B1 morphotype seems to represent some kind of "optimal" strategy for the survival of these genes, as this morphotype is by far the most frequent of all. While it is possible that this combination of genes arose only once (implying a monophyletic origin), it is just as possible that it arose several times (implying a polyphyletic origin) by convergent evolution.

Phages are often considered to be immensely conservative structures which resist change. In view of recent discoveries in microbial genetics we are inclined to invest this conservatism, not in the DNA of the whole phage but in its individual component modules or genes. This conservatism may be a feature of genes in general. A startling example has recently been provided by the discovery that the bacterium *E. coli* encodes a protein functionally similar to *human* insulin (Le Roith *et al.*, 1981). As Reanney (1978b) has suggested elsewhere, "recognizable ancestors for 90% of modern structural genes were already in existence 1.5 billion years ago." Subsequent evolution has largely been a matter of reshuffling these genes or their exon components.

What of the future? In one sense it is very bright. As this article has shown, in the case of intensively studied viruses such as ϕX174 and G4, it is possible to define evolutionary relationships with some precision. What is lacking to carry this process forward is not the technology but the incentive. The rise of recombinant DNA research has channeled much of the interest and associated grant finance in microbial genetics toward the cloning of eukaryote genes in bacterial plasmids and in phages such as λ. Phages have thus been reduced to the status of tools rather than objects which merit further study in their own right. The very success of microbial genetics in laying down the basic conceptual framework of molecular biology has diminished the desire to invest more time and energy in phage research per se. Present-day phage genetics is in part a victim of its own earlier triumphs.

In our view it would be unfortunate if interest in phage research were to wane at this point in time. We say that not as phage specialists but as evolutionary biologists. Many of the most fundamental questions in evolution have still to be answered. The same advantages that made phages popular genetic tools several years ago remain today; for example, the recent analysis of natural selection among prebiotic RNA molecules by Eigen *et al.* (1981) used phage Qβ as a model. The *in vitro* adaptation of Qβ remains virtually the only documented case in which the large-scale macroevolution of a self-replicating system has been studied in detail under defined conditions (Spiegelman, 1970). Phage genomes are probably easier to sequence than those of many plant or animal viruses because phages can be so easily cultivated. It would therefore be easier to build up a library of sequenced phage genomes for comparative study than any other set of natural genomes. The availability of such a library would allow evolutionists, for the first time, to construct a detailed genealogy for a major group of biounits. Information gained from such a study would have implications not just for phage taxonomy but for biology in general.

ACKNOWLEDGMENTS

We want to thank Dr. John M. Pemberton, St. Lucia, Queensland, for critically reading a draft version of the manuscript. Dr. Laurent Berthiaume and Dr. Sorin Sonea, Montreal, Canada, provided valuable suggestions. The *Pseudomonas* phage depicted in Fig. 2a was received from Dr. Odette Santos-Ferreira, Lisbon, Portugal. The aid of Mario Martin, Berthe Tessier, and Marie-Claude Jouvet, Quebec, Canada, in editing and typing is deeply appreciated.

REFERENCES

Accolas, J.-P., and Spillmann, H. (1979). *J. Appl. Bacteriol.* **47,** 309.
Ackermann, H.-W. (1974). *Pathol. Biol.* **22,** 909.
Ackermann, H.-W. (1975). *Pathol. Biol.* **23,** 247.
Ackermann, H.-W. (1976). *Pathol. Biol.* **24,** 359.
Ackermann, H.-W. (1978a). *In* "CRC Handbook of Microbiology" (A. I. Laskin and H. A. Lechevalier, eds.), 2nd Ed., Vol. 2, pp. 639–642. CRC Press, Cleveland, Ohio.
Ackermann, H.-W. (1978b). *In* "CRC Handbook of Microbiology" (A. I. Laskin and H. A. Lechevalier, eds.), 2nd Ed., Vol. 2, pp. 673–682. CRC Press, Cleveland, Ohio.
Ackermann, H.-W. (1978c). *Pathol. Biol.* **26,** 507.
Ackermann, H.-W. (1982). *In* "Virologie médicale" (J. Maurin, ed.). Flammarion, Paris, in press.
Ackermann, H.-W., and Brochu, G. (1978). *In* "CRC Handbook of Microbiology" (A. I. Laskin and H. A. Lechevalier, eds.), 2nd Ed., Vol. 2, pp. 691–695. CRC Press, Cleveland, Ohio.
Ackermann, H.-W., and Eisenstark, A. (1974). *Intervirology* **3,** 201.
Ackermann, H.-W., and Gauvreau, L. (1972). *Zentralbl. Bakteriol. Parasitenk. Infektionskr. Hyg. Abt. Orig.* **221,** 196.
Ackermann, H.-W., Brochu, G., and Cherchel, G. (1973). *J. Microsc. (Paris)* **16,** 215.
Ackermann, H.-W., Caprioli, T., and Kasatiya, S. S. (1975). *Can. J. Microbiol.* **21,** 571.
Ackermann, H.-W., Berthiaume, L., Sonea, S., and Kasatiya, S. S. (1976). *J. Virol.* **18,** 619.
Ackermann, H.-W., Roy, R., Martin, M., Murthy, M. R. V., and Smirnoff, W. A. (1978a). *Can. J. Microbiol.* **24,** 986.
Ackermann, H.-W., Fredette, T. V., and Vinet, G. (1978b). *Rev. Can. Biol.* **37,** 43.
Adams, M. H., ed. (1959). "Bacteriophages." Wiley (Interscience), New York.
Agol, V. I. (1976). *Orig. Life* **7,** 119.
Alexander, J. L., and Jollick, J. D. (1977). *J. Gen. Microbiol.* **99,** 325.
Allet, B., and Rochaix, J-D. (1979). *Cell* **18,** 55.
Amati, P., and Favre, R. (1968). *Cold Spring Harbor Symp. Quant. Biol.* **33,** 371.
Anderson, E. S. (1966). *Nature (London)* **209,** 637.
Anderson, T. F., and Stephens, R. (1964). *Virology* **23,** 113.
Arber, W., and Linn, S. (1969). *Annu. Rev. Biochem.* **38,** 467.

Artymiuk, P. J., Blake, C. C. F., and Sippel, A. E. (1981). *Nature (London)* **290,** 287.

Asbeck, F., Beyreuther, K., Köhler, H., von Wettstein, G., and Braunitzer, G. (1969). *Hoppe-Seylers Z. Physiol. Chem.* **350,** 1047.

Atkins, J. F., Steitz, J. A., Anderson, C. W., and Model, P. (1979). *Cell* **18,** 247.

Auling, G. (1978). *J. Gen. Virol.* **40,** 615.

Balch, W. E., Fox, G. E., Magrum, L. J., Woese, C. R., and Wolfe, R. S. (1979). *Microbiol. Rev.* **43,** 260.

Baltimore, D. (1980). *Ann. N.Y. Acad. Sci.* **345,** 491.

Barksdale, L., and Arden, S. B. (1974). *Annu. Rev. Microbiol.* **28,** 265.

Beaudoin, J., Henry, T. J., and Pratt, D. (1974). *J. Virol.* **13,** 470.

Beck, E., Sommer, R., Auerswald, E. A., Kurz, C., Zink, B., Osterburg, G., Schaller, H., Sugimoto, K., Sugisaki, H., Okamoto, T., and Takanami, M. (1978). *Nucleic Acids Res.* **5,** 4495.

Beier, H., and Hausmann, R. (1973). *J. Virol.* **12,** 417.

Bellett, A. J. D. (1967). *J. Mol. Biol.* **27,** 107.

Bendis, I., and Shapiro, L. (1970). *J. Virol.* **6,** 847.

Benveniste, R., and Davies, J. (1973). *Proc. Natl. Acad. Sci. U.S.A.* **70,** 2276.

Beremand, M. N., and Blumenthal, T. (1979). *Cell* **18,** 257.

Berg, D. E., Davies, J., Allet, B., and Rochaix, J. D. (1975). *Proc. Natl. Acad. Sci. U.S.A.* **72,** 3628.

Berkowitz, S. A., and Day, L. A. (1976). *J. Mol. Biol.* **102,** 531.

Berthiaume, L., and Ackermann, H.-W. (1977). *Pathol. Biol.* **25,** 195.

Blake, C. C. F. (1979). *Nature (London)* **277,** 598.

Blumenthal, T., Landers, T. A., and Weber, K. (1972). *Proc. Natl. Acad. Sci. U.S.A.* **69,** 1313.

Bogush, V. G., Zvenigorodsky, V. I., Rebentish, B. A., Smirnova, T. A., Permogorov, V. I., and Alikhanian, S. I. (1978). *Genetika* **14,** 867.

Botstein, D. (1980). *Ann. N.Y. Acad. Sci.* **354,** 484.

Botstein, D., and Herskowitz, I. (1974). *Nature (London) New Biol.* **251,** 584.

Boy de la Tour, E., and Kellenberger, E. (1965). *Virology* **27,** 222.

Bradley, D. E. (1964). *J. Gen. Microbiol.* **35,** 471.

Bradley, D. E. (1967). *Bacteriol. Rev.* **31,** 230.

Bradley, D. E. (1971). *In* "Comparative Virology" (K. Maramorosch and E. Kurstak, eds.), pp. 207–253. Academic Press, New York.

Bradley, D. E. (1973a). *J. Gen. Virol.* **20,** 249.

Bradley, D. E. (1973b). *J. Virol.* **12,** 1139.

Bradley, D. E. (1974). *Biochem. Biophys. Res. Commun.* **57,** 893.

Bradley, D. E. (1981). *In* "Molecular Biology, Pathogenicity and Ecology of Bacterial Plasmids" (S. B. Levy, R. C. Clowes, and E. L. Koenig, eds.), pp. 217–226. Plenum, New York.

Bradley, D. E., and Dewar, C. A. (1967). *J. Gen. Virol.* **1,** 179.

Bradley, D. E., and Pitt, T. L. (1974). *J. Gen. Virol.* **23,** 1.

Brinton, C. C. (1971). *CRC Crit. Rev. Microbiol.* **1,** 105.

Bruce, J., Gourlay, R. N., Hull, R., and Garwes, D. J. (1972). *J. Gen. Virol.* **16,** 215.

Brunovskis, I., Hyman, R. W., and Summers, W. C. (1973). *J. Virol.* **11,** 306.

Brutlag, D., and Kornberg, A. (1972). *J. Biol. Chem.* **247,** 241.

Buchanan, R. E., and Gibbons, N. E., eds. (1974). "Bergey's Manual of Determinative Bacteriology," 8th Ed. Williams & Wilkins, Baltimore, Maryland.

Buck, G., Groman, N., and Falkow, S. (1978). *Nature (London)* **271,** 683.

Bukhari, A. I., and Ambrosio, L. (1978). *Nature (London)* **271,** 575.

Bukhari, A. I., Shapiro, J. A., and Adhya, S. L., eds. (1977). "DNA Insertion Elements, Plasmids and Episomes." Cold Spring Harbor Laboratory, Cold Spring Harbor, New York.

Burgess, A. B. (1969). *Proc. Natl. Acad. Sci. U.S.A.* **64,** 613.

Campbell, A. M. (1972). *In* "Evolution of Genetic Systems" (H. H. Smith, ed.), pp. 543–562. Gordon & Breach, London.

Campbell, A., Heffernan, L., Hu, S.-L., and Szybalski, W. (1977). *In* "DNA Insertion Elements, Plasmids and Episomes" (A. I. Bukhari, J. A. Shapiro, and S. L. Adhya, eds.), pp. 375–379. Cold Spring Harbor Laboratory, Cold Spring Harbor, New York.

Calos, M. P., and Miller, J. H. (1980). *Cell* **20,** 579.

Casjens, S., and King, J. (1975). *Annu. Rev. Biochem.* **44,** 555.

Catterall, J., O'Malley, B., Robertson, M., Staden, R., and Tanaka, Y. (1978). *Nature (London)* **275,** 510.

Champe, S. P. (1974). *In* "CRC Handbook of Microbiology" (A. I. Laskin and H. A. Lechevalier, eds.), Vol. 4, pp. 605–608. CRC Press, Cleveland, Ohio.

Chang, S., and Cohen, S. N. (1977). *Proc. Natl. Acad. Sci. U.S.A.* **74,** 4811.

Chesney, R. H., and Scott, J. R. (1975). *Virology* **67,** 375.

Chilton, M. D., Drummond, M. H., Merlo, D. J., Schiahy, D., Montoya, A. L., Gordon, M. P., and Nester, E. W. (1977). *Cell* **11,** 99.

Chow, L. T., and Bukhari, A. I. (1977). *Virology* **74,** 242.

Chow, L. T., and Bukhari, A. I. (1978). *Gene* **3,** 333.

Chow, L. T., Boice, L. B., and Davidson, N. (1972). *J. Mol. Biol.* **68,** 391.

Clark, A. J., and Warren, G. J. (1979). *Annu. Rev. Genet.* **13,** 99.

Clements, J. B., and Sinsheimer, R. L. (1975). *J. Virol.* **15,** 151.

Cloud, P. E. (1968). *Science* **160,** 729.

Coetzee, J. N. (1974). *J. Gen. Microbiol.* **84,** 285.

Coetzee, J. N. (1975). *J. Gen. Microbiol.* **87,** 173.

Coetzee, W. F., and Bekker, P. J. (1979). *J. Gen. Virol.* **45,** 195.

Coetzee, J. N., Lecatsas, G., Coetzee, W. F., and Hedges, R. W. (1979). *J. Gen. Microbiol.* **110,** 263.

Coetzee, J. N., Sirgel, F. A., and Lecatsas, G. (1980). *J. Gen. Microbiol.* **117,** 547.

Coetzee, J. N., Bradley, D. E., Sirgel, F. A., and Bothma, T. (1981). *Int. Congr. Virol., 5th,* p. 279 (Abstr.).

Cohen, S. N. (1976). *Nature (London)* **263,** 731.

Cole, A., Dodds, J. A., and Hamilton, R. I. (1980). *Virology* **100,** 166.

Cole, R. M., Tully, J. G., and Popkin, T. J. (1974). *Coll. Inst. Natl. Santé Rech. Med.* **33,** 125.

Collins, J., and Bruening, J. J. (1978). *Gene* **4,** 85.

Collins, J., and Hohn, B. B. (1978). *Proc. Natl. Acad. Sci. U.S.A.* **75,** 4242.

Compton, S. C., Mayo, J. A., Ehrlich, M., Ackermann, H.-W., Tremblay, L., Cords, C. E., and Scaletti, J. V. (1979). *Can. J. Microbiol.* **25,** 1027.

Courvalin, P., Weisblum, B., and Davies, J. (1977). *Proc. Natl. Acad. Sci. U.S.A.* **74,** 999.

Cowrie, D. B., Avery, R. J., and Champe, S. P. (1971). *Virology* **45,** 30.

Crick, F. (1979). *Science* **204,** 264.

Cummings, D. J., and Bolin, R. W. (1975). *Bacteriol. Rev.* **40,** 314.

Cummings, D. J., Chapman, V. A., DeLong, S. S., and Couse, N. L. (1973). *Virology* **54,** 245.

Dai, H., Chiang, K.-S., and Kuo, T.-T. (1980). *J. Gen. Virol.* **46,** 277.

Darnell, J. E. (1978). *Science* **202,** 1257.

Datta, J. (1979). *In* "Plasmids of Medical, Environmental and Commercial Importance" (K. N. Timmis and A. Pühler, eds.), pp. 3–12. Elsevier, Amsterdam.

Davey, R. B., and Reanney, D. C. (1980). *Evol. Biol.* **13,** 113.

Davis, R. W., and Hyman, R. W. (1971). *J. Mol. Biol.* **62,** 287.

Dawkins, R., ed. (1976). "The Selfish Gene." Oxford Univ. Press, London and New York.

Dayhoff, M. O., ed. (1972, 1976). "Atlas of Protein Sequence and Structure," Vols. 2 and 5. National Biomedical Research Foundation, Silver Spring, Maryland.

De Ley, J. (1968). *In* "Evolutionary Biology" (T. Dobzhanski, M. K. Hecht, and W. C. Steere, eds.), Vol. 2, pp. 103–154. Appleton, New York.

De Ley, J., Gillis, M., Pootjes, C. F., Kersters, K., Tytgat, R., and Van Braekel, M. (1972). *J. Gen. Virol.* **16,** 199.

Denhardt, D. T. (1975). *CRC Crit. Rev. Microbiol.* **4,** 161.

Denhardt, D. T. (1977). *In* "Comprehensive Virology" (H. Fraenkel-Conrat and R. R. Wagner, eds.), Vol. 7, pp. 1–104. Plenum, New York.

Dhaese, P., Vandekerckhove, J. S., and van Montagu, M. C. (1979). *Eur. J. Biochem.* **94,** 375.

Dhaese, P., Lenaerts, A., Gielen, J., and van Montagu, M. (1980). *Biochem. Biophys. Res. Commun.* **94,** 1394.

Dhillon, T. S., Dhillon, E. K. S., Toyama, S., and Linn, S. (1980). *Microbiol. Immunol.* **24,** 515.

Dodds, J. A., and Cole, A. (1980). *Virology* **100,** 156.

Doermann, A. H., Eiserling, F. A., and Boehner, L. (1973). *J. Virol.* **12,** 374.

Dolman, C. E., and Chang, E. (1972). *Can. J. Microbiol.* **18,** 67.

Domingo, E., Sabo, D., Taniguchi, T., and Weissmann, C. (1978). *Cell* **13,** 735.

Doolittle, W. F. (1978). *Nature (London)* **272,** 581.

Doolittle, W. F., and Sapienza, C. (1980). *Nature (London)* **284,** 601.

Dove, W. F. (1968). *Annu. Rev. Genet.* **2,** 305.

Drake, J. W. (1974). *Symp. Soc. Gen. Microbiol.* **24,** 41.

Dubnau, E., and Stocker, B. A. D. (1964). *Nature (London)* **204,** 1112.

Earnshaw, C. W., and Casjens, S. R. (1980). *Cell* **21,** 319.

Edlin, G., Lin, L., and Kudrna, R. (1975). *Nature (London)* **255,** 735.

Edlin, G., Lin, L., and Bitner, R. (1977). *J. Virol.* **21,** 560.

Eigen, M., Gardiner, W., Schuster, P., and Winkler-Oswatitsch, R. (1981). *Sci. Am.* **244,** 88.

Eisenstark, M., Maaløe, O., and Birch-Andersen, A. (1961). *Virology* **15,** 56.

Eiserling, F. A. (1963). *Ph.D. Thesis,* University of California, Los Angeles.

Eiserling, F. A., and Dickson, R. C. (1972). *Annu. Rev. Biochem.* **41,** 467.

Eklund, M. W., and Poysky, F. T. (1974). *Appl. Microbiol.* **27,** 251.

Eklund, M. W., Poysky, F. T., Reed, S. M., and Smith, C. A. (1971). *Science* **172,** 480.

Elliott, S., and Stanisich, V. A. (1981). Personal communication.

Elton, R. A., Russell, G. J., and Subak-Sharpe, H. (1976). *J. Mol. Evol.* **8,** 117.

Enea, V., and Zinder, N. D. (1975). *Virology* **68,** 105.

Engelberg, H., and Schoulaker, R. (1976). *J. Mol. Biol.* **106,** 709.

Enger, M. D., and Kaesberg, P. (1965). *J. Mol. Biol.* **13,** 260.

Espejo, R. T., and Canelo, E. S. (1968). *Virology* **34,** 738.

Falkow, S., ed. (1975). "Infectious Multiple Drug Resistance." Pion, London.

Fenner, F. (1973). *Search* **4,** 477.

Fiandt, M., Hradecna, Z., Lozeron, H. A., and Szybalski, W. (1971). *In* "The Bacteriophage Lambda" (A. D. Hershey, ed.), pp. 329–354. Cold Spring Harbor Laboratory, Cold Spring Harbor, New York.

Fiddes, J. C., and Godson, G. N. (1979). *J. Mol. Biol.* **133,** 19.

Fiddes, J. C., Barrell, B. G., and Godson, G. N. (1978). *Proc. Natl. Acad. Sci. U.S.A.* **75,** 1081.

Fiers, W., Contreras, R., Duerinck, F., Haegeman, G., Iserentant, D., Merregaert, J., Min Jou, W., Molemans, F., Raeymaekers, A., van den Berghe, A., Volckaert, G., and Ysebaert, M. (1976). *Nature (London)* **260,** 500.

Fitch, W. M., and Margoliash, E. (1967). *Science* **155,** 279.

Fox, G. E., Stackebrandt, E., Hespell, R. B., Gibson, J., Maniloff, J., Dyer, T. A., Wolfe, R. S., Balch, W. E., Tanner, R. S., Magrum, L. J., Zablen, L. B., Blakemore, R., Gupta, R., Bonen, L., Lewis, B. J., Stahl, D. A., Luehrsen, K. R., Chen, K. N., and Woese, C. R. (1980). *Science* **209,** 457.

Frédéricq, P. (1963). *J. Theor. Biol.* **4,** 159.

Freeman, V. J. (1951). *J. Bacteriol.* **61,** 675.

Freeman, V. J., and Morse, I. U. (1952). *J. Bacteriol.* **63,** 407.

Furuse, K., Hirashima, A., Harigai, H., Ando, A., Watanabe, K., Kurosawa, K., Inokuchi, Y., and Watanabe, I. (1979). *Virology* **97,** 328.

Garro, A. J., and Marmur, J. (1970). *J. Cell. Physiol.* **76,** 253.

Geller, A., and Rich, A. (1980). *Nature (London)* **283,** 41.

Gibbs, A. J. (1980). *Intervirology* **14,** 101.

Gilbert, W. (1978). *Nature (London)* **271,** 501.

Gilbert, W. (1979). *ICN-UCLA Symp. Mol. Cell. Biol.* **14,** 1.

Godson, G. N. (1973). *J. Mol. Biol.* **77,** 467.

Godson, G. N. (1974). *Virology* **58,** 272.

Godson, G. N., Barrell, B. G., Staden, R., and Fiddes, J. C. (1978). *Nature (London)* **276,** 236.

Goldstein, R., Lengyel, J., Pruss, G., Barrett, K., Calendar, R., and Six, E. (1974). *Curr. Top. Microbiol. Immunol.* **68,** 59.

Greer, B. S., Hsiang, W., Musil, G., and Zinner, D. D. (1971). *J. Dent. Res.* **50,** 1594.

Griffith, J., and Kornberg, A. (1974). *Virology* **59,** 139.

Grimont, F., and Grimont, P. A. D. (1981). *Curr. Microbiol.* **6,** 65–69.

Gromov, B. V., and Mamkaeva, K. A. (1979). *Izvest. Akad. Nauk S.S.S.R., Ser. Biol.* **2,** 181.

Guay, R., Murthy, M. R. V., and Ackermann, H.-W. (1981). *In* "CRC Handbook of Microbiology" (A. I. Laskin and H. A. Lechevalier, eds.), 2nd Ed., Vol. 3, pp. 731–750. CRC Press, Cleveland, Ohio.

Haas, M., and Yoshikawa, H. (1969). *J. Virol.* **3,** 233.

Hall, T. C. (1979). *Int. Rev. Cytol.* **60,** 1.

Hartl, D., and Dykhuizen, D. (1979). *Nature (London)* **281,** 230.

Hedges, R. W., Jacob, A. E., Barth, P. T., and Grinter, N. J. (1975). *Mol. Gen. Genet.* **141,** 263.

Hemphill, H. E., and Whiteley, H. R. (1975). *Bacteriol. Rev.* **39,** 257.

Hirokawa, H., and Kadlubar, F. (1969). *J. Virol.* **3,** 205.

Hoffmann-Berling, H., Marvin, D. A., and Dürwald, H. (1963). *Z. Naturforsch., Teil B* **18,** 876.

Hoffmann-Berling, H., Kaerner, H. C., and Knipper, R. (1966). *Adv. Virus Res.* **12,** 329.

Holloway, B. W. (1979). *Plasmid* **2,** 1.

Holloway, B. W., and Cooper, G. N. (1962). *J. Bacteriol.* **84,** 1321.

Homma, J. Y., and Shionoya, H. (1967). *Jpn. J. Exp. Med.* **37,** 395.

Homma, J. Y., Goto, S., and Shionoya, H. (1967). *Jpn. J. Exp. Med.* **37,** 373.

Hopfield, J. J. (1974). *Proc. Natl. Acad. Sci. U.S.A.* **71,** 4135.

Huang, W. M., and Marmur, J. (1970). *J. Virol.* **5,** 237.

Hyman, R. W., Brunovskis, I., and Summers, W. C. (1973). *J. Mol. Biol.* **77,** 189.

Hyman, R. W., Brunovskis, I., and Summers, W. C. (1974). *Virology* **57,** 189.

Imaeda, T., and Rieber, M. (1968). *J. Bacteriol.* **96,** 557.

Inoue, M., and Mitsuhashi, S. (1975). *Virology* **68,** 544.

Ionesco, H. (1953). *C.R. Hebd. Seances Acad. Sci., Ser. D (Paris)* **237,** 1794.

Jaspars, S. (1974). *Adv. Virus Res.* **19,** 37.

Jawetz, E., Melnick, J. L., and Adelberg, E. A., eds. (1970). "Medical Microbiology," 9th Ed. Lange, Los Altos, California.

Jayawardene, A., and Farkas-Himsley, H. (1969). *Microbios* **1B,** 87.

Johnson, R. C., Wood, N. B., and Ely, B. (1977). *J. Gen. Virol.* **37,** 323.

Joklik, K. (1974). *Symp. Soc. Gen. Microbiol.* **24,** 293.

Jones, D., and Sneath, P. H. A. (1970). *Bacteriol. Rev.* **34,** 40.

Jones, L. A., Hahn, M. L., and Taylor, A. R. (1974). *J. Virol.* **14,** 493.

Josse, J., Kaiser, A. D., and Kornberg, A. (1961). *J. Biol. Chem.* **236,** 864.

Kageyama, M., Shinomiya, T., Aihara, Y., and Kobayashi, M. (1979). *J. Virol.* **32,** 951.

Kaiser, D., and Dworkin, M. (1975). *Science* **187,** 653.

Kameda, M., Harada, K., Suzuki, M., and Mitsuhashi, S. (1965). *J. Bacteriol.* **90,** 1174.

Kamiunten, H., and Wakimoto, S. (1979). *Ann. Phytopathol. Soc. Jpn.* **45,** 174.

Kamp, D., Kahmann, R., Zipser, D., Broker, T. R., and Chow, L. T. (1978). *Nature (London)* **271,** 577.

Kandler, O. (1981). *Naturwissenschaften* **68,** 183.

Kchromov, I. S., Sorotchkina, V. V., Nigmatullin, T. G., and Tikhonenko, T. I. (1980). *FEBS Lett.* **118,** 51.

Keegstra, W., Godson, G. N., Weisbeek, P. J., and Jansz, H. S. (1979). *Virology* **93,** 527.

Khatoon, H., Iyer, R. V., and Iyer, V. N. (1972). *Virology* **48,** 145.

Kim, J.-S., and Davidson, N. (1974). *Virology* **57,** 93.

Kleckner, N. (1977). *Cell* **11,** 11.

Klipikova, F. S., Belyaeva, N. N., Zhdanov, V. G., and Tikhonenko, A. S. (1978). *Mikrobiologiya* **47,** 312.

Klein, J. P., and Frank, R. M. (1973). *J. Biol. Buccale* **1,** 79.

Kondo, E., and Mitsuhashi, S. (1964). *J. Bacteriol.* **88,** 1266.

Konigsberg, W., Maita, T., Katze, J., and Weber, K. (1970). *Nature (London)* **227,** 271.

Konkel, D. A., Maizel, J. V., and Leder, P. (1979). *Cell* **18,** 865.

Korsten, K. H., Tomkiewicz, C., and Hausmann, R. (1979). *J. Gen. Virol.* **43,** 57.

Krämer, J., and Lenz, W. (1975). *Zentralbl. Bakteriol. Parasitenk. Infektionskr. Hyg. Abt. Orig.* **231,** 421.

Kretschmer, P. J., and Egan, J. B. (1975). *J. Virol.* **16,** 642.

Krieg, A., ed. (1973). "Arthropodenviren." Thieme, Stuttgart.

Krylov, V. N., Bogush, V. G., Yanenko, A. S., and Kirsanov, N. B. (1980). *Genetika* **16,** 975.

Krzywy, T. (1972). *Arch. Immunol. Ther. Exp.* **20,** 621.

Kuo, T.-T., Huang, T.-C., and Chow, T.-Y. (1969). *Virology* **39,** 548.

Kuroda, K., and Kageyama, M. (1979). *J. Biochem. (Tokyo)* **85,** 7.

Kuroda, K., Kageyama, M., Maeda, T., and Fujime, S. (1979). *J. Biochem. (Tokyo)* **85,** 21.

Kushner, D. J. (1969). *Bacteriol. Rev.* **33,** 302.

Lacey, R. W. (1975). *Bacteriol. Rev.* **39,** 1.

Lamb, R. A., and Lai, C.-J. (1980). *Cell* **21,** 475.

Landy, A., and Ross, W. (1977). *Science* **197,** 1147.

Lau, A. H. S., Hawirko, R. Z., and Chow, C. T. (1974). *Can. J. Microbiol.* **20,** 385.

Lee, C. S., Davis, R. W., and Davidson, N. (1970). *J. Mol. Biol.* **48,** 1.

Lemke, P. A. (1976). *Annu. Rev. Microbiol.* **30,** 105.

Lengyel, J. A., Goldstein, R. N., Marsh, M., and Calendar, R. (1974). *Virology* **62,** 161.

Le Roith, D., Shiloach, J., Roth, J., and Leanink, M. (1981). *Proc. Natl. Acad. Sci. U.S.A.* **77,** 6184.

Lin, J.-Y., Tsung, C. M., and Fraenkel-Conrat, H. (1967). *J. Mol. Biol.* **24,** 1.

Lin, J.-Y., Wu, C.-C., and Kue, T.-T. (1971). *Virology* **45,** 38.

Lin, L., Bitner, R., and Edlin, G. (1977). *J. Virol.* **21,** 554.

Lindquist, B. (1974). *Proc. Natl. Acad. Sci. U.S.A.* **71,** 2752.

Liss, A., and Maniloff, J. (1973). *Virology* **55,** 118.

Liss, A., Ackermann, H.-W., Meyer, L. W., and Zierdt, H. (1981). *Intervirology* **15,** 71.

Lopez, R., Ronda, C., Tomasz, A., and Portoles, A. (1977). *J. Virol.* **24,** 201.

Lotz, W. (1976). *Prog. Mol. Subcell. Biol.* **4,** 53.

Lotz, W., and Mayer, F. (1972). *J. Virol.* **9,** 160.

Lotz, W., and Pfister, H. (1975). *J. Virol.* **16,** 725.

Lowe, D. R. (1980). *Nature (London)* **284,** 441.

Luria, S. E., and Darnell, J. E., eds. (1968). *In* "General Virology," 2nd Ed., pp. 439–454. Wiley, New York.

Lwoff, A. (1953). *Bacteriol. Rev.* **17,** 269.

Lwoff, A. (1957). *J. Gen. Microbiol.* **17,** 239.

McCorquodale, D. H. (1975). *CRC Crit. Rev. Microbiol.* **4,** 101.

MacHattie, L. A., and Shapiro, J. A. (1978). *Proc. Natl. Acad. Sci. U.S.A.* **75,** 1490.

Mahler, D. H., and Perlman, P. S. (1979). *ICN-UCLA Symp. Mol. Cell. Biol.* **15,** 11.

Maramorosch, K. (1972). *In* "A Symposium on Ecosystematics" (R. T. Allen and F. C. James, eds.), pp. 194–217. Univ. of Arkansas Museum, Fayetteville.

Margulis, L. (1972). *In* "Exobiology" (C. Ponnamperuma, ed.), pp. 342–386. North-Holland Publ., Amsterdam.

Martin, D. F., and Godson, G. N. (1977). *J. Mol. Biol.* **117,** 321.

Marvin, D. A., and Hohn, B. (1969). *Bacteriol. Rev.* **33,** 172.

Matthews, B. W., Grütter, M. G., Anderson, W. F., and Remington, S. J. (1981). *Nature (London)* **290,** 334.

Matthews, R. E. F. (1978). *Int. Congr. Virol., 4th,* p. 622 (Abstr.). Centre for Agricultural Publishing and Documentation, Wageningen.

Matthews, R. E. F. (1979). *Intervirology* **12,** 129.

Meijer, M., Beck, E., Hansen, F. G., Bergmans, H. E. N., Messer, W., von Meyenburg, K., and Schaller, H. (1979). *Proc. Natl. Acad. Sci. U.S.A.* **76,** 580.

Meynell, G. G., and Lawn, A. M. (1968). *Nature (London)* **217,** 1184.

Mise, K. (1976). *Virology* **71,** 531.

Mise, K., and Arber, W. (1976). *Virology* **69,** 191.

Moazamie, N., Ackermann, H.-W., and Murthy, M. R. V. (1979). *Can. J. Microbiol.* **25,** 1063.

Modak, M. J., and Notani, G. W. (1969). *Experientia* **25,** 1027.

Moore, G. W., and Goodman, M. (1977). *J. Mol. Evol.* **9,** 121.

Mosig, G., Carnighan, J. R., Bibring, J. B., Cole, R., Bock, H.-G. O., and Bock, S. (1972). *J. Virol.* **9,** 857.

Moskovets, S. N., Mendzhul, M. I., Zhigir, V. V., Nesterova, N. V., and Khil, O. S. (1970). *Vopr. Virusol.* **16,** 98.

Mount, D. W. A., Harris, A. W., Fuerst, C. R., and Siminovitch, L. (1968). *Virology* **35,** 134.

Muir, M. D., and Hall, D. O. (1974). *Nature (London)* **252,** 376.

Murialdo, H., and Becker, A. (1978). *Microbiol. Rev.* **42**, 529.

Murooka, Y., and Harada, T. (1979). *Appl. Environ. Microbiol.* **38**, 754.

Nagy, E. (1974). *Acta Microbiol. Acad. Sci. Hung.* **21**, 257.

Nagy, E. (1981). *Int. Congr. Virol., 5th,* p. 282 (Abstr.).

Nagy, E., Prágai, B., and Ivánovics, G. (1976). *J. Gen. Virol.* **32**, 129.

Nahmias, A. J., and Reanney, D. C. (1977). *Annu. Rev. Ecol. Syst.* **8**, 29.

Nakashima, Y., Wiseman, R. L., Konigsberg, W., and Marvin, D. A. (1975). *Nature (London)* **258**, 68.

Nauman, R. K., and Wilkie, E. F. (1974). *J. Virol.* **13**, 1151.

Nishihara, T., Haruna, I., Watanabe, I., Nozu, Y., and Okada, Y. (1969). *Virology* **37**, 153.

Nishihara, T., Nozu, Y., and Okada, Y. (1970). *J. Biochem. (Tokyo)* **67**, 403.

Nishioka, Y., and Leder, P. (1979). *Cell* **18**, 875.

Normore, W. M. (1981). *In* "CRC Handbook of Microbiology" (A. I. Laskin and H. A. Lechevalier, eds.), 2nd Ed., Vol. 3, pp. 529–730. CRC Press, Cleveland, Ohio.

Notani, W. D., Engelhardt, D. L., Konigsberg, W., and Zinder, N. D. (1965). *J. Mol. Biol.* **12**, 260.

Novick, R. (1967). *Virology* **33**, 155.

Ogata, S., Mihara, O., Ikeda, Y., and Hongo, M. (1972). *Agric. Biol. Chem.* **36**, 1413.

Ohsumi, M., Vovis, G. F., and Zinder, N. D. (1978). *Virology* **89**, 438.

Ohtsubo, H., and Ohtsubo, E. (1978). *Proc. Natl. Acad. Sci. U.S.A.* **75**, 615.

Okamoto, K., Mudd, J. A., Mangan, J., Huang, W. M., Subbaiah, T. V., and Marmur, J. (1968). *J. Mol. Biol.* **34**, 413.

Olsen, R. H., and Shipley, P. (1973). *J. Bacteriol.* **113**, 772.

Orgel, L. E., and Crick, F. (1980). *Nature (London)* **284**, 604.

Ortel, S. (1965). *Zentralbl. Bakteriol. Parasitenk. Infektionskr. Hyg. Abt. Orig.* **198**, 423.

Overby, L. R., Barlow, G. H., Doi, R. H., Jacob, M., and Spiegelman, S. (1966). *J. Bacteriol.* **92**, 739.

Pate, J. L., Petzold, S. J., and Umbreit, T. H. (1979). *Virology* **94**, 24.

Pemberton, J. M., and Tucker, W. T. (1977). *Nature (London)* **266**, 50.

Piffaretti, J.-C., and Pitton, J.-S. (1976). *J. Virol.* **20**, 314.

Poljak, R. J., and Suruda, A. J. (1969). *Virology* **39**, 145.

Pollock, T. J., Tessman, I., and Tessman, E. S. (1978). *Nature (London)* **274**, 34.

Porter, R. D., Shoemaker, R. B., Rampe, G., and Guild, W. R. (1979). *J. Bacteriol.* **137**, 556.

Pruss, G., Goldstein, R. N., and Calendar, R. (1974). *Proc. Natl. Acad. Sci. U.S.A.* **71**, 2367.

Rautenstein, Ya. I., Solovieva, N. Ya., Moskalenko, L. N., and Filatova, A. D. (1971). *Mikrobiologiya* **40**, 1094.

Ray, D. S. (1977). *In* "Comprehensive Virology" (H. Fraenkel-Conrat and R. R. Wagner, eds.), Vol. 7, pp. 105–178. Plenum, New York.

Reanney, D. C. (1974). *J. Theor. Biol.* **48**, 243.

Reanney, D. C. (1976a). *In* "Yearbook of Science and Technology" (D. N. Lapedes, S. P. Parker, and J. Weil, eds.), pp. 243–245. McGraw-Hill, New York.

Reanney, D. C. (1976b). *Bacteriol. Rev.* **40**, 552.

Reanney, D. C. (1977). *Brookhaven Symp. Biol.* **29**, 248.

Reanney, D. C. (1978a). *ICN-UCLA Symp. Mol. Cell. Biol.* **11**, 311.

Reanney, D. C. (1978b). *Int. Rev. Cytol. (Suppl.).* **8**, 1.

Reanney, D. C. (1979). *Nature (London)* **277**, 598.

Reanney, D. C. (1981). *In* "The Human Herpesviruses" (A. J. Nahmias, W. R. Dowdle, and R. Schinazi, eds.), pp. 519–536. Elsevier, Amsterdam.

Reanney, D. C., and Ackermann, H.-W. (1981). *Intervirology* **15**, 190.

Reilly, B. E., Nelson, R. A., and Anderson, D. L. (1977). *J. Virol.* **24**, 363.

Richmond, M. H., and Wiedeman, B. (1974). *Symp. Soc. Gen. Microbiol.* **24**, 59.

Roberts, R. J. (1976). *CRC Crit. Rev. Biochem.* **4**, 123.

Robinson, J. W. (1972). *Biochem. J.* **128**, 481.

Rudinski, M. S., and Dean, D. H. (1979). *Virology* **99**, 57.

Rutberg, L., Armentrout, R. W.., and Jonasson, J. (1972). *J. Virol.* **9**, 732.

Sakaki, Y., and Oshima, T. (1976). *Virology* **75**, 256.

Sakaki, Y., Yamada, K., Oshima, M., and Oshima, T. (1977). *J. Biochem. (Tokyo)* **82**, 1451.

Salivar, W. O., Henry, T. J., and Pratt, D. (1967). *Virology* **32**, 41.

Sandoval, H. K., Reilly, H. C., and Tandler, B. (1965). *Nature (London)* **205**, 522.

Sanger, F., Air, G. M., Barrell, B. G., Brown, N. L., Coulson, A. R., Fiddes, J. C., Hutchison, C. A. III, Slocombe, P. A., and Smith, M. (1977). *Nature (London)* **265**, 687.

Savage, D. C. (1977). *Annu. Rev. Microbiol.* **31**, 107.

Saxelin, M.-L., Nurmiaho, E.-L., Korshola, M. P., and Sundman, V. (1979). *Can. J. Microbiol.* **25**, 1182.

Schäfer, R., Huber, U., and Franklin, R. M. (1977). *Eur. J. Biochem.* **73**, 239.

Schekman, R., Weiner, J. H., Weiner, A., and Kornberg, A. (1975). *J. Biol. Chem.* **250**, 5859.

Schell, J., and van Montagu, M. (1977). *Brookhaven Symp. Biol.* **29**, 36.

Schildkraut, C. L., Wierzchowski, K. L., Marmur, J., Green, D. M., and Doty, P. (1962). *Virology* **18**, 43.

Scholl, D. R., and Jollick, J. D. (1980). *J. Virol.* **35**, 949.

Schopf, W. (1972). *In* "Exobiology" (C. Ponnamperuma, ed.), pp. 16–61. North-Holland Publ., Amsterdam.

Schopf, J. W. (1978). *Sci. Am.* **239**, 110.

Schreier, M. H., Staehelin, T., Gesteland, R. F., and Spahr, P. F. (1973). *J. Mol. Biol.* **75**, 575.

Scott, J. R. (1973). *Virology* **53**, 327.

Scott, J. R., and Zinder, N. D. (1967). *In* "The Molecular Biology of Viruses" (J. S. Colter and W. Paranchych, eds.), pp. 211–218. Academic Press, New York.

Seaman, E., Tarmy, E., and Marmur, J. (1964). *Biochemistry* **3**, 607.

Sharp, P. A. (1981). *Cell* **23**, 643.

Shinomiya, T., and Shiga, S. (1979). *J. Virol.* **32**, 958.

Sik, T., and Orosz, L. (1971). *Plant Soil* (special vol.), p. 57.

Simon, M. N., and Davidson, N. (1969). *Fed. Proc., Fed. Am. Soc. Exp. Biol.* **28**, 531.

Simon, M. N., Davis, R., and Davidson, N. (1971). *In* "The Bacteriophage Lambda" (A. D. Hershey, ed.), pp. 313–328. Cold Spring Harbor Laboratory, Cold Spring Harbor, New York.

Sinsheimer, R. L. (1968). *Prog. Nucleic Acid Res. Mol. Biol.* **8**, 115.

Sirgel, F. A., Coetzee, J. N., Hedges, R. W., and Lecatsas, G. (1981). *J. Gen. Microbiol.* **122**, 155.

Sladkova, I. A., Chinenova, T. A., Lomovskaya, N. D., and Mkrtumyan, N. M. (1979). *Genetika* (transl.) **15**, 1305.

Smit, J. A., Hugo, N., and de Klerk, H. C. (1969). *J. Gen. Virol.* **5**, 33.

Smith, H. W. (1972). *Nature (London) New Biol.* **238**, 205.

Smith, R. A., and Parkinson, J. S. (1980). *Proc. Natl. Acad. Sci. U.S.A.* **77**, 5370.

Snell, D. T., and Offord, R. E. (1972). *Biochem. J.* **127**, 167.

Sonea, S., and Panisset, M., eds. (1980). "Introduction à la nouvelle bactériologie." Press of Univ. of Montreal, Montreal.

Spiegelman, S. (1970). *In* "The Neurosciences" (G. C. Quarton, T. Melnechuk, and Adelman, G., eds.), pp. 927–945. Rockefeller Univ. Press, New York.

Steensma, H. Y., and Robertson, L. A. (1978). *FEMS Microbiol. Lett.* **3**, 313.

Steensma, H. Y., Robertson, L. A., and van Elsas, J. D. (1978). *Antonie van Leeuwenhoek* **44**, 353.

Steven, A., and Wurtz, M. (1977). *Recherche* **8**, 26.

Stiube, P., and Dimitriu, C. (1969). *Arch. Roum. Pathol. Exp. Microbiol.* **28**, 809.

Streisinger, G., Emrich, J., and Stahl, M. M. (1967). *Proc. Natl. Acad. Sci. U.S.A.* **57**, 292.

Studier, F. W. (1979). *Virology* **95**, 70.

Subak-Sharpe, H., Burk, R. R., Crawford, L. W., Morrison, J. M., Hay, J., and Keir, H. M. (1966). *Cold Spring Harbor Symp. Quant. Biol.* **31**, 737.

Subak-Sharpe, H., Elton, R. A., and Russell, G. J. (1974). *Symp. Soc. Gen. Microbiol.* **24**, 131.

Sun, S. M., Slightom, J. L., and Hall, T. C. (1981). *Nature (London)* **289**, 37.

Susskind, M. M., and Botstein, D. (1978). *Microbiol. Rev.* **42**, 385.

Szybalski, W., and Szybalski, E. H. (1974). *In* "Viruses, Evolution and Cancer" (E. Kurstak and K. Maramorosch, eds.), pp. 563–580. Academic Press, New York.

Takano, T., and Ikeda, S. (1976). *Virology* **70**, 198.

Takeya, K., and Amako, K. (1966). *Virology* **28**, 163.

Takeya, K., Minamishima, Y., Ohnishi, Y., and Amako, K. (1969). *J. Gen. Virol.* **4**, 145.

Thomas, C. A., and MacHattie, L. A. (1967). *Annu. Rev. Biochem.* **29**, 485.

Thurm, P., and Garro, A. J. (1975). *J. Virol.* **16**, 184.

Tikhonenko, A. S., and Bespalova, I. A. (1961). *Mikrobiologiya* **30**, 867.

Tikhonenko, A. S., and Zavarzina, N. B. (1966). *Mikrobiologiya* **35**, 848.

Tikhonenko, A. S., Bespalova, I. A., and Belyaeva, N. N. (1971). *Mikrobiologiya* **40**, 128.

Tikhonenko, T. I. (1978). *In* "Comprehensive Virology" (H. Fraenkel-Conrat and R. R. Wagner, eds.), Vol. 12, pp. 235–269. Plenum, New York.

Tikhonenko, T. I., Velikodvorskaya, G. A., Bobkova, A. F., Bartoshevich, Yu. E., Lebed, E. P., Chaplygina, N. M., and Maksimova, T. S. (1974). *Nature (London)* **249**, 454.

Tohyama, K., Sakurai, T., Arai, H., and Oda, A. (1972). *Jpn. J. Microbiol.* **16**, 385.

Tomashow, M. F., Nutter, R., Montoya, A. L., Gordon, M. P., and Nester, E. W. (1980). *Cell* **19**, 729.

Torsvik, T., and Dundas, I. D. (1974). *Nature (London)* **248**, 680.

Toussaint, A., Lefebvre, N., Scott, J. R., Cowan, J. A., de Bruijn, F., and Bukhari, A. I. (1978). *Virology* **89**, 146.

Truffaut, N., Revet, B., and Soulie, M.-O. (1970). *Eur. J. Biochem.* **15**, 391.

Tschopp, J., and Smith, P. R. (1977). *J. Mol. Biol.* **114**, 281.

Tucker, W. T., and Pemberton, J. M. (1978). *J. Bacteriol.* **135**, 207.

Tucker, W. T., and Pemberton, J. M. (1980). *J. Bacteriol.* **143**, 43.

Uhlenhopp, E. L., Zimm, B. H., and Cummings, D. J. (1974). *J. Mol. Biol.* **89**, 689.

Van Wezenbeek, P. M. G. F., Hulsebos, T. J. M., and Schoenmakers, J. G. G. (1980). *Gene* **11**, 129.

Vervliet, G., Holsters, M., Teuchy, H., van Montagu, M., and Schell, J. (1975). *J. Gen. Virol.* **26**, 33.

Vidaver, A. K., Koski, R. K., and van Etten, J. L. (1973). *J. Virol.* **11**, 799.

Vieu, J.-F., Croissant, O., and Hamon, Y. (1967). *C.R. Hebd. Seances Acad. Sci., Ser. D (Paris)* **264**, 181.

Wahba, J. J., Miller, M. J., Niveleau, A., Landers, T. A., Carmichael, G. G., Weber, K., Hawley, D. A., and Slobin, L. I. (1974). *J. Biol. Chem.* **249**, 3314.

Wais, A. C., Kon, M., MacDonald, R. E., and Stollar, B. D. (1975). *Nature (London)* **256,** 134.

Wakimoto, S., Katsuki, Y., Matsuo, N., and Kamiunten, H. (1979). *Ann. Phytopathol. Soc. Jpn.* **45,** 168.

Walker, D. H., and Anderson, T. F. (1970). *J. Virol.* **5,** 765.

Walter, M. R. (1977). *Am. Sci.* **65,** 563.

Walter, M. R., Buick, R., and Dunlop, J. S. R. (1980). *Nature (London)* **284,** 443.

Wandersman, C., and Yarmolinsky, M. (1977). *Virology* **77,** 386.

Wang, J. C., Martin, K. V., and Calendar, R. (1973). *Biochemistry* **12,** 2119.

Warren, R. A. J. (1980). *Annu. Rev. Microbiol.* **34,** 137.

Weber, K., and Konigsberg, W. (1967). *J. Biol. Chem.* **242,** 3563.

Willetts, N., and Skurray, R. (1980). *Annu. Rev. Genet.* **14,** 41.

Wimmer, E., ed. (1979). "The Molecular Biology of Picornaviruses," pp. 175–190. Plenum, New York.

Wiseman, R. L., Dunker, A. K., and Marvin, D. A. (1972). *Virology* **48,** 230.

Wittman-Liebold, B., and Wittman, H. G. (1967). *Mol. Gen. Genet.* **100,** 358.

Wood, W. B., and King, J. (1979). *In* "Comprehensive Virology" (H. Fraenkel-Conrat and R. R. Wagner, eds.), Vol. 13, pp. 581–633. Plenum, New York.

Yasunaka, K. (1977). *Fukuoka Acta Med.* **68,** 425.

Yokokura, T., Kodaira, S., Ishiwa, H., and Sakurai, T. (1974). *J. Gen. Microbiol.* **84,** 277.

Younghusband, H. B., and Inman, R. B. (1974). *Virology* **62,** 530.

Yun, T., and Vapnek, D. (1977). *Virology* **77,** 376.

Zechel, K., Bouche, J. P., and Kornberg, A. (1975). *J. Biol. Chem.* **250,** 4684.

Zinder, N. D. (1980). *Intervirology* **13,** 257.

ADVANCES IN VIRUS RESEARCH, VOL. 27

MECHANISMS OF VIRAL TUMORIGENESIS

Raymond V. Gilden and Harvey Rabin

Biological Carcinogenesis Program
NCI-Frederick Cancer Research Facility
Frederick, Maryland

I. INTRODUCTION

The mechanism or mechanisms by which viruses cause tumors represents a complex *in vivo* problem. Clues to what are presumably critical events in this process have been sought for many years in tissue culture using what have come to be known as acute transforming viruses. The interactions of viruses such as simian virus 40 (SV40) and Rous sarcoma virus (RSV) with sensitive target cells have shown that integration into host cell DNA of specific viral genes is essential for initiation of cellular transformation. However, most viruses with known oncogenic potential either lack genes which code directly for products with transforming function or, like Epstein–Barr virus (EBV), exhibit the capacity to immortalize certain cell types without actually inducing malignant changes. Thus, for these so-called subacute viruses, analyses *in vitro* of their interactions with cellular targets are limited in so far as their applicability to cell transformation is concerned. By combining findings relative to transformation *in vitro* with observations on experimental tumors, exciting new findings are emerging which suggest that some subacute transforming avian retroviruses may induce tumors via a mechanism not unlike that ascribed to the acutely transforming viruses, but using a cellular gene.

281

For this reason, immediately after a brief introduction to the retroviruses, we begin with a review of acute transforming viruses, thereby introducing the newer concepts and terminology relative to transformation early in the chapter. We then proceed to a discussion of the subacute oncogenic retroviruses and broaden our discussion to include both molecular mechanisms based on what has been learned from acute transforming viruses and alternate models where the virus may act as an initiator of transformation via immunological or other mechanisms. We conclude the first part of the review with a discussion of cellular transforming genes and their products as a possible unifying mechanism for carcinogenesis over a wide range of carcinogens, viral and nonviral.

The lymphotropic herpesviruses are associated not only with experimental disease, but also with several forms of human disease. In man, EBV is associated with monoclonal tumors such as Burkitt's lymphoma which arise long after initial infection and also with acute primary diseases such as infectious mononucleosis and immunoblastic sarcoma. To a large extent, the outcome of EBV infection and that of its counterpart viruses in other species may be a result of the functional integrity of the host's immune system. Recently described conditions, such as the X-linked lymphoproliferative syndrome, may serve to increase our awareness of the role played by EBV in the etiology of lymphoproliferative diseases. It is not yet possible to construct molecular models by which EBV and EBV-like viruses may immortalize cells, but the genomes of these large DNA viruses have been physically mapped, several viral proteins have been isolated and characterized, and we can anticipate the mapping of several viral genes in the relatively near future.

Finally, we return to the intriguing examples of both experimental and naturally occurring tumors or transformed cells which have clear associations with certain viruses, but which fail to yield evidence for retention of viral genes or gene products. Such tumors raise the possibilities of transient virus–cell interactions which nevertheless are sufficient for tumorigenesis.

II. RNA Tumor Viruses

During the last decade, information on the various aspects of RNA tumor virus biology has been forthcoming at an ever-increasing rate. Reviews on the numerous characteristics of these viruses appear at regular intervals and, as such, we will select only those areas which most clearly relate to the question of oncogenesis for detailed consideration.

It is now recognized that the family Retroviridae includes a large number of unique agents with diverse pathogenic potential united by their common feature of reproduction via a DNA intermediate which integrates in the host cell genome. The family includes members which spread horizontally within a species, termed exogenous, and also those that are inherited within the cell genome, termed endogenous. Activation of endogenous viral genomes by a variety of chemical and physical treatments, often eventuating in the isolation of oncogenic agents, has provided the basis for theories of carcinogenesis involving an interplay of factors resembling the multihit nature of most malignant diseases in man (Bentvelzen, 1972; Heubner and Todaro, 1969; Todaro and Heubner, 1972). More recently, it has been established that the oncogenic potential of acute acting RNA tumor viruses has arisen by recombination of helper viruses and genes present in the host cell genome (reviewed by Bishop, 1981). The helper viruses contain the genes for the virion structural components and the rescued cellular genes presumably code for transformation-related proteins (Stephenson, 1980; reviewed by Hayman, 1981). The latter have been described for several transforming viruses and, at present, the evidence suggests that they are membrane proteins. As may be predicted from the presence of these transforming genes in the normal cell, low amounts of the "transforming protein" have also been found in uninfected cells (Collett et al., 1978; Brugge et al., 1979; Oppermann et al., 1979), most notably in the case of the avian Rous sarcoma virus, designated pp60^{src} (a phosphoprotein of 60,000 MW produced from a sarcoma-specific gene) (Brugge and Erickson, 1977; Collett and Erickson, 1978). Thus, current interest centers on a complete definition of the kinase–substrate system(s) with the hope that it will clearly define the important parameters in malignant transformation (reviewed by Hunter, 1980). Along with the concept that the virus adds a transforming gene to newly infected cells, other mechanisms of virus involvement in cancer induction have also been suggested. As discussed below, these mechanisms involve, among other possibilities, the insertion of viral promotor sequences at crucial sites in the cell genome and an immunological means for the expansion of target cell populations.

The presence of viral genes in host cells has suggested that these should serve a normal physiological function. Many studies have emphasized the evolutionary divergence patterns of certain classes of endogenous viruses suggesting long-term residence in the host cell genome (reviewed by Todaro, 1980). While evidence of the interspecies transfer of retroviruses has now been obtained, until recently this was considered to be a historically remote event measured in millions of years. There are now increasing examples of the variability of integration sites of endogenous viruses within a species, as dramatized by the ability to employ selective breeding for the derivation of a line of chickens which lack any evidence of endogenous viral genomes (Astrin *et al.*, 1980). When such animals are obtained in quantity, one should be able to evaluate whether or not the endogenous viruses play any role in development (it is already clear that they are not essential for production of viable individuals) and oncogenesis or simply carry out one phase of a complex life cycle which permits a safe parasitic existence. Recent evidence that the integrated proviral DNA has a structure similar to that of bacterial transposable elements (Dhar *et al.*, 1980; Sutcliffe *et al.*, 1980; Shimotono *et al.*, 1980; Ju and Skalka, 1980) clearly raises the possibility of their independent existence and we have speculated (Gilden *et al.*, 1981) that this is perhaps the prime example of "selfish" DNA (Doolittle and Sapienza, 1980; Orgel and Crick, 1980).

Theories suggesting normal function for virus structural genes, "virogenes," would seem to require their presence in an orderly fashion in all members of a species and in all species. This is clearly not the case (Astrin *et al.*, 1980; reviewed by Weinberg, 1980), as there is no compelling evidence for any natural function, although their involvement in oncogenesis is established. The ability of endogenous viruses to recombine with either exogenous agents or "transforming genes" may lead to the generation of transmissible (both intra- and interspecies) oncogenic agents with pathogenic significance for the host species. Such events, by analogy with the generation of virulent influenza viruses via recombination or reassortment, should not be dismissed lightly. To date, the clearest evidence for transspecies creation of transforming viruses has come from laboratory experiments; however, the single isolate of a sarcomagenic virus from a Woolly monkey (Theilen *et al.*, 1971) where the virogenes are clearly related to a group of viruses isolated from gibbons (Todaro *et al.*, 1975) and the transforming gene, "oncogene," was derived from the Woolly monkey (Wong-Staal *et al.*, 1981), does offer one possible example of a naturally occurring recombination. Recent evidence also suggests that reticuloendotheliosis virus, an acute avian transforming virus, is a recombinant between an avian transforming gene (Simek and Rice, 1980) and a primate-related helper virus (Oroszlan and Gilden, 1980; Oroszlan *et al.*, 1981a; Rice *et al.*, 1981). In addition, most naturally occurring subacute oncogenic RNA viruses are not endogenous (i.e., not represented in normal cellular

DNA of their current host), thus leading to the speculation that they have arisen in another species (Gilden, 1977).

In attempting this review, one knows that some portions will be obsolete prior to publication, thus giving striking testimony to the capabilities of the contemporary techniques in biology. While the impact of the results of research on these viruses to human disease has yet to be felt, the confidence that this will occur is certainly increasing. In fact, Gallo and colleagues (Poiesz *et al.*, 1980a, 1981; Rho *et al.*, 1981; Kalyanaraman *et al.*, 1981a) have recently described the isolation of a unique retrovirus from a subclass of human adult T-cell lymphomas. The virus, designated HTLV, is clearly exogenous and early serologic surveys have yielded results consistent with it being the etiologic agent of these lymphomas (Kalyanaraman *et al.*, 1981b; Posner *et al.*, 1981). HTLV appears to be distantly related to the bovine leukemia virus based on protein-sequencing studies (Oroszlan *et al.*, 1982) and, in fact, there are many similarities in the biology of these viruses.

Recently, another laboratory (Miyoshi *et al.*, 1981) described a retrovirus in cultures of an adult T-cell leukemia. Cocultivation of these cells from a female patient with human cord leukemia of male origin yielded a virus-positive established male cell line with T-cell markers. This line did not contain EBV nuclear antigen. The serologic data reported were very similar, although using different methodology, to those of Gallo and colleagues, suggesting that a similar agent has been isolated in both laboratories. The observations of potential transformation are noteworthy and clearly deserve repetition and extensive analysis.

A. *Acute Transforming Viruses—src Gene and Analogous Transforming Genes*

There are approximately one dozen distinctive retroviruses which represent five species sharing the property of directly transforming cell cultures *in vitro* and inducing tumors rapidly *in vivo*. These include three distinct sarcoma viruses (distinct *onc* genes) of birds (Rous, Fujinami, and Y-73 prototype strains), four acute leukemia viruses of birds (avian myeloblastosis, erythroblastosis, myelocytomatosis, and reticuloendotheliosis), three sarcoma viruses with related rat transforming genes (Harvey, Kirsten, and Rasheed), several mouse sarcoma viruses (Moloney, FBJ, AT-8), three characterized sarcoma viruses of cats (Gardner–Arnstein, Snyder–Theilen, McDonough), and one sarcomagenic isolate from a Woolly monkey. A number of these transforming genomes have now been molecularly cloned, the transforming gene identified by molecular hybridization and transfection experiments, and the product of the gene characterized. The use of transformation-defective mutants for biochemical mapping studies has produced further evidence that the transforming function resides within these genes, variously designated as *src* for the Rous sarcoma virus or *onc* as a more general term (for a detailed nomenclature, see Coffin *et al.*, 1981). The following generalizations apply to these viruses, although with certain exceptions: the viral *src* (*v-src*) has a normal cell equivalent (designated *sarc* or more recently *c-src*) which is present in low copy number and highly conserved in an evolutionary sense. For example, both avian (Stehelin *et al.*, 1976) and mammalian DNA contain sequences which hybridize to a cDNA copy of the RSV *src* gene (Spector *et al.*, 1978a) and the product of the rat *c-src* gene, p21, can be readily detected in many vertebrate cells (Langbeheim *et al.*, 1980). The normal *c-src* gene homologs of several transforming viruses have been isolated by cloning procedures. These do not transform cells directly; however, when mixed or ligated with that portion of the helper virus containing promotor-like sequences, they are biologically active (Blair *et al.*, 1981). Thus, there can be no doubt concerning the transforming potential of these normal genes in all cellular DNA. There is also variability in the structure of *sarc* genes; for example, the MSV *sarc* contains no introns (Oskarsson *et al.*, 1980), while in the rat, two variants of the Harvey *c-src* gene have been isolated, one with three putative introns and one with none (DeFeo *et al.*, 1981). Biological activity in this case has been shown for both genes. Intervening sequences have also been noted in the cat cell homolog (*c-fes*) of feline sarcoma virus (Franchini *et al.*, 1981) and the latter publication refers to similar observations for RSV, AEV, and SSV (see also Favera *et al.*, 1981). The Abelson virus *c-src* gene also has been shown to contain intervening sequences (Goff *et al.*, 1980). The complete DNA sequence of the RSV *src* gene has been recently published (Czernilofsky *et al.*, 1980); the deduced amino acid sequence now provides opportunity for synthesis of specific regions which may be useful in the study of direct effects on cultured cells. Given the occurrence of a small number of *onc*-containing viruses, there is an obvious question of the number of corresponding *c-oncs* and their interrelationships. These questions are still open, awaiting detailed sequence analysis. The recent findings of cross-hybridization between strains of feline sarcoma virus and a distinctive avian sarcoma virus (Fujinami strain) (Shibuya *et al.*, 1980) and sequence homology between RSV and MSV *v-srcs* (Van Beveren *et al.*, 1981) suggest a finite number. While the majority of viruses clearly carry one gene of cellular origin, avian erythroblastosis virus (AEV) possibly contains two or has a gene with two separable domains. This acute acting virus transforms both fibroblasts and erythroid precursors. The virus contains a large cellular insert (Lai *et al.*, 1979), codes for two distinctive proteins (Lai *et al.*, 1980), and transformation-defective mutants are available which have lost the capability to transform one cell type and not the other (Beug *et al.*, 1980). In one such case (Beug *et al.*, 1980), loss of erythroblast transformation was found to correlate with a deletion in one of the proteins. Clearly the one approach to determination of potential involvement of the current array of *onc* genes in human DNA requires isolation by cloning and preparation of specific probes and hybridization to cell mRNA. Such studies are now in progress in several laboratories (Eva *et al.*, 1981; Westin *et al.*, 1981).

The product of the RSV *src* gene has been characterized in detail and found to be a phosphoprotein of 60,000 MW designated pp60^{src} (Brugge *et al.*, 1978a,b). This protein also possesses kinase activity which has the distinctive property of phosphorylating at tyrosine residues (Collett and Erickson, 1978; Hunter and Sefton, 1980; Collett *et al.*, 1980). This property is shared by other products of acute transforming retroviruses, including those of the Abelson virus of mice (Witte *et al.*, 1980a) and the Fujinami sarcoma virus of chickens (Feldman *et al.*, 1980). In these cases, transformation-defective mutants also are deficient in kinase activity (Collett *et al.*, 1979; Witte *et al.*, 1980b; Pawson *et al.*, 1980). Among the group, the Harvey (HaSV) and Kirsten (KiSV) *src* product, p21, possesses binding activity for guanine nucleotides (Scolnick *et al.*, 1979). A unique feature of the acute viruses is that the majority are replication-defective, requiring a helper virus for progeny formation but not transformation. In the majority of the cases, the cellular insert is in the *gag* region of the genome as opposed to the situation with RSV where *src* occurs 3′ in the RNA genome to the known structural genes, thus allowing replication competence. The protein product of the majority of the defective acute transforming viruses appears to be initially a fusion product containing a variable portion of the *gag* gene product (Hayman, 1981). This may provide a membrane anchor function and thus, in some cases, be crucial for the transformation effect. In contrast to RSV pp60^{src}, the product of the rat HaSV and KiSV *src* genes, p21, does not appear to have a *gag*-derived NH$_2$ terminus (Shih *et al.*, 1979a,b), while a separate rat-transforming virus, the Rasheed strain, does yield a p29 which contains NH$_2$-terminal sequences of the rat helper virus *gag* product (Young *et al.*, 1979, 1981). This protein is cross-reactive with p21s from the other viruses. The history and structure of the defective Harvey and Kirsten genomes indicate a complex involvement of mouse helper virus, sequences derived from a retrovirus-like element in normal rat cells

(30 S) and the rat c-src gene (Shih et al., 1978; Ellis et al., 1980; Young et al., 1980). It seems possible that the 30 S sequences 5' to the src gene might provide an anchor function and, thus, p21 might still be a fusion product of a 30 S product and the rat src gene. These possibilities are amenable to direct experimental test as 30 S, cellular c-src, and viral src are available for sequence analysis.

Irrespective of their origin as fusion products or free of helper virus contribution, the src products appear to localize in the cell membrane (Hynes, 1980); Abelson src has at least one region exposed to the external environment (Witte et al., 1979), whereas both rat p21 (Willingham et al., 1980) and RSV pp60^{src} (Willingham et al., 1979) have been localized to the inner surface of the cell membrane. These conclusions are limited at present by the available reagents and more refinement can be expected. The normal c-src counterparts have not been studied as extensively but at least for pp60^{c-src}, convincing evidence has been presented for a similar membrane location.

Given the generality of membrane localization, a key role in the multiple membrane alterations suggested for transformation seems logical. Thus, extensive analysis of src (v and c) function as well as interactions with membrane-localized molecules are being made. A number of src products appear to possess protein kinase activity, the common striking characteristic being phosphorylation at tyrosine residues. This property is also shared with the middle T antigen of polyoma virus (Eckhart et al., 1979) which plays a key role in transformation by that DNA virus. The potential substrates for pp60^{src} kinase have been studied by several laboratories. As reviewed by Hunter (1980) and Hynes (1980), there are highly suggestive interactions with elements of the cytoskeleton also including some proteins with undetermined function. The generalizations emerging are thus a spectrum of activities related to function of the cytoskeletal elements and perhaps second messenger functions. These clearly are within the predicted sites of action of transformation-specific products. The expectation should not be for a single common function for all the transforming genes; rather one might expect them to represent key factors in a multicomponent system controlling growth and differentiation. The process of transformation thus results from a noncontrolled, overproduction of one of a number of factors, the gene for which has been usurped from its normal position in the cell genome. While current data do not allow a precise delineation of the uniqueness of each of the known oncs, within the chicken there are at least seven known distinct genes which, even if distantly related, may be assumed to perform different functions.

The known *onc* genes and the discovery of normal counterparts have provided important new insights and reagents for a variety of studies in cell biology and carcinogenesis research. Currently there is no evidence of a regular direct involvement of any of these genes in naturally occurring malignancies; in fact, to date the evidence is to the contrary (however, see Section II,C). Such experiments have generally consisted of examination of spontaneous or nonviral induced tumors for elevated levels of *onc*-specific RNA (Fischinger, 1980). This indicates that either these genes are irrelevant or, more logically, they represent a minimal sample of a much larger potential population of genes. Given the rare occurrence of natural isolates of acute transforming viruses and the recent evidence of a highly conserved normal cellular function for *c-src* genes, we presume that such genes must be relatively well "protected." It should also be emphasized that a complete inventory of human homologs of known transforming genes/products is not yet available and, thus, a meaningful evaluation of their potential role in human malignancy has not yet been made.

B. Acute Transforming Viruses —env Gene in Recombinant and Defective Viruses

The *onc* gene-containing viruses provide relatively straightforward conceptual and technical approaches to a detailed understanding of viral transformation. Yet, the attention given these viruses should not overshadow the generality that most naturally occurring virus-related malignancies do not involve these viruses, but rather are associated with those with no direct transforming potential and no demonstrable *onc*-like genes. These are discussed in Section II, C. In this section we will consider an intermediate virus, one which acts in acute fashion *in vivo* but does not appear to transform cells *in vitro* nor carry cellular genes. The prototype of this group is the Friend virus of mice whose target cell *in vivo* is an erythroid precursor. The history and molecular biology of Friend virus have recently been reviewed (Troxler *et al.,* 1980) and we will summarize major recent findings here. There are stable variants of this virus (MacDonald *et al.,* 1980) which induce either anemia (FV-A) (Friend, 1957) or polycythemia (FV-P) (Mirand *et al.,* 1968). These viruses are distinguished from other murine leukemia viruses on the basis of short latent periods (10–30 days for disease induction), pathogenicity for both adult and newborn mice, lack of involvement of lymph nodes or thymus, and their specific effects on erythroid precursors. Virus preparations (Friend virus complex) contain a replication-defective component designated spleen focus-forming virus (SFFV) (Axelrod and Steeves, 1964) quantitated by its ability to induce macroscopic foci of transformed cells in adult mouse spleens and a replication-competent virus (F-MuLV) which produces erythro-leukemia only in newborn mice (Troxler and Scolnick, 1978). As SFFV is a defective virus whose genome can be rescued by other helper MuLVs, it is not surprising to find multiple helper viruses in animal passaged preparations, some of which are nonpathogenic or induce lymphoid leukemia (Steeves *et al.*, 1971). The availability of cloned SFFV and F-MuLV makes it possible to refer to the specific entity under discussion. SFFV has been isolated free of helper in tissue culture despite the lack of transforming potential by cloning cells immediately after infection using a stock of virus with equivalent helper and SFFV titers (Troxler *et al.*, 1977). Clonal lines would thus have F-MuLV alone, F-MuLV plus SFFV, neither virus, or SFFV alone. Spleen focus-forming virus-containing cells free of helper were detected by addition of F-MuLV and inoculation of adult mice in which F-MuLV alone has no effect. In this manner, even in the absence of morphological changes *in vitro,* a pathogenic genome could be identified. This gave formal proof that SFFV was replication-defective. The same procedure of assay was used for analysis of molecularly cloned SFFV which has resulted in isolation of a focus-forming genome integrated into the plasmid pBR322 (Linemeyer *et al.*, 1980). Similarly, a subgenomic fragment of F-MuLV possessing erythroleukemia-inducing potential has also been molecularly cloned (Oliff *et al.*, 1980). Thus, it is now possible to specify the transforming gene of the two viruses and its product. Different SFFV isolates code for a variable portion of the retrovirus *gag* gene, but all yield a partial *env* gene product of 52,000 to 55,000 daltons (Ruscetti *et al.*, 1979; Dresler *et al.*, 1979; MacDonald *et al.*, 1980). The p15(E) component of the *env* gene primary product is not found in nonproducer cells containing SFFV, so that a deletion in the 3' region of *env* is indicated (Schultz *et al.*, 1980). The subgenomic active fraction of F-MuLV (Oliff *et al.*, 1980) also contains, in a 4.1 kb fragment, *env*, the terminal repeat and the C region. Spleen focus-forming virus gp52 appears to have determinants cross-reactive with gp70 of MCF viruses (Ruscetti *et al.*, 1979), a class of MuLVs which appear to have arisen by recombination in the *env* gene between xenotropic and ecotropic MuLVs. This is consistent with finding of cross-hybridization between SFFV and xenotropic MuLVs (Troxler *et al.*, 1977). Spleen focus-forming virus gp52, while cross-reactive with MCF gp70s, can also be distinguished from these products by loss of certain group and interspecies determinants (Ruscetti *et al.*, 1979). The possibility of origin of SFFV by recombination between F-MuLV and xeno-MuLV in the *env* gene region is thus raised. Gp52 is not incorporated into virions, appears at the cell surface of all SFFV nonproducer lines, and is produced *in vivo* by rescued SFFV (Dresler *et al.*, 1979; Ruscetti *et al.*, 1979). This correlation has suggested that the *env* gene is, in fact, the transforming gene. A similar suggestion can be made for F-MuLV: namely, that the *env* gene carries the transforming function.

At present there is no direct explanation for transforming potential residing in an *env* gene and attempts to form a unifying hypothesis with the *onc*-containing viruses are difficult. There is much data gathering to be done on the possible function of *env* molecules, an interesting area for speculation based on widespread distribution of related molecules in nonvirus-producing normal cells of the mouse. If progressive differentiation requires external signals (hormones, etc.), then one could imagine the disruption of a normal receptor or the nonfunctional binding of an external effector molecule which could lead to differentiation arrest. The molecular definition of F-MuLV and SFFV transforming genes and their mechanism of action are important questions deserving intensive investigation.

C. Subacute Oncogenic Viruses

A large number of the known oncogenic RNA viruses, including avian leukosis, murine leukemia, mouse mammary tumor, feline leukemia, bovine leukemia, and gibbon ape leukemia viruses, induce disease only after long latent periods ranging from several months to perhaps several years. Despite intensive attempts, these viruses have not been found to transform cells *in vitro* nor do they appear to contain transforming genes. In a number of cases it is established that persistence of an intact viral genome is not required and in the case of lymphosarcoma in cats, fully one-third of the tumors linked to FeLV exposure fail to show any evidence of viral antigens or specific viral sequences (Hardy *et al.*, 1980). Among the alternatives suggested to account for the activity of these slowly acting oncogenic agents are insertion of viral promotor sequences adjacent to a *c-onc* gene and indirect effects involving intercellular messages (e.g., lymphokines, hormones) in which the virus plays a role in initiation but is not the ultimate carcinogen. Evidence for these two mechanisms will be summarized in this section keeping in mind that they are not mutually exclusive nor is it necessary for one mechanism to apply in all cases.

1. The Promotor Insertion Model

Avian leukosis viruses (ALV) generally cause B-cell lymphomas with a latent period of up to 1 year postinfection. Other pathologic manifestations of viral infection at lower frequency are osteopetrosis and erythroblastosis. Independent examination of a large number of ALV-induced tumors has revealed that most have defective exogenous proviruses and do not express the virus-specific 35 and 21 S mRNAs coding for the structural genes of the virion (Neel *et al.,* 1981; Payne *et al.,* 1981). This provides initial evidence that the virus per se is not required for maintenance of transformation. Earlier studies had established that a region of the viral genome 3' to the structural genes, termed *c,* controlled the efficiency of replication and oncogenicity in ALVs (Crittenden *et al.,* 1980; Tsichlis and Coffin, 1980). Infected cells in all the retrovirus systems reported to date integrate virus at multiple random sites (Steffen and Weinberg, 1978; Shimotohno *et al.,* 1980). This general finding contrasts sharply with the situation in ALV lymphomas in which the exogenous provirus is found only in a limited number of sites (Neel *et al.,* 1981; Payne *et al.,* 1981). Utilizing probes specific for the sequences at the termini of integrated retroviruses (LTRs) and those representative of the structural genes (*rep* probe) of the virus, mRNA species at high levels containing LTR sequences covalently attached to a nonviral RNA were detected. The size of this new RNA species was 2.5 kb in a number of tumors but slightly larger and heterogeneous in others. As expected, some virus-producing tumors were encountered. These produce 35 and 21 S viral RNAs positive with the LTR and *rep* probes. The new RNA species were then assayed with cDNA probes specific for the known avian *onc* genes. In a large number of tumors (85%) the new RNA species hybridized with a probe specific for the transforming gene of the MC-29 strain of myelocytomatosis virus (the gene is termed *myc*), thus providing strong evidence for a specific insertion of ALV next to *c-myc* as a crucial event in ALV oncogenesis (Hayward *et al.,* 1981). Ordinarily *myc*-specific RNA can be detected at a few copies per cell (consistent with evidence that all *c-oncs* have a function in normal cells); however, the expression in tumor cells is 30- to 100-fold higher. The *myc*-negative tumors did show enhanced expression of LTR sequences, thus it seems likely that another *c-onc* may be activated in these tumors.

Direct evidence of integration of the ALV LTR adjacent to *c-myc* in cellular DNA has also been obtained (Hayward *et al.,* 1981). In normal cells, it has been previously found (Sheiness *et al.,* 1980) that *c-myc* sequences were contained within a 14-kb *Eco*RI restriction fragment. This enzyme cuts at a specific site within the ALV LTR leaving about 150 bases adjacent to cellular DNA. Tumors contain a new RI fragment smaller than the normal which contains both ALV 5'-specific sequences and *c-myc* sequences. This fragment has been cloned from two tumors and the presence of both sequences has been confirmed, thus providing compelling evidence for covalent linkage (referred to as Neel *et al.*, unpublished, in Hayward *et al.*, 1981). The essential factors are continuous transcription into adjacent cellular sequences from promotor sequences in either left or right LTRs. Heterogeneity in size of new mRNA species might reflect use of the left LTR, some structural sequences (variable), and the *c-myc* gene whereas the 2.5-kb RNA is most logically derived from a single LTR (no structural sequences) immediately adjacent to *c-myc*. Creation of a defective provirus thus seems to be a potential selective step leading to oncogenesis. This selection could be based on loss of virus-specified cell surface antigens or loss of proviral sequences which otherwise would prevent downstream transcription into the cell genome. The presence of identical structures including promotors and cap sites, at both ends of the integrated provirus, has raised the possibility that integration could lead to promotion of normal cellular genes. Suggestive evidence has been obtained that this might occur (Quintrell *et al.*, 1980; referred to as unpublished in Hayward *et al.*, 1981); however, this may have no untoward consequence except in the case in which a specific clonal expansion occurs (as is true of ALV-induced tumors). Presumably, then, cloned cells from a randomly infected population would show different patterns of new mRNA species depending on the site of integration.

The surprising feature of these data is that MC-29 is not a leukosis-inducing virus, but rather transforms macrophages. It seems possible that other factors might influence the cell tropism of viruses for transformation and that endogenous genes may be expressed totally independent of such restrictions. These hypothetical factors could range from viral envelope receptor interactions to the state of the cellular DNA at the time of infection (e.g., integration next to *c-myc* can occur only at certain times in the cell cycle or cell lineage).

One prediction of the activated *c-onc* hypothesis is that transfection of tumor DNA should select for activated *c-oncs*. In studies to date where indeed enhanced transformation capability has been obtained with avian lymphoma DNA, no evidence for the involvement of viral or known *c-onc* sequences has been obtained (Cooper and Neiman, 1980). This general topic will be discussed in Section II,D.

The dramatic findings of Hayward and colleagues (1981) compel a detailed examination of other tumor systems. One obvious situation, referred to above, is that of virus-free lymphosarcomas in the cat where previous exposure to feline leukemia virus is implicated (Hardy *et al.*, 1980). In the single study known to the authors (J. Casey, J. Mullins, and N. Davidson, personal communication) where probes specific for the promotor region of the viral genome have been utilized, no evidence for such sequences in cellular DNA of virus-free tumors was found. This would seem to indicate that an alternate mechanism for viral involvement in these tumors is operative; however, these experiments do not rule out activation by endogenous promotors. We note here that in the case of avian lymphomas the findings of Hayward *et al.* (1981) have been confirmed by other laboratories (Payne *et al.*, 1981; Fung *et al.*, 1981), so that the activation of *c-myc* by promotor insertion by ALV is a reproducible finding.

2. Subacute Retroviruses and the Immune System

The predominant neoplasm induced by avian leukosis virus is a B-cell lymphoma originating in the Bursa of Fabricius. ALVs grow readily in B cells and one can easily imagine a succession of new viral integrations, generation of defective genomes, leading finally to the insertion adjacent to the *c-myc* gene. If integration can occur at numerous sites, the chance of a hit at any one site would be a rare event, thus accounting for long latency and monoclonality of resulting tumors.

In the case of murine leukemia viruses of the subacute variety, the predominant neoplasm induced is a T-cell leukemia. MuLVs do not appear to infect normal T cells (Horak *et al.*, 1981), thus many of the viral isolates from tumor cells would appear to be secondary to the disease process. Included in this category are the so-called mink cell focus (MCF) isolates (Hartley *et al.*, 1977) which are thought to be recombinants between normal ecotropic and xenotropic viruses (Hartley *et al.*, 1977; Elder *et al.*, 1977; Rommerlaere *et al.*, 1978). The following discussion refers to the AKR mouse which has been purposefully bred for high leukemia incidence. Early in life considerable levels of poorly leukemogenic ecotropic virus are found in these mice; at this stage, passive immunization with antiviral antibody can prevent disease (Heubner *et al.*, 1976; Schwarz *et al.*, 1979). At roughly 6 months of age, first xenotropic and then MCF classes of MuLV can be detected in the thymus, thus leading to the prediction that generation of recombinant viruses was a crucial event in the disease process (Staal *et al.*, 1977). Consistent with this, transplantation of infected thymocytes prior to the onset of neoplastic disease has shown that these are not leukemic cells per se (Metcalf, 1966; Haran-Ghera, 1977), but they do accelerate leukemia in the host. A number of these MCFs and yet another class of MuLVs which are called SL viruses (isolated from spontaneous leukemia; these do not form XC plaques characteristic of ecotropic viruses nor do they form foci in mink cells characteristic of MCF viruses) are positive in leukemia acceleration tests (Nowinski *et al.*, 1977; Nowinski and Hays, 1978), yet clearly these are not acute transforming viruses.

A number of studies established that early viremia was a crucial component of the disease process (Lilly *et al.*, 1975; Lee and Ihle, 1979), a characteristic shared with other subacute agents such as feline leukemia virus and mouse mammary tumor viruses (in the latter case, virus in milk is the key parameter). While availability of one parental virus in a requisite recombinational process might be envisioned to explain the viremia requirement, alternate hypotheses were suggested by studies of immune function. In AKR mice and also in BALB/c mice injected with Moloney leukemia virus (MoLV), development of leukemia is associated with a chronic blastogenic response to viral antigens, especially the envelope glycoprotein, gp71 (Ihle *et al.*, 1980). These findings and related studies of virus binding to lymphoma cells (McGrath and Weissman, 1978 and 1979) suggest that chronic immune stimulation might be a crucial factor in tumor development. In a survey of a number of mouse strains, one strain, CBA/N, was found to exhibit high levels of viremia with no development of leukemia after MoLV inoculation (Lee and Ihle, 1981). These mice failed to develop a blastogenic response to the virus, thus establishing that viremia itself was not sufficient to induce disease. CBA/N mice have an X-chromosome-linked immunodeficiency syndrome (Berning *et al.*, 1980) which involves certain T-independent antigens. In the case of MoLV-inoculated mice, early exposure may induce tolerance, although this remains speculative at present. The dissociation of infectivity and immune response would seem to argue against a promotor insertion model, as viral integrations occur in both cases. In addition, the MoLV stocks used in these experiments contain viruses of the MCF class, thus these viruses alone do not appear to account for tumor development. If chronic immune stimulation is the key factor, then one possibility would be that receptor-positive cells would ultimately transform *in vitro*. The results of McGrath and Weissman (1979) indicate that the normal thymus contains a low proportion of such cells, suggesting that these are the precursors to the eventual leukemic cells which, themselves, are characterized by the presence of specific receptors. Indeed, it has proven possible to maintain T cell lines for at least 1 year in the continuous presence of virus or viral gp70; however, these remain normal (Ihle *et al.*, 1980). These initial studies have led to a further examination of the blastogenic response itself. One key result was that the majority of the response is from cells other than those immediately responsive to viral antigens (Ihle *et al.*, 1981a). These are recruited by lymphokines released by the antigen-stimulated cells. A novel lymphokine, termed interleukin-3 (IL-3) (Ihle *et al.*, 1981b), was described which could be successfully used to grow cells with the surface markers of helper T cells (Hapel *et al.*, 1981) and which is assayed by converting 20α-hydroxy steroid dehydrogenase-negative nude mouse lymphocytes to an enzyme-positive state. This conversion is felt to be an early step in T-cell differentiation. Thus one sees that viral antigen-driven immune responses can result in stimulation of differentiation as well as expansion of lymphocyte populations. The potential targets for transformation can thus be far removed from the point of initial contact with virus. What then is the mechanism of transformation by this category of viruses? One cannot yet rule out viral promotor insertion, but the circumstantial evidence makes it seem unlikely as an explanation for the activity of all subacute acting viruses.

It may well be that expanded or particularly responsive cell populations are made available and thus the frequency of the rare transforming event may be increased. This could involve activation of *c-oncs* by a broad category of molecular events (e.g., mutation, transposition) and even a late-occurring viral insertion. The lack of transforming genes in the subacute viruses places them in a general category of environmental agents which could ultimately impinge on a common genetic pathway. Klein (1979) has reviewed data leading to the proposal that convergent cytogenetic evolution can result from a variety of initiating causes. In this context, AKR leukemias, C57L Black leukemias induced by radiation leukemia virus, or dimethylbenzanthracene show a high frequency of trisomy for chromosome 15. These events have parallels with the oncogenic herpesviruses as discussed in Section III.

D. Detection of Cellular Transforming Genes

As described in Section II,A, cellular equivalents of *v-onc* genes designated *c-onc* or protooncogenes (Bishop, 1981) exist in cellular DNA and are expressed at relatively low levels. Ligation of *v-* or *c-onc* to viral promotor sequences results in enhanced malignant transformation (Blair *et al.*, 1980, 1981; Oskarsson *et al.*, 1980). Dosage effects resulting in enhanced product formation may thus provide the simplest explanation for *v-onc* activity. The natural extension of this concept to malignancy resulting from nonviral (or subacute viral) causes (i.e., chemical carcinogens, radiation, and genetic factors) would require enhanced expression of one of the protooncogenes where the gene is defined in terms of its ability to induce transformation. At present, detailed knowledge of such genes and their products is restricted to *v-onc* genes. Any factors which promote permanent enhanced expression of protooncogenes might be considered as targets for the transformation process; these might include mutations in positive or negative control elements, transpositional effects (e.g., the chance insertion of an active promotor), or changes within the gene itself. The case of ALV-induced lymphomas provides one example of oncogenesis mediated by promotor insertion. A logical prediction of the activated oncogene hypothesis is the ability to obtain transformation with DNA from malignant cells. Two laboratories have reported considerable success in such efforts, using the NIH mouse 3T3 cell as recipient. These cells can be transformed directly but at low efficiency with subgenomic fragments of both MSV (Andersson *et al.*, 1979) and RSV (Copeland *et al.*, 1979). Transfection was attempted with DNA from methylcholanthrene-transformed quail, BALB 3T3 mouse, and normal chicken and mouse fibroblasts. Both high-molecular-weight ($>20 \times 10^6$) and sonicated DNA ($0.3-3 \times 10^6$) preparations were used. A soft agarose growth assay was used to measure transformation. High-molecular-weight preparations did not increase the frequency of background transformants whereas sonicated DNA from both normal and the two MC-transformed cells gave a 10-fold increase over background. This was similar to the results obtained with subgenomic fragments of avian sarcoma virus-infected cells. The efficiency of secondary transfections was dramatically increased, ca. 100-fold, and now positive results were obtained with unsonicated DNA ($0.1-1.5$ colonies/μg DNA were obtained). Restriction endonuclease digestions were performed with three of these DNAs and specific but different patterns were seen, two of the lines differing from the third. These experiments indicate that secondary transfections were mediated by specific donor DNA segments. One interpretation of these results (Cooper *et al.*, 1980) is that endogenous transforming genes are under the control of flanking sequences removed by sonication in the first transfection step. Activity results from a low probability insertion into a region influenced by an active promotor; this active region can then be transferred in a second round of infection by high-molecular-weight DNA. This model is in essence the reciprocal of promotor insertion and perhaps may be termed protooncogene insertion. It was of obvious interest to perform similar experiments on avian B-cell lymphomas (Cooper and Neiman, 1980) where one might expect to transfer the activated *c-myc* gene (Section I,C,1). High-molecular-weight DNA of seven tumors, but not ALV infected or uninfected tissue DNAs, induced transformation in the range of 0.02–0.9 transformants/μg DNA (3T3 cells). Secondary transfections were again positive with a similar high frequency. Six transformed cell lines were analyzed for ALV-related sequences including those found in the long terminal repeat of proviral DNA. These yielded negative results and thus the authors concluded that oncogenesis by ALV may involve indirect activation of cellular transforming genes in distinction to the promotor insertion model. Expansion of a target cell population by ALV (see Section II,C,2) at an early stage (preneoplastic) in the total disease process was considered a plausible role for this virus. The activation of *c-myc* might be a crucial factor in this expansion. Krontiris and Cooper (1981) have also reported that 2 of 26 human tumors and cell lines yielded high-molecular-weight DNA capable of transforming 3T3 cells. These results suggest that at least in some human tumors dominant mutations or DNA rearrangements activate protooncogenes. The negative cases suggest that the more common changes in human malignancy are much more complex, involving "epigenetic changes, recessive mutations, or mutations in multiple unlinked genes" (Krontiris and Cooper, 1981).

Weinberg and colleagues have performed similar experiments also using NIH 3T3 cells as the recipient of high-molecular-weight DNA. In initial experiments (Shih *et al.*, 1979c), 5 of 15 chemically transformed murine cell line DNAs gave foci at high efficiency (0.1–0.2 focus/μg DNA). These could also be passaged serially indicating a stable genetic element. In one case, chromosomal transfer from a donor cell whose DNA was negative in transfection attempts yielded a significant number of transformants. One presumes that the greater number of cell lines used by Shih *et al.* (1979c) explains the negative results of Cooper *et al.* (1980) with the two MC-transformed mouse and chicken cells. Following these initial results, Shilo and Weinberg (1981) have now studied the DNA of four methylcholanthrene-transformed mouse cell DNAs by cleavage with five restriction endonucleases. Each of the DNAs gave a similar pattern (i.e., *Eco*RI and *Hin*dIII completely abolished activity whereas *Bam*HI, *Xho*I, and *Sal*I did not). The authors conclude "that the transforming genes of the four mouse lines have a high probability of being associated with an identical set of nucleotide sequences." As the authors point out, this does not mean a single gene for all MC-transformed lines since transfection-negative lines have been described; however, a finite number of such genes is suggested. In addition, transfection-positive DNA from a rat neuroblastoma gave a different pattern of sensitivity to restriction enzymes. Shih *et al.* (1981) have also recently reported successful transfection of transforming DNA from human, rabbit, and mouse bladder carcinoma lines, a lung carcinoma line, and a rat neuroblastoma and mouse glioma cell line. In several of the above reports attempts were made to detect or rescue

nucleic acid hybridization which has revealed multiple copies of the EBV genome, by the presence of the EBV nuclear antigen, EBNA, which is always expressed in EBV-infected cells, and by the recovery of the virus, itself, from cultured tumor cells (for review see Epstein and Achong, 1979a,b). This association between EBV and BL holds for the great majority of BL cases which occur in Africa. Outside the BL endemic areas of Africa and New Guinea the obverse is true, with the majority of BL cases being EBV negative. This observation suggests that there may be two etiologically different forms of BL, one form associated with EBV and the other not. The association between EBV and BL is fairly specific as other types of B-cell lymphomas such as Hodgkin's disease tumors were negative for the EBV genome even when arising in serologically positive patients (zur Hausen, 1976).

Besides virological evidence, serological studies indicated a relationship between EBV and BL. These results have been extensively reviewed (Henle and Henle, 1979; Pearson, 1980). Some early studies suggested that EBV might be related to several diseases—IM, BL, NPC, Hodgkin's Disease, and chronic lymphocytic leukemia, on the basis of an increased percentage of patients with elevated antibody titers, in this case antiviral capsid antigen or VCA, compared to matched controls. Tests for other viral antigens, early antigen (EA), and EBNA indicated more specific relationships with BL, NPC, and IM. Burkitt's lymphoma serology is characterized by high IgG antibody to VCA, antibody to the R (restricted) component of EA, and moderate titers to EBNA. Nasopharyngeal carcinoma patients characteristically show high antibody titers to most EBV-associated antigens and have both IgA and IgG antibody to VCA and EA. Antibody to the D (diffuse) component of EA is also characteristic of NPC. Moreover, there is evidence that throat washings from NPC patients contain IgA as well as IgG antibodies to EBV-associated antigens and that IgA-containing throat washings have EBV-neutralizing activity (Desgranges et al., 1977). Patients with IM show IgM and IgG antibody to VCA and have antibody to the D component of EA. Moreover, there is evidence that antibodies to EA (R) have prognostic value in BL. Similarly, antibodies to both VCA and EA (D) are related to the stage of disease in NPC patients. More recently, antibody to EBV-associated membrane antigens (MA) have shown direct relationships in both NPC and BL (Pearson et al., 1978, 1979).

Thus, these results indicate the presence of relatively specific antibody patterns for BL, NPC, and IM. It is of interest that the antibody patterns for BL tend to be similar regardless of whether the disease occurs in endemic or nonendemic areas (Judson et al., 1977). In terms of etiology, IM was shown to be the result of a primary EBV infection by serological conversion and virological studies. Hodgkin's disease patients may be free of EBV or may become infected during disease so the antibody patterns in this disease and presumably some other neoplastic conditions may only monitor the disease (Lange *et al.*, 1978). In the areas of endemic BL and NPC it has been established that EBV infection occurs early in childhood but the onset of disease occurs much later in life. In the case of BL a large prospective study was undertaken involving the collection of serum samples from about 42,000 children in an attempt to determine if the cases of BL resulted from rare instances of primary infection in older individuals or as a secondary event in already infected individuals and to distinguish among several possible hypotheses regarding the etiologic role of EBV in BL. Among the possibilities were that EBV plays no causal role in BL, that BL is the result of a primary EBV infection, or that BL develops in patients following severe infection or heavy exposure to EBV and that some secondary factors might be necessary for the disease to occur.

In the study group, 14 cases of BL were diagnosed but of these one was later listed as an unclassified lymphoma. This tumor and one other were found to be free of the EBV genome by nucleic acid hybridization. A third case was clinically and serologically atypical. Six of the other tumors were tested for EBV genome and found to be positive with high levels of genome equivalents. Antibodies to VCA in the BL patients' sera collected prior to disease were significantly higher than those in matched controls with geometric mean titers being higher by a ratio of about 3 to 1. Individually, the VCA titers of the BL patients tended to be higher than any of their controls or as high as the highest control. Further analysis comparing serology before and after BL showed that the anti-VCA titers were stable in most cases. It was also shown that anti-EA titers, especially to the R component, increased after disease. Antibodies to herpes simplex virus, cytomegalovirus, and measles virus showed no disease association (de-The *et al.*, 1978). These results have been interpreted to suggest that high anti-VCA titers are associated with a greatly increased risk of developing EBV-positive BL in the endemic area. The time interval found between initial serum collection and development of BL, 7 to 54 months, also suggested that BL may not be the sole cause of the disease, as it arises long after initial infection.

There is recent evidence that IgA antibodies to EBV-associated antigens may provide a means of early detection for NPC. Ho and colleagues described four individuals who had IgA anti-VCA antibody detectable in their sera several months prior to the onset of NPC (Ho *et al.,* 1978a,b). Lanier and associates (Lanier *et al.,* 1980) described the presence of IgA anti-VCA titers in three Alaskan patients prior to the diagnosis of NPC. In a large-scale serological survey of 56,584 individuals aged 30 years or older in South China, Zeng *et al.* (1980) found 117 sera positive for IgA anti-VCA. Of those positive individuals, 20 were subsequently diagnosed as having NPC. de-The *et al.* (1981a) described 56 individuals from China who had persistent IgA antibody levels to VCA. Four of these individuals were diagnosed clinically or histologically as having NPC, five others showed the presence of EBV DNA and EBNA, and four had EBV DNA without EBNA in biopsy tissue (de-The *et al.,* 1981b).

Thus, in BL the presence of virus in the tumor cells, characteristic antibody profiles in the patients, and the high risk associated with long-standing, stable, high anti-VCA titers may implicate a direct role for EBV in BL. This contention is supported by two biological properties of the virus, its ability to transform or immortalize lymphocytes (reviewed by Pope, 1979), and its ability to induce B-cell lymphomas in cotton-topped marmosets (Shope *et al.,* 1973). There are two other cellular characteristics of BL which need to be mentioned. First, the tumor cells show a specific marker chromosome, 14q+, (Manolov and Manolova, 1972) in which a specific portion of chromosome 8 is translocated to chromosome 14 (Zech *et al.,* 1976). Analysis at early metaphase and phophase has shown that this translocation is reciprocal (Manolova *et al.,* 1979). Second, BL has a monoclonal origin as indicated by the presence of single glucose-6-phosphate dehydrogenase isozymes in tumors of heterozygous females (Fialkow *et al.,* 1970).

Klein (1979) has presented the hypothesis that BL develops in a three-step process. First EBV infection leads to the stable infection and immortalization of B cells. In endemic areas, the level of such immortalized cells may be high compared to general levels in nonendemic areas. The second stage is related to an environmental factor which would lead to persistent proliferation of EBV-infected cells and which might also enhance their ability to grow through the immunosuppression of the host. The possible significance of a relationship between BL, EBV, and malaria was suggested several years ago (O'Conor, 1970). It is interesting in this regard that *Plasmodium berghei yoelii* infection in mice was reported to enhance the tumorigenicity of Moloney leukemia virus (Wedderburn, 1970) and that plasma from owl monkeys infected with *Plasmodium falciparum* inhibited several responses *in vitro* (mitogenic response, one-way mixed lymphocyte reaction) of normal primate peripheral blood lymphocytes (Taylor and Siddiqui, 1978). The third stage would occur when the specific chromosome translocation took place, thereby establishing the truly neoplastic clone. Thus, EBV-positive BL and EBV-negative BL, while similar in their final stages, would have quite different initiating and promoting steps.

3. Experimental Infection by EBV in Primates

It is of interest to contrast what seem to be the factors which lead to BL in humans with the pathogenic potential of EBV in various primate hosts. This subject has been recently reviewed (Miller, 1979), but a few points should be restated and emphasized in the present context. At least three species have responded to EBV inoculation by the development of lymphoproliferative conditions: cotton-topped marmosets, owl monkeys, and common marmosets. In the most susceptible species, cotton-topped marmosets, both lymphoma and lymphoid hyperplasia have been seen (Miller et al., 1977). In some cases the lymphoreticular hyperplasia seen in biopsies of lymph nodes relatively early after virus inoculation disappeared on reexamination at necropsy while at least one other case resulted in fatal hyperplastic disease after 38 days. In some instances both lymphoid hyperplasia and lymphoma were present in the lymph nodes of the same marmoset. Evidence for monoclonality of at least one marmoset tumor was obtained by analysis of surface immunoglobulins and karyotypes (Rabin et al., 1977). In this study EBV-positive B cell lines established from separate tumors which had occurred in the liver, spleen, and mesenteric lymph nodes all had the same immunoglobulin light chain, λ, and heavy chains, μ and γ, and had the same karyotype showing the loss of one homolog of chromosome 2. These cell lines were established from tissue taken at necropsy nearly 5 months after virus inoculation. In general, inoculated marmosets produced antibodies to EBV-associated antigens. Antibody to VCA was the most common but antibody to EA and even to EBNA have been observed (Miller et al., 1977; Rabin et al., 1977; Werner et al., 1975; Deinhardt et al., 1975). In some marmosets tumors arose rapidly (1–2 months) and the animals died without having made a detectable antibody response (Miller et al., 1977). Thus, in marmosets a range of results from no tumor induction to both transient and progressive hyperplasia to fatal lymphoma induction has been observed. In some instances strong antibody responses have been demonstrated but in most cases antibody titers are moderate while in some instances there has been a failure to produce antiviral antibodies even in the presence of apparently virus-induced disease. Latent periods have been relatively short but have extended to nearly 5 months. Importantly, the EBV genome has been consistently associated with tumors in cotton-topped marmosets and has been demonstrated by both molecular hybridization and by immunofluorescence staining for EBNA (Miller *et al.,* 1977; Werner *et al.,* 1975). While the level of EBNA-positive cells and the number of EBV genome equivalents per cell were lower than in tumor cells, lymph node cells from an animal with EBV-associated hyperplasia were positive for the EBV genome.

Inoculation of common marmosets with EBV has not induced lymphomas but one report describes the induction of lymphoproliferative disease in seven of eight inoculated animals (Falk *et al.,* 1976). This disease was characterized by a diffuse hyperplasia comprised of small lymphocytes, lymphoblasts, immunoblasts, and plasmacytoid cells. The disease involved to varying degrees the lymph nodes, thymus, spleen, kidney, salivary glands, and tissues surrounding bronchi, bronchioles, and pulmonary vessels. Impression smears of several tissues from three marmosets were stained for EBNA with negative results and no continuous cell lines were established from thymus, spleen, and lymph nodes from five marmosets. Six of the eight marmosets developed low levels of antibody to EBV VCA. In another study involving five common marmosets (Ablashi *et al.,* 1978) no tumors were induced although as judged from antibody response all the marmosets were infected, with two of them developing moderately high titers to both VCA and early antigen. Attempts to establish continuous lymphoid cell lines from the inoculated marmosets failed. Inoculation of owl monkeys resulted in the production of a reticuloproliferative disease in one of three injected animals (Epstein *et al.,* 1973a). This disease was characterized by a heterogeneous cell population and in this respect resembled certain types of human and experimental lymphomas. A continuous lymphoid cell line was established from an involved lymph node of this monkey (Epstein *et al.,* 1973b). These cells were virus producers as determined by electron microscopic analysis and this virus transformed human cord lymphocytes *in vitro.* Nucleic acid hybridization analysis showed the presence of multiple copies of the EBV genome in these cells (Epstein *et al.,* 1975). These cells were clearly B cells as shown by the presence of surface immunoglobulin and receptors for activated complement. The immunoglobulin was $\mu\kappa$ and was detected on 1–6% of the cells (Rabin *et al.,* 1973). These cells showed two distinct populations cytogenetically (Jarvis, 1974). One of these populations had 54 chromosomes while the other had 55. The general karyotype of both types conformed to owl monkey karyotype I (Ma *et al.,* 1976); the proportion of the two cell types in the population varied to a degree but the cells with 55 chromosomes predominated. While the origin of the extra chromosome is unclear, it does suggest a level of heterogeneity in the transformed cell population. It is of interest in this regard that Miller (1979) pointed out that by ultrastructural observation experimental marmoset tumors appeared to be of two types although only one type may have persisted in cultures established from such tumors. The owl monkey in which disease was induced as well as several other owl monkeys developed antibodies to several EBV-associated antigens (Rabin *et al.*, 1976) including several monkeys which produced antibodies to EA and EBNA as well as VCA. However, of the nine owl monkeys inoculated with transforming strains of EBV, only the one already described developed a lymphoproliferative disease.

Thus, a range of conditions has been produced experimentally with EBV. The most sensitive host is the cotton-topped marmoset where both lymphomas and hyperplastic conditions have been produced in a relatively high proportion of animals. These conditions seem definitely associated with the EBV genome. However, attempts at neutralization *in vivo* yielded somewhat incomplete results (Shope, 1975). In these experiments antibody-positive serum blocked infection of marmosets but marmosets which received virus plus antibody-negative serum while developing antibody failed to come down with progressive disease. Due to the heterogeneous populations of cells observed histologically it may be that many of these diseases are polyclonal, although in one case the tumor may well have been uniclonal as indicated by studies on tumor-derived cell lines. The conditions in common marmosets seem more hyperplastic than neoplastic and evidence indicating the association of virus with the proliferating cells is lacking. In owl monkeys, only a single case of progressive disease has been reported although these monkeys seem to be quite susceptible to virus infection. The cell line established from this diseased owl monkey was cytogenetically heterogeneous. All progressive diseases induced in primates have resulted from the inoculation of transforming strains of EBV. Nontransforming virus of the P3HR-1 strain has not induced disease in limited numbers of cotton-topped marmosets (Deinhardt *et al.*, 1975) or owl monkeys (Rabin *et al.*, 1976).

4. EBV and Acute Lymphoproliferative Disease in Man

In addition to its association with BL, NPC, and IM, EBV has been associated in varying degrees with several other lymphoproliferative diseases in man. In at least some of these conditions, the evidence linking EBV to the disease is both relatively strong and direct. Virelizier *et al.* (1978) described a case of persistent EBV infection which was associated with a fatal case of polyclonal lymphoid hyperplasia in a 5-year-old girl. The disease was characterized in part by enlargement of spleen, liver, and lymph nodes. Biopsy specimens of lymph nodes showed infiltration of large, basophilic immunoblasts. Many mitotic figures were seen and a large number of the cells showed various immunoglobulin classes as determined by immunofluorescence. Similar cells were seen in a lung biopsy taken after X-ray examination revealed interstitial pneumonitis. Blast cells were also seen in the peripheral blood and bone marrow. The blast cells in the peripheral blood increased toward the end of the disease and appeared to be polyclonal B cells. These cells as well as those from bone marrow had normal karyotypes and were EBNA negative. All serum immunoglobulin classes except for IgE were elevated. There was no electrophoretic evidence of monoclonal serum immunoglobulins and the ratio of κ to λ light chains was normal.

The patient had exceptionally high IgG antibodies to VCA which tended to increase with time (1 : 20,000 to 1 : 160,000 over the course of about 1 year). IgM antibodies to this same antigen were present but declined over the same time period from 1 : 320 to 1 : 20 suggesting that the infection was primary. Antibodies to EA were present at unusually high titers, 1 : 5000 to 1 : 80,000, but did not show a temporal trend. Antibodies to EBNA were present throughout the same period but at relatively normal levels of 1 : 40 to 1 : 1280. About 1% of lymphocytes from the axillary lymph node and from peripheral blood were EBNA positive. Lymphoid cell lines were repeatedly established from the peripheral blood and once from a lymph node over the course of 1 year without the use of exogenously supplied EBV.

Although certain aspects of cellular immunity in this patient appeared normal, background levels of thymidine incorporation by peripheral leukocytes were high (4 to 40 times above the normal range). In spite of these high levels in unstimulated cells, proliferative responses to EBV-associated antigens *in vitro* were found. Although it was possible to generate *in vitro* cytotoxic lymphocytes to allogeneic lymphocytes, including Raji cells (a BL-derived, EBV-positive, lymphoid cell line), there was no background of cells which were spontaneously cytotoxic for Raji cells even though normal proportions of various peripheral lymphocyte types were found as judged by Fc receptor, complement receptor, and surface immunoglobulin analysis. Although peripheral leukocytes made approximately normal levels of leukocyte interferon (after treatment with Newcastle disease virus) they failed to produce immune interferon following mixed lymphocyte cultivation with Raji cells.

Another case with more direct evidence on the role of EBV in inducing primary lymphoproliferative disease was reported by Robinson *et al.* (1980). This case was that of a 4-year-old girl who died of a rapidly developing disease characterized by extensive infiltrations of poorly differentiated lymphoplasmacytic cells. The infiltrate involved lymph nodes, spleen, liver, bone marrow, lungs, stomach, and submandibular glands, and, in addition, the peripheral leukocyte counts had been high prior to death. The majority of the leukocytes in the peripheral blood were B cells as determined by surface marker analysis. The population was polyclonal as shown by the presence of several immunoglobulin classes including both κ and λ light chains. The majority of these B cells were EBNA positive, had EBV DNA at about 20 genome copies per cell, and could induce the transformation of human cord leukocytes on cocultivation. Similarly, EBNA-positive cells were demonstrated in lymph nodes, spleen, liver, lung, and submaxillary gland. Virus was demonstrated in saliva and isolated from salivary gland extract. While it was possible to transform cells with peripheral blood lymphocytes, only one lymphoid line was established from this source by mass culture, presumably because the cultured cells entered a virus-productive cycle. Seven lines from both peripheral blood and spleen were established by direct cloning in agarose. Among these seven cell lines, five immunoglobulin types were evident and two were heteroploid with one line having a trisomy of chromosome 11 and the other having a 17p+ marker chromosome.

The patient's lymphocytes, even after T-cell enrichment, responded poorly to phytohemagglutinin (PHA). Her medical history showed indications of possible defects in cell-mediated immune function. That her EBV infection was a primary one was inferred from her serology. IgM as well as IgG antibodies to VCA were detected, with the IgM titer declining with time. Neutralizing antibody also was detected in serum taken after the onset of the disease.

In a third instance, Bornkamm *et al.* (1976) investigated the presence of EBV-specific DNA in a variety of tumors including a case of immunoblastic lymphadenopathy with excessive plasmacytosis. This condition arose in a 66-year-old female. The patient had generalized lymph node enlargement and fever and died 4 weeks after the onset of the disease. Her serum showed a moderate (1:40) antibody level to VCA but antibody to EA was not detected. Molecular hybridization showed two to three genome equivalents of EBV per cell.

A fourth case occurred in a 16-year-old male who developed a rapidly progressing disease and died 18 days after symptoms first appeared (Bar *et al.*, 1974). The disease was characterized by infiltration of atypical, polyclonal B cells, some of which had a degree of plasmacytic differentiation. These cells were present in peripheral blood, all lymphoid tissues, liver, small intestine, and brain. Peripheral blood lymphocytes were nonresponsive to both PHA and pokeweed mitogen (PWM) and had abnormally high backgrounds of thymidine uptake. EBV DNA was found by hybridization of DNA from the spleen with EBV-specific complementary RNA at a level of six genome equivalents per cell. Transforming virus, neutralizable by anti-EBV serum, was recovered from spleen extracts; electron microscopic observation of a cervical lymph node (taken at biopsy prior to death) showed infiltration of atypical mononuclear cells and the presence of herpesvirus particles in lymphoid cells. This patient had no detectable antibodies to EBV but was positive for heterophile antibody. This patient just described had two male cousins who died of a similar condition, one at 17 years of age and the other at 22 years of age. In one case, that of the 17-year-old, actual tumor formation was described in several organs. A brother of the 22-year-old had died at 7 years of age from a disease diagnosed as leukemia, but without the benefit of a postmortem examination.

Further indications of a familial association with severe and even fatal lymphoproliferative disease (fatal IM and malignant lymphoma) are documented in reports by Falletta et al. (1973), Provisor et al. (1975), and Maurer et al. (1976). Falletta and his associates described a family in which 17 preschool-age boys, over a span of two generations, died of a disease characterized by its rapid course (average duration of 2–3 days), fever, hepatosplenomegaly, enlarged lymph nodes, purpura, hyperglobulinemia, and anemia. Histologic examination of 12 cases showed widespread infiltration of immature mononuclear cells and mature plasma cells. All the cases were maternally related and the condition was described as X-linked recessive reticuloendotheliosis. More detailed study of one case suggested the presence of a chromosomal abnormality (partial deletion of the short arm of an A-group chromosome), and a failure to give positive skin tests to a series of recall antigens. Although specific tests for EBV were not performed on this patient, it was apparently possible to obtain replicating lymphocytes from peripheral blood after 30 days in culture. The peripheral blood count showed 83% lymphocytes many of which were atypical.

Provisor and associates described three cases of a disease resembling IM which occurred in two brothers and their maternal male cousin. Agammaglobulinemia developed in two of the cases and the third patient died during the acute phase of the disease. In the two cases examined, antibodies to EBV intracellular antigens were found at low titer (1 : 10–1 : 50). Maurer and colleagues described four cases of malignant lymphoma in three brothers and the son of their sister. The tumors arose in the small intestine in three cases and in the retroperitoneum in the fourth case. Microscopically the tumors were diffuse, and of lymphocytic or mixed lymphocytic–histiocytic type. Depressed immunoglobulin levels, low plasma cell levels in a bone marrow aspirate, and a lack of germinal centers in a lymph node biopsy were found in one patient and a second patient had negative skin test reactions to recall antigens during a recurrence of lymphoma. No specific assays for EBV were performed.

Purtilo *et al.* (1975) described an additional family in which 6 males out of 18 died of a lymphoproliferative disease. Fatal IM occurred in three brothers and two half-brothers who were maternal cousins of the first three patients. All the patients had signs of lymphoid proliferation including two with lymphoma. A sixth cousin developed various signs of immunodeficiency following IM and subsequently died of septicemia. A seventh male cousin also showed clinical signs of immunodeficiency. Subsequently Purtilo *et al.* (1977) described another family in which a large number of boys showed various proliferative B-cell conditions such as American BL, immunoblastic sarcoma, plasmacytoma, and fatal IM. Nonproliferative conditions such as acquired agammaglobulinemia, agranulocytosis, and aplastic anemia were seen in this kindred. In several of the individual patients, more than one of the phenotypes were noted concurrently. Because of the mode of inheritance and the various types of conditions seen, the name X-linked recessive lymphoproliferative syndrome or more simply X-linked lymphoproliferative (XLP) syndrome was proposed for these cases and it was postulated that a single locus, the lymphoproliferative control locus (X_{LC}), when defective, was a basic element of the disease.

In 1978 the XLP Registry (Hamilton *et al.*, 1980) was established in order to better identify and investigate XLP. Purtilo (1981) recently summarized the variety of phenotypes represented in XLP. Data on 100 patients showed that approximately 70% had fatal IM, 15% had fatal IM plus immunoblastic sarcoma, 10% survived IM but developed agammaglobulinemia or dysgammaglobulinemia, 25% had malignant lymphoma, and 15% had aplastic anemia. Two phenotypes often occurred together but aplastic anemia and agammaglobulinemia never occurred concurrently. Only 20 of the 100 patients survived. Thus, while there is apparently a variety of phenotypes exhibited in XLP, the majority of cases show severe, often fatal IM which outside of XLP families is considered a rare (Penman, 1970) but not unknown (Crawford *et al.*, 1979) occurrence. Because of the general acceptance of EBV as the etiologic agent of IM it was reasonable to propose that EBV was playing a major role in XLP. In fact, as mentioned, virus was demonstrated in one case of fatal IM (Bar *et al.*, 1974) and a second case was described (Purtilo, 1976) in which EBNA-positive cells were cultivated *in vitro* from the patient. In both of these cases, antibodies to EBV were not detected. Preliminary data also indicated the presence of EBV genome in a benign lymphoid tumor in a third patient (Hamilton *et al.*, 1980). More recent findings indicate that presumably abnormal tissues from seven XLP patients were positive for the EBV genome (Purtilo *et al.*, 1981).

Many of the XLP patients showed evidence of immunodepression as did two of the non-XLP cases with lymphoproliferative disease. In XLP, there appears to be a defect in natural killer (NK) cell activity in which of the 12 patients tested, 10 had levels of NK activity below that of the lowest levels found in normal controls, XLP carrier females, and patients with acute IM. While NK activity of XLP patients was enhanced by interferon, normal levels were never achieved (Sullivan *et al.*, 1980). X-linked lymphoproliferative patients show diverse antibody patterns to EBV. Of 15 patients studied, 6 were without antibodies to VCA, EA, or EBNA, 6 others had antibody to VCA but not to the other 2 antigens, and 3 had antibodies to VCA and EA. The geometric mean titers (GMT) to VCA (1 : 110 and 1 : 93) of these two groups were in the same range as those of a control group of healthy individuals (1 : 100). Carrier females were uniformly positive for anti-VCA and had a very high GMT of 1 : 350. Four carrier females were also positive for anti-EA and all had antibodies to EBNA (Sakamoto *et al.*, 1980). Carrier females have been shown to have IgM and IgA as well as IgG antibodies to VCA which are not seen in XLP patients or controls (Purtilo *et al.*, 1981). The anti-EA antibodies in patients and carriers suggest active infection in those individuals. Because it is recognized that individuals with either hereditary (Spector *et al.*, 1978a) or acquired (Gatti and Good, 1971) immunodeficiency have a higher incidence of cancer, particularly lymphoreticular neoplasms, it is not surprising to see the same type of correlation occur in the XLP and other cases which apparently involve EBV. What is of great potential importance is the extent to which EBV may play a critical role in the etiology of lymphoid neoplasms in individuals with either hereditary or acquired immunodeficiency. In this context it is of interest to consider the types of immunity which seem to be involved in recovery from acute EBV infection and in the maintenance of long-term immunity to EBV-transformed cells. Aspects of specific cytotoxic immunity will be stressed. For a comprehensive review of humoral immunity, see Pearson (1980).

Lymphocytes with preferential cytotoxicity for EBV-positive as opposed to EBV-negative cell lines were initially demonstrated by Svedmyr and Jondal (1975). These authors showed that if the background natural killer cells were removed from peripheral blood lymphocyte populations of IM patients, but not of normal individuals, a subpopulation remained which was cytotoxic for a range of EBV-positive cell lines but not for EBV-negative cell lines. Bakacs *et al.* (1978) subsequently showed that this cytotoxic cell was T cell-negative for the Fc receptor of IgG. In addition to being cytotoxic for established EBV-positive cell lines, T cells from IM patients can block the outgrowth of EBV-infected fetal lymphocytes (Rickinson *et al.*, 1977).

A second type of specific inhibitory cell was demonstrated in the peripheral blood of normal adults subsequent to observations on EBV transformation of adult compared to fetal lymphocytes (Thorley-Lawson *et al.*, 1977). It was shown that unfractionated fetal lymphocytes transformed with the same efficiency as determined by the time required for outgrowth as did purified, fetal B cells. Unfractionated adult lymphocytes, however, transformed much less efficiently than purified, adult B cells. Moreover, adult B cells transformed with the same efficiency as did fetal B cells, indicating that the lower efficiency of transformation of adult cells was not directly attributable to the B cells themselves. Autologous reconstruction experiments indicated that T cells present in the non-B-cell fraction of the peripheral blood were responsible for the inhibition of B-cell outgrowth. In a subsequent paper (Thorley-Lawson, 1980) evidence was presented that T cells suppress the EBV-induced proliferation of B cells, which is a measure of the immortalization phenomenon. The T-cell inhibition was best demonstrated when the T cells were added at the time of or within a few hours of virus infection. Delay in the addition of T cells by 24 hours resulted in a greatly reduced degree of inhibition. When added at the time of virus infection, the T cells could be removed after as little as 3 hours and still induce inhibition. The T-cell inhibitory activity was partially resistant to X-irradiation (1000 R) and the T cells had no effect on the transforming capacity of the virus itself. Thus, there seems to be some T-cell capacity for inhibition of an early stage in transformation.

Working along these same lines Moss *et al.* (1978) showed that if unfractionated lymphocytes from EBV-seropositive donors were infected with virus and then incubated at relatively high cell concentrations ($\sim 10^6$ cells/ml) there was a high degree of regression of cells which initially appeared to be undergoing transformation. This type of regression was progressively less apparent as the cell concentration was reduced to values of 2.5×10^5/ml. No such regression was observed with either fetal lymphocytes or lymphocytes from EBV-seronegative donors. T-cell depletion and reconstitution experiments showed that the regression was T-cell dependent. Further analysis of this T-cell-dependent inhibition showed that viral antigen apparently was not the target for the T cells and that the inhibition was not mediated through antiviral antibody or through a soluble factor produced by infected cultures from seropositive donors (Rickinson *et al.*, 1979). In analyzing the kinetics of the response it was demonstrated that the number of T cells in infected cultures from EBV-seropositive donors increased during the first 2 to 3 weeks of culture. EBNA-positive cells increased initially but by 1 to 2 weeks began to decrease and by 3.5 weeks were essentially eliminated. T cells harvested from unfractionated cultures at 11–14 days after infection were highly toxic for autologous EBV-transformed B cells. The cytotoxic activity of these T cells was much more pronounced on autologous target cells compared to allogeneic cells (Moss *et al.*, 1979) indicating a possible HLA restriction, a point on which supporting data were subsequently obtained (Misko *et al.*, 1980). In a study designed to gain information on the time required to obtain a normal complement of the apparent T memory cells, assays were performed with lymphocytes from patients with IM from the acute stage of the illness to 83 weeks after onset (Rickinson *et al.*, 1980). In a series of normal seropositive donors it was determined that while there was a degree of individual variation, an average of 4.6 $\times$ 10^5 lymphocytes/ml was required to induce 50% regression in EBV-infected cultures. In contrast, 3.7×10^6 lymphocytes/ml were required to achieve the same degree of regression in cultures prepared from IM patients during the acute stage. These figures presumably are related to the concentration of specific memory T cells present in the peripheral blood. In following IM patients it was found that the lymphocyte concentration required for regression dropped slightly between 5 and 23 weeks but normal levels were not generally reached until about 30 weeks after onset.

Thus, several cellular mechanisms seem to operate against EBV-immortalized B cells. In IM there seems to be an activated cytotoxic cell which is effective against EBV-positive cells in a non-HLA-restricted fashion. In normal seropositive individuals and in individuals who have recovered from IM there is a memory T cell which can expand on exposure to EBV-positive cells and which seems to have an HLA-restricted inhibitory action. In addition, there is evidence for a T cell which can inhibit outgrowth of newly infected B cells. Nonspecific inhibitors such as NK cells are apparently important as inferred from XLP and other immunodeficient patients. Humoral mechanisms such as antibody-determined cellular cytotoxicity (ADCC) may be important as in both BL and NPC. Antibody-determined cellular cytotoxicity titers were lower in patients with a poorer survival pattern than in patients with a better survival pattern (Pearson, 1980). Although the antigens of importance in humoral immunity may well be different than those of importance in strictly cellular immune reactions, it is not clear whether or not there are some forms of immunity which are more critical than others or whether the host must be immunologically intact overall both in order to prevent disease on primary exposure to EBV or its reactivation. Without sufficiently intact immunological function the immortalization of host B cells may be sufficient to induce the rapidly fatal lymphoproliferative diseases seen in patients with hereditary or acquired forms of immunodeficiency. In this regard it is of interest to note that allograft recipients treated with the fungal peptide cyclosporin A which can inhibit T-cell-dependent immune responses *in vivo* (Borel *et al.*, 1977) and *in vitro* have shown a significant incidence of lymphomas (Calne *et al.*, 1979; Nagington and Gray, 1980). Nonhuman primate allograft recipients receiving cyclosporin A have also shown high levels of lymphoma development (Bieber *et al.*, 1980). A percentage of treated patients have shown increases in anti-VCA titers (Nagington and Gray, 1980) and one patient's lymphoma was shown to be EBNA positive (Crawford *et al.*, 1980). *In vitro*, lymphocytes from cyclosporin A-treated patients transformed more readily with EBV than did controls showing high levels of transformation even at high initial cell concentrations (Crawford *et al.*, 1981), suggesting an inhibition of specific memory cell proliferation. Cyclosporin added to lymphocyte cultures of normal, seropositive donors *in vitro* was found to enhance the establishment of lymphoid cell lines with or without the addition of EBV (Bird *et al.*, 1981), suggesting an effect on EBV-specific memory T cells. It seems clear that there are differences between cell lines arising as the result of immortalization, including lymphoid cell lines (LCL) established *in vitro* from normal donors, which may arise as a result of transformation *in vitro* from virus activated from latently infected cells (Epstein and Achong, 1979a,b), and those recovered from BL tumors. The LCL prior to becoming aneuploid have shown little capacity for growth in semisolid medium and for transplantation to athymic, nude mice by the subcutaneous route. Lines of BL origin, on the other hand, have the characteristic 14q+ chromosome marker, can grow in agarose, and can transplant subcutaneously in athymic mice (Nilsson *et al.*, 1977). It should be noted that cells and LCL transformed *in vitro* can transplant to athymic mice by the intracranial route (Rabin *et al.*, 1978; Schaadt *et al.*, 1979) which is a preferred site for the transplantation of lymphoid tumor cells (Epstein *et al.*, 1976). It would be of interest to determine the susceptibility levels to EBV of B cells from patients with XLP and other immunodeficiency diseases. The possibility exists that these cells may be more sensitive to virus and may have heightened growth or virus-producing properties after infection. It is also possible that factors which may effect the functional properties of cells which are of importance in controlling the growth of EBV immortalized cells, such as the IgG Fc minus killer cells present in acute IM, may be present and inhibit normal reactivity. The proposition that a compromised immunological makeup predisposes to EBV-induced disease may explain the high degree of susceptibility of cotton-topped marmosets to EBV. These marmosets show decreased antibody production to conventional antigens in assays both *in vivo* and *in vitro* as compared to white-lipped marmosets. Moreover, both cotton-topped and white-lipped marmosets show delayed allograft and even xenograft rejection compared with other mammalian species (Gengozian, 1978). An extensive exposition on EBV-associated lymphoproliferative disease in immunodeficient patients has recently been published in symposium form and should be consulted for further information (Klein and Purtilo, 1981).

While EBV seems confined to human beings in its natural host range, it is related to several other B-cell tropic viruses of Old World monkeys and apes. Viruses have been found in chimpanzees, baboons, orangutans, gorillas (reviewed by Deinhardt and Deinhardt, 1979), and African green monkeys (Bocker *et al.*, 1980). These simian viruses tend to have biological properties similar to EBV and to be related to EBV both antigenically and by DNA sequence homology. These viruses become latent in B cells *in vivo* and can immortalize B cells *in vitro*. Where studied, the structure of their genomes is also similar to that of EBV, and is collinear with it (Lee *et al.*, 1981a,b; Heller *et al.*, 1981; Heller and Kieff, 1981).

Of all these viruses, the isolate from hamadryas baboons, designated as Herpesvirus papio (HVP), has shown some evidence of disease relationship (Lapin, 1973, 1975, 1976). Baboons in the research colony at the Institute of Experimental Pathology and Therapy of the USSR Academy of Medical Science, Sukhumi, have had a continuing high incidence of spontaneous lymphoma since 1967. Affected animals show splenomegaly, lymphadenopathy, and other symptoms such as gingivitis and dermatitis. Tumor growth was described as mainly diffuse but nodular tumors also occurred. Analysis of 55 cases (Yakovleva *et al.*, 1980) indicated that the majority of tumors were non-Hodgkins lymphomas of the lymphoid type. These cases included lymphoblastic lymphosarcomas of the BL type and prolymphocytic lymphosarcomas.

Single cases of lymphoplasmacytic and immunoblastic lymphomas were seen and some cases of reticulosarcoma and lymphogranulomatosis were also noted. Among the lymphomas most were B-cell tumors but T-cell and null-cell neoplasms were apparently also identified. Cytogenetic studies have indicated a high frequency of chromosome breaks and structural rearrangements in cells from lymph node, spleen, and bone marrow of lymphomatous baboons (Markaryan, 1980). Chromosomes 20 and 1 were found to be affected more than other chromosomes. Peripheral blood lymphocytes from baboons with advanced disease were found to be hyporesponsive to phytohemagglutin and concanavalin A (Neubauer et al., 1980).

The tumor cases are exclusively confined to the main colony of about 900 animals. There is a second, smaller colony of about 100 baboons which is kept in a geographically separate site under seminatural conditions. No lymphoid disease has been observed in this so-called forest colony. Results of a serological study showed that baboons with clinical signs of disease had, as a group, antibodies to more HVP-associated antigens (VCA, EA, soluble, and nuclear) and at higher titers than did baboons living in contact with them or than baboons from the forest colony (Neubauer et al., 1979). No similar disease-related antibody patterns were seen with two other viruses originally isolated from the baboons of the main colony: a foamy virus and a cytomegalovirus. It has been possible to recover virus regularly from oral swabs taken from animals of the main colony while it has proved difficult to do so from baboons of the forest colony (Agrba et al., 1980). Preliminary results have indicated that the HVP genome is present in multiple copies in tumor tissue and at low or undetectable levels in normal tissues (Neubauer et al., 1980). However, extensive studies on this important point have not been reported.

For the most part, inoculation of HVP or HVP-containing cell lines into various species has not resulted in tumor induction although the recipients generally seroconverted and many shed virus following inoculation (Gerber et al., 1977; Rabin, unpublished results). Species used in these studies were cotton-topped marmosets, yellow baboons (Papio cynocephalus), and rhesus monkeys. Deinhardt et al. (1978) did achieve induction of an acute, widely disseminated, lymphoproliferative condition in adult cotton-topped marmosets but not in white-lipped marmosets on inoculation of large numbers ($1.3–5 \times 10^8$) of HVP-producer baboon cells. Similar inoculations in newborn marmosets induced only minimal signs of disease. It was apparently difficult to characterize the lesions in the adult marmosets precisely and they were described as being similar to a strong reactive response, to IM, or to immunoblastic lymphoadenopathy, and in some organs to lymphoma (Deinhardt and Deinhardt, 1979). Affected tissues were tested for viral genome using an EBV DNA probe with negative results. Several lymphoid cell lines were established from inoculated marmosets all of which appeared to be B cell in type and of marmoset origin. At least one cell line showed a low level of genome by EBV cRNA : cell DNA hybridization.

Thus, the Sukhumi colony shows a great deal of clinical disease which histologically resembles various types of lymphoid neoplasms. An EBV-related virus is highly prevelant in the colony where it can readily be isolated from oral swabs of diseased animals and their contacts. Diseased baboons have antibodies to several virus-associated antigens at relatively high titers. There is a degree of immunosuppression associated with the disease and there is some evidence that the viral genome is present in tumor cells in multiple copies. Finally, under certain circumstances the virus has shown a limited ability to induce a lymphoproliferative condition in adult cotton-topped marmosets.

This spontaneous and ongoing outbreak of lymphoid disease would seem to represent a nearly ideal situation in which to study the mechanism involved in the expression of what are apparently a diverse series of neoplasms. While there is too little information on hand to speculate as to these possible mechanisms, the roles played by possible inbreeding (Lapin, 1973), crowding, stress, and viruses such as HVP, would seem to be leading candidates for study. The role of HVP might be amenable to rather clear-cut evaluation by the use of a vaccination study employing a membrane antigen vaccine of the type suggested by Thorley-Lawson (1979).

While the simian EBV-related viruses have mainly been recovered from normal animals, including several isolates made in the United States from *Papio hamadryas, Papio anubis, Papio papio,* and *Papio cynocephalus* (Deinhardt and Deinhardt, 1979), the virus of orangutan origin was isolated from a cell line established from an animal which died of monomyelocytic leukemia (Rasheed et al., 1977). The cell line, even early after its establishment in culture, showed multiple chromosomal abnormalities including a deletion of the short arm of chromosome 22 which was also seen in cultured skin fibroblasts of the animal. Several other abnormalities were seen only in the lymphoid cells including deletion of a large portion of the long arm of one homolog of the X-chromosome, an increased prevalence of monoscmy of chromosomes 12 and 7, and a high degree of polyploidy. These cells in contrast to LCL can form colonies in agarose medium and transplant efficiently to athymic nude mice by the subcutaneous route (Rabin *et al.*, 1978). Even though the animal did not have a lymphoma, the presence of a constitutive chromosomal abnormality or the ready development of chromosomal abnormalities upon virus-induced transformation may well have contributed to the phenotype exhibited by these cells. In this same vein, cells in an apparently benign lymph node from a patient with nasopharyngeal carcinoma was found to have the 14q+ chromosome marker (Mitelman *et al.*, 1979). A percentage of cells in the lymph node population also had trisomy of chromosome 17 and an additional marker chromosome was present in a proportion of the cells. This node in addition to being histologically benign was negative for the presence of the EBV genome. Thus, it may be that in certain cases of BL, the chromosomal changes may occur prior to virus infection. When immortalization of such already abnormal cells occurs, the resultant population may be much more neoplastic in phenotype than is the case when normal diploid cells are immortalized.

5. T-Cell Tumors Associated with Herpesviruses of New World Primates

In contrast to the B-cell tropic viruses of man, Old World monkeys, and apes, the lymphotropic herpesviruses of New World primates are T-cell tropic. Such viruses have been isolated from two species, HVS from squirrel monkeys and HVA from spider monkeys. Several reviews have been written on these viruses and the diseases which they induce (Fleckenstein, 1979; Neubauer and Rabin, 1979; Deinhardt and Deinhardt, 1979; Rangan and Gallagher, 1979) and we will not attempt to duplicate this material here except where necessary for clarity.

Herpesvirus saimiri is the more extensively studied virus of the two. Several strains of HVS exist which are closely related but which can be differentiated from each other by the cleavage patterns of their DNAs with restriction endonucleases (Fleckenstein, 1979). Both field strains and laboratory strains of HVS exist and exhibit different biological properties, especially in regard to oncogenicity. It has been possible to generate attenuated strains of HVS by various tissue culture techniques. These attenuated strains have been used as experimental vaccines (see Fleckenstein, 1979, for discussion).

Herpesvirus saimiri has a broad experimental host range regularly producing tumors in cotton-topped and white-lipped marmosets, and in owl monkeys and howler monkeys. Tumorigenicity in common marmosets and capuchin monkeys has been variable from laboratory to laboratory. The virus has not been shown to induce tumors in its natural host under natural conditions or after intentional inoculation even when combined with immunosuppressive treatment (Falk *et al.*, 1973; Martin and Allen, 1975).

In its natural host HVS is horizontally transmitted probably by way of oral secretions (Falk *et al.*, 1973). Virus can be transmitted from squirrel monkeys to owl monkeys by apparently the same means and can result in the development of lymphoma (Hunt *et al.*, 1973; Rabin *et al.*, 1975). Herpesvirus saimiri can be readily recovered from peripheral blood lymphocytes of squirrel monkeys where it is associated with the T-cell fraction (Wright *et al.*, 1976). In infected owl and marmoset monkeys the virus also exhibits strong tropism for T cells (Wallen *et al.*, 1973). There is no evidence that it can be transmitted among experimentally infected monkeys, other than squirrel monkeys.

In infected owl monkeys, while there is wide variation in the time course of the disease (deaths occurring from 60 to 330 days after inoculation and certain other parameters such as the levels of DNA synthesis of PBL and detectable levels of antiproliferative factors in the peripheral blood) several events occur on a consistent basis. These events are (1) the loss of lymphocyte responsiveness to general mitogens, (2) the appearance of high levels of lymphocyte infectious centers (IC), (3) the rise in titer of antibody to HVS-associated EA, (4) a rise in titer of antibody to HVS-associated membrane antigen as determined by ADCC, and (5) the appearance of a suppressor T-cell population which accompanies the loss of mitogen response and increase in ADCC titer. These signs appear in the 75% of owl monkeys which develop fatal disease. The other 25% of infected owl monkeys survive with a chronic HVS infection where there is apparently no immunosuppression nor dramatic increases in antibody titers. Neither chronically nor fatally infected owl monkeys show a cell-mediated immune response to HVS antigens in tests with peripheral blood lymphocytes while such responses are readily demonstrable in naturally infected squirrel monkeys (Neubauer *et al.*, 1980b). The susceptibility of owl monkeys to HVS does not seem to be related to any particular karyotype of the several types described for owl monkeys (Ma *et al.*, 1976, 1978; Neubauer *et al.*, 1980b).

Recent studies on infected owl monkeys in our laboratory (Neubauer *et al.*, 1981) have been concerned with the distribution of HVS among various cell types and its effect on the functional behavior of infected cells. In chronically infected monkeys and in a leukemic animal, HVS was found to be associated solely with T cells. Separation of T cells into those positive for the Fc receptor of IgG (Tγ+ cells) and those negative for this receptor (Tγ- cells) showed that essentially both types of T cells were infected and to about the same levels. In the chronically infected monkeys which had been thought to be immunologically normal, we found that the response of the Tγ+ cells to PHA was lost while that of the Tγ− cells was intact. During leukemia both Tγ+ and Tγ− cells increased in the peripheral blood. This observation tends to confirm previous results in infected marmosets that the tumor is polyclonal (Chu and Rabson, 1972; Rabin et al., 1973). Marmosets are hematologic chimaeras and cell lines established from tumors of marmosets which are bisexual chimaeras have shown both male and female cells. Moreover, in this particular owl monkey, the suppressor cell was shown to be an X-ray-resistant, Tγ− suppressor cell.

Thus, HVS-induced lymphoma seems to be a primary consequence of virus infection. In this respect it resembles the acute, polyclonal B-cell tumors associated with EBV. Similarly, HVS-induced tumors are associated with a high degree of immunological alteration in the host, although many of these alterations seem to be associated with the developing disease, at least in owl monkeys. The same arguments pertaining to a lack of full immunocompetency on the part of cotton-topped marmosets, which may explain the high degree of susceptibility of these animals to EBV-induced disease, may be made for their high level of susceptibility to HVS.

Little is known about the phenotypes of HVS tumor cells or cells transformed by HVS in vitro. Transformation by this virus in vitro has been extremely difficult to accomplish and cell lines have not been available for study. There are recent indications that the technical problems relative to HVS-induced transformation have been overcome (Fleckenstein, 1979) and we may expect to see studies on these cells reported in the near future. There are several cell lines available which were established from experimental HVS tumors. There are scattered observations on these cell lines which are of some interest. These lines generally maintain a mixed bisexual population if initiated from a bisexually chimaeric marmoset, the karyotypes seem to be normal diploid. The cells are functionally active producing interferon continuously over periods in culture of several months (Wright et al., 1974). An antiproliferation factor was described (Neubauer et al., 1975), and short-term cultures from owl monkeys were shown to release several lymphokines including lymphotoxin, leukocyte inhibitory factor, interferon, and antiproliferation factor (Neubauer and Rabin, 1979). The cells have shown little capacity for heterotransplantation, even by the intracranial route in nude mice, and generally fail to form colonies in agarose medium (Rabin and Hopkins, unpublished results). Those cell lines of marmoset origin replicate in culture without the need for added T-cell growth factor (TCGF). One of these lines, 70-N-2 (Falk et al.,

1974) which has a $T\gamma-$ phenotype, has recently been shown to exhibit helper cell function in a mixed-species assay (J. J. Marchalonis, personal communication). These cells also react with a monoclonal antibody which defines human helper cells (Neubauer, 1982; Ledbetter *et al.*, 1981). Interestingly, 70-N-2 cells and several other HVS or HVA tumor or transformed marmoset cells showed cytotoxic activity for a variety of human target cell types (Johnson and Jondal, 1981). Earlier studies had shown that lymphocytes from HVS-infected owl monkeys were cytotoxic for xenogeneic target cells (Wallen *et al.*, 1974). Owl monkey lines may soon become available for comparative study as G. Pearson (personal communication) has been able to establish a continuous line of owl monkey T cells from an HVS tumor with the help of added TCGF. Surface properties of the marmoset lines maintain T-cell characteristics, sheep cell rosette formation, and characteristic glycoprotein profile (Wallen *et al.*, 1973; Strnad *et al.*, 1979). There is recent evidence that neoplastic T cells can be cultured from humans with the aid of TCGF and without the need for prior activation *in vitro* (Poiesz *et al.*, 1980b). It would be of great interest to see if this same ability would be a consistent feature of the HVS tumor cells.

It is of interest to speculate on the polyclonal nature of HVS lymphoma. Little is known about HVS transformation or the properties of HVS-transformed cells. The tumor-derived cell lines have not shown evidence of a neoplastic phenotype by karyotype, transplantability, or growth in agarose. The disease is characterized by a cytotoxic/suppressor cell but at least one continuous cell line apparently has helper cell function in addition to cytotoxic activity. In chronically infected monkeys there is evidence for impairment of $T\gamma+$ cell function while $T\gamma-$ cells seem unaffected. While it is possible that all T-cell types are altered in growth and function by the virus, it may also be that alterations occur only in certain subpopulations which then maintain the other T cells in continuous growth via the release of factors, such as TCGF. Examination of tumor cell populations from several animals for evidence of clonal subpopulations should be of value in future studies. Attempts to develop cloned tumor cell lines for detailed characterization relative to neoplastic properties should also be of value.

Herpesvirus ateles has been less well studied than HVS but offers the potential value that it transforms lymphocytes readily *in vitro* (Falk *et al.*, 1974). Such transformed cells have the general properties of T lymphocytes but otherwise have been largely uncharacterized. They do not appear to require TCGF for growth. Studies *in vivo* with HVA for comparative purposes would seem to be long overdue.

At the present time HVS and HVA represent the only feasible means of regularly generating primate models of T-cell lymphoma. Unlike EBV which seems to reproduce a spectrum of EBV-associated human lymphoproliferative diseases, it is not yet clear what human neoplasms may be most closely paralleled by HVS- and HVA-induced disease. Human T-cell lymphomas are apparently less frequent than B-cell tumors but are found among 20–30% of acute lymphoblastic leukemias, a low percentage of chronic lymphocytic leukemias and non-Hodgkin's lymphomas, and in Sezary's syndrome and mycosis fungoides (see review by Moretta *et al.*, 1979). Examination of lymphocytes from four of five patients with Sezary's syndrome showed that these cells had helper cell function as indicated by their capacity to aid normal B lymphocytes synthesize immunoglobulins after treatment with pokeweed mitogen (Broder *et al.*, 1976). Work from this same group (Broder *et al.*, 1978) describes a case of acute lymphoblastic leukemia in which the leukemic cells suppressed the helper cell function of normal T cells in an immunoglobulin release helper cell assay. This result is similar to that described for HVS leukemia where in one case the suppressor cells were shown to inhibit normal T cell help in a mitogen-induced, proliferation assay (Neubauer *et al.*, 1981). Finally, there is at least one case of T-cell leukemia in man characterized by the presence of both suppressor and helper cells (Saxon *et al.*, 1979). A patient with ataxia telangiectasia was described who developed lymphoid leukemia. This patient's peripheral blood was comprised mainly of T cells with both $T\mu$ (Fc receptor for IgM) and $T\gamma$ cells present. In *in vitro* assays both radiation-resistant helper cell function and radiation-sensitive suppressor cell function (at high T-cell concentrations) were demonstrated. The $T\mu$ and $T\gamma$ populations were distinct, but both contained a characteristic 14q+ chromosome marker. Mitogen-stimulated bone marrow cells did not show this marker. It is conceivable in this case that a transformation occurred in a stem cell which then differentiated giving rise to the two populations of T cells found in the peripheral blood and that the leukemia was not the result of two separate transforming events. As discussed, resolution of the tumorigenesis mechanisms remains to be resolved in the cases of both HVS- and HVA-induced neoplasms.

6. Possible Mechanisms of Herpes Simplex Virus Transformation

A brief discussion of cell transformation by HSV is included in this review as illustrating a possible alternative means by which common viruses may alter cell phenotypes. This discussion will be limited to tumorigenic transformation and will not cover either biochemical or morphological transformation. For a comprehensive review of HSV-associated cell transformation, see Hampar (1981).

The HSV system suffers in comparison to similar studies with lymphotropic herpesviruses in that it is difficult to relate transformation *in vitro* to tumorigenesis *in vivo,* as HSV has not yet been shown to be tumorigenic. Moreover, tumors which on epidemiological grounds are thought to be associated with HSV are carcinomas, while the systems *in vitro* use fibroblasts. These factors make it difficult to assess results obtained *in vitro* in relationship to what may happen in actual tumorigenesis.

However, it is possible in the HSV system to begin with cells which are not tumorigenic, as defined by their being able to transplant to syngeneic or allogeneic hosts, and confer this property to them by exposure to inactivated virus, viral DNA, or DNA fragments. In the case of HSV, transformation is a relatively late event which occurs with low frequency. Therefore, it is difficult to determine if the viral genes are limited to initiation of the process and other factors account for the tumorigenic properties. This circumstance would be somewhat analogous to the acquisition of tumorigenic properties by EBV-lymphoid cell lines after prolonged periods of growth *in vitro* or even in the ideas relative to BL where EBV is thought to immortalize B cells while other events, such as acquisition of the specific chromosomal changes, fix the neoplastic state. A second feature of HSV transformation is the variable retention of viral genes. While it is clear that transformed cells of some species retain viral genes and express them, other cells tend to lose sequences on continued cultivation, while still others seem to lack detectable levels even when assayed early after transformation (see discussion by Hampar, 1981). The specific initiation of transformation without the necessary retention of viral genes, usually presumed to be critical for the maintenance of the transformed state, has been referred to as a "hit and run" mechanism (Skinner, 1976).

An illustration of how such a mechanism might apply to HSV tumorigenesis can be found in the recent experiments of Hampar *et al.* (1980). In these experiments, a cloned line of BABL/c mouse fibroblasts, 10E2 cells, was used. 10E2 cells at low passage levels (<p35) are uniformly flat, grow poorly in low serum medium, have a relatively low saturation density, do not form colonies in semisolid medium, and fail to transplant to nude mice. At higher passage levels 10E2 cells show increased growth capacity (higher saturation density, better growth on low serum medium) and become transplantable to nude mice. They also show foci of morphologically distinct but contact inhibited cells which when expanded are also transplantable. At relatively low pas-

sage levels (e.g., p17), ultraviolet light-inactivated HSV induced foci of cells with the same morphology as those spontaneously transformed foci which appear at later passage. Cells propagated from these HSV-induced foci transplant efficiently to both nude and intact BALB/c mice, as do cells grown from spontaneous foci. When assayed for a variety of properties including plating efficiency, growth in agarose and agar media, sensitivity to infection by HSV, and cell growth in 1 and 10% serum media, as well as transplantability, the two types of cells were indistinguishable from each other. Moreover, no evidence for HSV-specific DNA, RNA, or protein could be found in these cells. Similar results were reported by Bellett and Younghusband (1979) in which adenovirus type 5 increased the frequency of virus-negative, tumorigenic variants of C57 black mouse embryo cells in culture.

These experiments can be interpreted to mean that in this case, HSV potentiates what is essentially a cellular event and is not altering the cell phenotype by directly superimposing its genes on those of the cell. If such a situation actually exists under natural circumstances, common viruses may be more highly involved in tumor initiation than previously thought. A possible example of such a circumstance is represented by the apparently virus-negative, feline lymphosarcomas which are epidemiologically associated with exposure to feline leukemia virus (Hardy et al., 1980). Alternatively, of course, it is possible that small fragments of viral nucleic acid persist in the transformed cells and are critical to the maintenance of the transformed state, but current viral probes are incapable of detecting them.

IV. Conclusions

The extensive efforts in the study of transforming retroviruses have led to the discovery of the transforming genes of these viruses and their counterparts present in normal cells. In at least one system involving a nontransforming or subacute virus, evidence for such a cellular gene was found associated with viral LTR sequences in a high percentage of tumors. The ability to transfect cells successfully with cellular DNA from tumor, transformed, and even normal cells indicates the presence of specific transforming sequences in these cells, suggesting the possibility of a general mechanism of cell transformation. However, the possibility for multiple transforming genes not necessarily belonging to a single class is suggested by the idea that the env gene of SFFV may be the transforming gene of this virus.

In the case of EBV and of certain nontransforming retroviruses, the concept that the virus serves to initiate a multistage process eventually leading to neoplasia has been well developed for some time. The extreme cases of this type may be represented by viruses which maintain an impermanent association with parental cells whose progeny may transform (HSV) or which may act indirectly, possibly through lymphokines, on potentially transformable cells. In the case of the T-cell tropic herpesviruses, it is even difficult at the moment to know if there are multiple target cells for virus transformation or if the expanding T-cell population represents the results of both transformation and indirect immunologic events.

The capacity of EBV to stimulate polyclonal replication of B cells which, in an immunologically intact individual, is a benign process, would appear to be a deadly property of the virus in individuals with certain hereditary or acquired immune deficiencies. The nature of the dysfunctions which permit rapid primary EBV-induced proliferative disease to occur remains to be described. The always intriguing question of what controls EBV-positive cells in infected individuals has been answered in part with the finding of specific memory T cells. What happens to such cells in individuals developing EBV-associated tumors is an important area for study.

Thus, while much speculation on mechanisms of viral tumorigenesis remains, recent findings have clarified points relative to both transformation and tumor immunity and have helped to focus many efforts currently under way in the study of experimental and spontaneous tumors.

ACKNOWLEDGMENTS

The authors thank Margaret Fanning and Lindy Lawyer for assistance in the preparation of this manuscript.

This work was supported by Contract NO-1-CO-75380 with the National Cancer Institute, National Institutes of Health, Bethesda, Maryland.

REFERENCES

Ablashi, D. V., Pearson, G., Rabin, H., Armstrong, G., Easton, J., Valerio, M., and Cicmanec, J. (1978). Biomedicine 29, 7–10.

Agrba, V. Z., Lapin, B. A., Timanovskaya, V. V., Dzhachvliany, M. Ch., Kokosha, L. C., Chuvirov, G. N., and Djatchenko, A. G. (1980). Exp. Mol. Pathol. 18, 269–274.

Andersson, P., Goldfarb, M. P., and Weinberg, R. A. (1979). Cell 16, 63–75.

Astrin, S., Buss, E. G., and Hayward, W. S. (1980). Nature (London) 282, 339–340.

Axelrad, A. A., and Steeves, R. A. (1964). Virology 24, 513–518.

Bakacs, T., Svedmyr, E., Klein, E., Rombo, L., and Weiland, D. (1978). Cancer Lett. (Shannon, Irel.) 4, 185–189.

Bar, R. S., DeLor, C. J., Clausen, K. P., Hurtubise, P., Henle, W., and Hewetson, J. F. (1974). *N. Engl. J. Med.* **290**, 363–367.

Bellet, A. J. D., and Younghusband, H. B. (1979). *J. Cell. Physiol.* **101**, 33–48.

Bentvelzen, P. (1972). *In* "RNA Viruses and Host Genome in Oncogenesis" (P. Emmelot and P. Bentvelzen, eds.), pp. 309–337. Elsevier, Amsterdam.

Berning, A. K., Eicher, E. M., Paul, W. E., and Scher, I. (1980). *J. Immunol.* **124**, 1875–1877.

Beug, H., Kitchener, G., Doedeclein, G., Graf, T., and Hayman, M. J. (1980). *Proc. Natl. Acad. Sci. U.S.A.* **77**, 6683–6686.

Bieber, C. P., Reitz, B. A., Jamieson, S. W., Oyer, P. E., and Stinson, E. B. (1980). *Lancet* **1**, 43.

Bird, A. G., McLachlan, S. M., and Britton, S. (1981). *Nature (London)* **289**, 300–301.

Bishop, J. M. (1981). *Cell* **23**, 5–6.

Blair, D. G., McClements, W. L., Oskarsson, M. K., Fischinger, P. J., and Vande Woude, G. F. (1980). *Proc. Natl. Acad. Sci. U.S.A.* **77**, 3504–3508.

Blair, D. G., Oskarsson, M., Wood, T. G., McClements, W. L., Fischinger, P. J., and Vande Woude, G. F. (1981). *Science* **212**, 941–942.

Bocker, J. F., Tredemann, K.-H., Bornkamm, G. W., and zur Hausen, H. (1980). *Virology* **101**, 291–295.

Borel, J. F., Feurer, C., Magnee, C., and Stahelin, H. (1977). *Immunology* **32**, 1017–1025.

Bornkamm, G. W., Stein, H., Lennert, K., Ruggeberg, R., Bartels, H., and zur Hausen, H. (1976). *Int. J. Cancer* **17**, 177–181.

Broder, S., Edelson, R. L., Lutzner, M. A., Nelson, D. L., MacDermott, R. P., Durm, M. E., Goldman, C. K., Meade, B. D., and Waldmann, T. A. (1976). *J. Clin. Invest.* **58**, 1297–1306.

Broder, S., Poplack, D., Whang-Peng, J., Durm, M., Goldman, C., Muul, L., and Waldmann, T. A. (1978). *N. Engl. J. Med.* **298**, 66–72.

Brugge, J. S., and Erikson, R. L. (1977). *Nature (London)* **269**, 346–348.

Brugge, J. S., Erikson, E., Collett, M. S., and Erikson, R. L. (1978a). *J. Virol.* **26**, 773–782.

Brugge, J. S., Steinbaugh, P. J., and Erikson, R. L. (1978b). *Virology* **91**, 130–140.

Brugge, J. S., Collett, M. S., Siddiqui, A., Marczynka, B., Deinhardt, F., and Erikson, R. L. (1979). *J. Virol.* **29**, 1196–1203.

Calne, R. Y., Rolles, K., White, D. J. G., Thiru, S., Evans, D. B., McMaster, P., Dunn, D. C., Craddock, G. N., Henderson, R. G., Aziz, S., and Lewis, P. (1979). *Lancet* **2**, 1033–1036.

Chu, E. W., and Rabson, A. S. (1972). *J. Natl. Cancer Inst.* **48**, 771–775.

Coffin, J. M., Varmus, H. E., Bishop, J. M., Essex, M., Hardy, W. D., Jr., Martin, G. S., Rosenberg, N. E., Scolnick, E. M., Weinberg, R. A., and Vogt, P. K. (1981). *J. Virol.* **40**, 953–957.

Collett, M. S., and Erikson, R. L. (1978). *Proc. Natl. Acad. Sci. U.S.A.* **75**, 2021–2024.

Collett, M. S., Brugge, J. S., and Erikson, R. L. (1978). *Cell* **15**, 1363–1369.

Collett, M. S., Erikson, E., and Erikson, R. L. (1979). *J. Virol.* **29**, 770–781.

Collett, M. S., Purchio, A. F., and Erikson, R. L. (1980). *Nature (London)* **85**, 167–169.

Cooper, G. M., and Neiman, P. E. (1980). *Nature (London)* **287**, 656–659.

Cooper, G. M., Okenquist, S., and Silverman, L. (1980). *Nature (London)* **284**, 418–421.

Copeland, N. G., Zelenetz, A. D., and Cooper, G. M. (1979). *Cell* **17**, 993–1002.

Crawford, D. H., Epstein, M. A., Achong, B. G., Finertz, S., Newman, J., Liversedge, S., Tedder, R. S., and Steward, J. W. (1979). *J. Infect.* **1**, 37–48.

Crawford, D. H., Thomas, J. A., Janossy, G., Sweny, P., Fernando, O. N., Moorhead, J. F., and Thompson, J. H. (1980). *Lancet* **1**, 1355–1356.

Crawford, D. H., Sweny, P., Edwards, J. M. B., Janossy, G., and Hoffbrand, A. V. (1981). *Lancet* **1**, 10–12.

Crittenden, L. B., Hayward, W. S., Hanafusa, H., and Fadley, A. M. (1980). *J. Virol.* **33**, 915–919.

Czernilofsky, A. P., Levinson, A. D., Varmus, H. E., Bishop, J. M., Tischer, E., and Goodman, H. M. (1980). *Nature (London)* **287**, 198–203.

DeFeo, D., Gonda, M. A., Young, H. A., Chang, E. H., Lowy, D. R., Scolnick, E. M., and Ellis, R. W. (1981). *Proc. Natl. Acad. Sci. U.S.A.* **78**, 3328–3332.

Deinhardt, F., and Deinhardt, J. (1979). *In* "The Epstein-Barr Virus" (M. A. Epstein and B. G. Achong, eds.), pp. 373–415. Springer-Verlag, Berlin and New York.

Deinhardt, F., Falk, L., Wolfe, L. G., Paciga, J., and Johnson, D. (1975). *In* "Oncogenesis and Herpesviruses II" (G. de-The, M. A. Epstein, and H. zur Hausen, eds.), Part 2, pp. 161–168. International Agency for Research on Cancer, Lyons, France.

Deinhardt, F., Falk, L., Wolfe, L. G., Schudel, A., Nonoyama, M., Lai, P., Lapin, B., and Yakovleva, L. (1978). *Primatol. Med.* **10**, 163–170.

Desgranges, C., de-The, G., Ho, J. H. C., and Ellouz, R. (1977). *Int. J. Cancer* **19**, 627–633.

de-The, G., Zeng, Y., and Desgranges, C. (1981a). *Proc. Int. Symp. Comp. Res. Leukemia Related Dis., 10th, Los Angeles,* p. 165.

de-The, G., Desgranges, C., Bornkamm, G. W., and Zeng, Y. (1981b). *Int. Workshop Herpesviruses,* p. 241.

de-The, G., Geser, A., Day, N. E., Tukei, P. M., Williams, E. H., Beri, D. P., Smith, P. G., Dean, A. G., Bornkamm, G. W., Feorino, P., and Henle, W. (1978). *Nature (London)* **274**, 756–761.

Dhar, R., McClements, W. L., Enquist, L. W., and Vande Woude, G. F. (1980). *Proc. Natl. Acad. Sci. U.S.A.* **77**, 3937–3941.

Doolittle, W. F., and Sapienza, C. (1980). *Nature (London)* **284**, 601–603.

Dresler, S., Ruta, M., Murray, M. J., and Kabat, D. (1979). *J. Virol.* **30**, 564–575.

Eckhart, W., Hutchinson, M. A., and Hunter, T. (1979). *Cell* **18**, 925–933.

Elder, J., Gautsch, J., Jensen, F., Lerner, R., Hartley, J., and Rowe, J. P. (1977). *Proc. Natl. Acad. Sci. U.S.A.* **74**, 4676–4680.

Ellis, R. W., DeFeo, D., Maryak, V. M., Young, H. A., Shih, T. Y., Chang, E. H., Lowy, D. R., and Scolnick, E. M. (1980). *J. Virol.* **36**, 408–420.

Epstein, M. A., and Achong, B. G. (1979a). *In* "The Epstein-Barr Virus," (M. A. Epstein and B. G. Achong, eds.), pp. 1–22. Springer-Verlag, Berlin and New York.

Epstein, M. A., and Achong, B. G. (1979b). *In* "The Epstein-Barr Virus" (M. A. Epstein and B. G. Achong, eds.), pp. 321–337. Springer-Verlag, Berlin and New York.

Epstein, M. A., Hunt, R. D., and Rabin, H. (1973a). *Int. J. Cancer* **12**, 309–318.

Epstein, M. A., Rabin, H., Ball, G., Rickinson, A. B., Jarvis, J., and Melendez, L. V. (1973b). *Int. J. Cancer* **12**, 319–332.

Epstein, M. A., zur Hausen, H., Ball, G., and Rabin, H. (1975). *Int. J. Cancer* **15**, 17–22.

Epstein, A. L., Herman, M. M., Kim, H., Dorfman, R. F., Path, M. R. C., and Kaplan, H. S. (1976). *Cancer* **37**, 2158–2176.

Eva, A., Robbins, K., Andersen, P., Srinivasan, A., Papas, T., Tronick, S., Reddy, E. P., Westin, E., Wong-Staal, E., Ellmore, N., Gallo, R., and Aaronson, S. A. (1981). *Proc. Int. Symp. Comp. Res. Leukemia Related Dis., 10th, Los Angeles.*

Falk, L. A., Nigida, S., Deinhardt, F., Cooper, R. W., and Hernandez-Camacho, J. I. (1973). *J. Natl. Cancer Inst.* **51**, 1987–1989.

Falk, L. A., Wright, J., Wolfe, L., and Deinhardt, F. (1974). *Int. J. Cancer* **14**, 244–251.

Falk, L., Deinhardt, F., Wolfe, L., Johnson, D., Hilgers, J., and de-The, G. (1976). *Int. J. Cancer* **17**, 785–788.

Falletta, J. M., Fernbach, D. J., Singer, D. B., South, M. A., Landing, B. H., Heath, C. W., Jr., Shore, N. A., and Barrett, F. F. (1973). *J. Pediatr.* **83**, 549–556.

Favera, R. D., Gelmann, E. P., Gallo, R. C., and Wong-Staal, F. (1981). *Nature (London)* **292**, 31–35.

Feldman, R. A., Hanafusa, T., and Hanafusa, H. (1980). *Cell* **22**, 757–765.

Fialkow, P. J., Klein, G., Gartler, S. M., and Clifford, P. (1970). *Lancet* **1**, 384–386.

Fischinger, P. J. (1980). *In* "Molecular Biology of RNA Tumor Viruses" (J. R. Stephenson, ed.), pp. 163–198. Academic Press, New York.

Fleckenstein, B. (1979). *Biochim. Biophys. Acta* **560**, 301–342.

Franchini, G., Even, J., Sherr, C. J., and Wong-Staal, F. (1981). *Nature (London)* **290**, 154–157.

Friend, C. (1957). *J. Exp. Med.* **105**, 307–318.

Fung, Y. T., Fadly, A. M., Crittenden, L. B., and Kung, H. J. (1981). *Proc. Natl. Acad. Sci. U.S.A.* **78**, 3418–3422.

Gatti, R. A., and Good, R. A. (1971). *Cancer* **28**, 89–98.

Gengozian, N. (1978). *Primatol. Med.* **10**, 173–183.

Gerber, P., Kalter, S. S., Schidlovsky, G., Peterson, W. D., Jr., and Daniel, M. D. (1977). *Int. J. Cancer* **20**, 448–459.

Gilden, R. V. (1977). *In* "Molecular Biology of Animal Viruses" (D. P. Nayak, ed.), pp. 435–542. Dekker, New York.

Gilden, R. V., Oroszlan, S., Young, H. A., Rice, N. R., Gonda, M. A., Cohen, M., and Rein, A. (1981). *In* "Frontiers in Immunogenetics" (W. H. Hildemann, ed.), pp. 191–223. Elsevier, Amsterdam.

Goff, S. P., Gibbon, E., Witte, O. N., and Baltimore, D. (1980). *Cell* **22**, 777–785.

Hamilton, J. K., Paquin, L. A., Sullivan, J. L., Maurer, H. S., Cruzi, F. G., Provisor, A. J., Steuber, C. P., Hawkins, E., Yawn, D., Cornet, J. A., Clausen, K., Finkelstein, G. Z., Landing, B., Grunnet, M., and Purtilo, D. T. (1980). *J. Pediatr.* **96**, 669–673.

Hampar, B. (1981). *Adv. Cancer Res.* **35**, 27–47.

Hampar, B., Boyd, A. L., Derge, J. G., Zweig, M., Eader, L., and Showalter, S. D. (1980). *Cancer Res.* **40**, 2213–2222.

Hapel, A. J., Lee, J. C., Farrar, W. L., and Ihle, J. N. (1981). *Cell* **25**, 179–186.

Haran-Ghera, N. (1977). *In* "Symposium on Radiation Induced Leukemogenesis and Related Virus" (J. F. Duplan, ed.), pp. 79–89. Elsevier, Amsterdam.

Hardy, W. D., Jr., McClelland, A. J., Zuckerman, E. E., Snyder, H. W., Jr., MacEwan, E. G., Francis, D., and Essex, M. (1980). *Nature (London)* **288**, 90–92.

Hartley, J. W., Wolford, N. K., Old, L. J., and Rowe, W. P. (1977). *Proc. Natl. Acad. Sci. U.S.A.* **74**, 789–792.

Hayman, M. J. (1981). *J. Gen. Virol.* **52**, 1–4.

Hayward, W. S., Neel, B. G., and Astrin, S. M. (1981). *Nature (London)* **209**, 475–480.

Heller, M., and Kieff, E. (1981). *J. Virol.* **37**, 821–826.

Heller, M., Gerber, P., and Kieff, E. (1981). *J. Virol.* **37**, 698–709.

Henle, W., and Henle, G. (1979). *In* "The Epstein-Barr Virus" (M. A. Epstein and B. G. Achong, eds.), pp. 61–78. Springer-Verlag, Berlin and New York.

Heubner, R. J., and Todaro, G. J. (1969). *Proc. Natl. Acad. Sci. U.S.A.* **64**, 1087–1094.

Heubner, R. J., Gilden, R. V., Toni, R., Hill, R. W., Trimmer, R. W., Fish, D. C., and Sass, B. (1976). *Proc. Natl. Acad. Sci. U.S.A.* **73**, 4633–4635.

Ho, H. C., Kwan, H. C., Mun, H. Ng, and de-The, G. (1978a). *Lancet* **1**, 436.

Ho, H. C., Mun, H. Ng, and Kwan, H. C. (1978b). *Br. J. Cancer* **37**, 356–362.

Horak, I., Enjuanes, L., Lee, J. C., and Ihle, J. N. (1981). *J. Virol.* **37**, 483–487.

Hughes, S. H., Shank, P. R., Spector, D. H., Kung, H.-J., Bishop, J. M., Varmus, H. E., Vogt, P. K., and Breitman, M. L. (1978). *Cell* **15**, 1397–1410.

Hunt, R. D., Garcia, F. G., Barahona, H. H., King, N. W., Fraser, C. E. O., and Melendez, L. V. (1973). *J. Infect. Dis.* **127**, 723–725.

Hunter, T. (1980). *Cell* **22**, 647–648.

Hunter, T., and Sefton, B. (1980). *Proc. Natl. Acad. Sci. U.S.A.* **77**, 1311–1315.

Hynes, R. O. (1980). *Cell* **21**, 601–602.

Ihle, J. N., Lee, J. C., Enjuanes, L., Cicurel, L., Horak, I., and Pepersack, L. (1980). *In* "Viruses in Naturally Occurring Cancers" (M. A. Essex, G. J. Todaro, and H. zur Hausen, eds.), Vol. 7, pp. 1049–1064. Cold Spring Harbor Laboratory, Cold Spring Harbor, New York.

Ihle, J. N., Lee, J. C., and Rebar, L. (1981a). *J. Immunol.* **127**, 2565–2570.

Ihle, J. N., Pepersack, L., and Rebar, L. (1981b). *J. Immunol.* **126**, 2184–2189.

Jarvis, J. E. (1974). *Br. J. Cancer* **30**, 164–167.

Johnson, D. R., and Jondal, M. (1981). *Nature (London)* **291**, 81–83.

Ju, G., and Skalka, A. M. (1980). *Cell* **22**, 379–386.

Judson, S. C., Henle, W., and Henle, G. (1977). *N. Engl. J. Med.* **297**, 464–468.

Kalyanaraman, V. S., Sarngadharan, M. G., Poiesz, B. J., Ruscetti, F. W., and Gallo, R. C. (1981a). *J. Virol.* **38**, 906–915.

Kalyanaraman, V. S., Sarngadharan, M. G., Bunn, P. A., Minna, J. D., and Gallo, R. C. (1981b). *Nature (London)* **294**, 271–273.

Klein, G. (1979). *Proc. Natl. Acad. Sci. U.S.A.* **76**, 2442–2446.

Klein, G., and Purtilo, D. T. (1981). *Cancer Res.* **41**, 4209–4304.

Krontiris, T. G., and Cooper, G. M. (1981). *Proc. Natl. Acad. Sci. U.S.A.* **78**, 1181–1184.

Lai, M. M. C., Hu, S. S. F., and Vogt, P. K. (1979). *Virology* **97**, 366–377.

Lai, M. M. C., Neil, J. C., and Vogt, P. K. (1980). *Virology* **100**, 475–483.

Lange, B., Arbeter, A., Hewetson, J., and Henle, W. (1978). *Int. J. Cancer* **22**, 521–527.

Langebeheim, H., Shih, T. Y., and Scolnick, E. M. (1980). *Virology* **106**, 292–300.

Lanier, A. P., Henle, W., Bender, T. R., Henle, G., and Talbot, M. L. (1980). *Int. J. Cancer* **26**, 133–137.

Lapin, B. A. (1973). *In* "Unifying Concepts of Leukemia, Bibl. Haemat." (R. M. Dutcher and L. Chieco-Bianchi, eds.), No. 39, pp. 263–268. Karger, Basel.

Lapin, B. A. (1975). *In* "Comparative Leukemia Research 1973, Leukemogenesis, Bibl. Haemat." (Y. Ito and R. M. Dutcher, eds.), No. 40, pp. 75–84. Karger, Basel.

Lapin, B. A. (1976). *In* "Comparative Leukemia Research 1975, Bibl. Haemat." (J. Clemmesen and D. S. Yohn, eds.), No. 43, pp. 212–215. Karger, Basel.

Ledbetter, S. A., Evans, R. L., Lipinski, M., Cunningham-Rundles, C., Good, R. A., and Herzenberg, L. A. (1981). *J. Exp. Med.* **153**, 310–323.

Lee, J. C., and Ihle, J. N. (1979). *J. Immunol.* **123**, 2351–2358.

Lee, J. C., and Ihle, J. N. (1981). *Nature (London)* **289**, 407–409.

Lee, Y. S., Tanaka, A., Lau, R. Y., Nonoyama, M., and Rabin, H. (1981a). *J. Virol.* **37**, 710–720.

Lee, Y. S., Nonoyama, M., and Rabin, H. (1981b). *Virology* **110**, 248–252.

Lilly, F., Duran-Reynals, M. L., and Rowe, W. P. (1975). *J. Exp. Med.* **141**, 882–889.

Linemeyer, D. L., Ruscetti, S. K., Menke, J. G., and Scolnick, E. M. (1980). *J. Virol.* **35**, 710–721.

Ma, N. S. F., Jones, T. C., Miller, A. C., Morgan, L. M., and Adams, E. A. (1976). *Lab. Anim. Sci.* **26**, 1022–1036.

Ma, N. S. F., Rossan, R. N., Kelley, S. T., Harper, J. S., Bedard, M. T., and Jones, T. C. (1978). *J. Med. Primatol.* **7**, 146–155.

MacDonald, M. E., Reynolds, F. H., Jr., Van de Ven, W. J. M., Stephenson, J. R., Mak, T. W., and Bernstein, A. (1980). *J. Exp. Med.* **151**, 1477–1492.

McGrath, M. S., and Weissman, I. L. (1978). *Cold Spring Harbor Conf. Cell Proliferation* **5**, Book B, 577–589.

McGrath, M. S., and Weissman, I. L. (1979). *Cell* **17**, 65–75.

Manolov, G., and Manolova, Y. (1972). *Nature (London)* **237**, 33–34.

Manolova, Y., Manolov, G., Kieler, J., Levan, A., and Klein, G. (1979). *Hereditas* **90**, 5–10.

Markaryan, D. S. (1980). *In* "Advances in Comparative Leukemia Research 1979" (D. S. Yohn, B. A. Lapin, and J. R. Blakeslee, eds.), pp. 415–416. Elsevier, Amsterdam.

Martin, L. N., and Allen, W. P. (1975). *Infect. Immun.* **12**, 528–535.

Maurer, H. S., Gotoff, S. P., Allen, L., and Bolan, J. (1976). *Cancer* **37**, 2224–2231.

Metcalf, D. (1966). *J. Natl. Cancer Inst.* **37**, 425–442.

Miller, G. (1979). *In* "The Epstein-Barr Virus" (M. A. Epstein and B. G. Achong, eds.), pp. 351–372. Springer-Verlag, Berlin and New York.

Miller, G., Shope, T., Coope, D., Waters, L., Pagano, J., Bornkamm, G. W., and Henle, W. (1977). *J. Exp. Med.* **145**, 948–967.

Mirand, E. A., Steeves, R. A., Avila, L., and Grace, J. J. (1968). *Proc. Soc. Exp. Biol. Med.* **127**, 900–904.

Misko, I. S., Moss, D. J., and Pope, J. H. (1980). *Proc. Natl. Acad. Sci. U.S.A.* **77**, 4247–4250.

Mitelman, F., Klein, G., Andersson-Anvret, M., Forsby, N., and Jóhansson, B. (1979). *Int. J. Cancer* **23**, 32–36.

Miyoshi, I., Kubonsihi, I., Yoshimoto, S., Akagi, T., Ohtsuki, Y., Shiraishi, Y., Nagata, K., and Hinuma, Y. (1981). *Nature (London)* **294**, 770–771.

Moretta, L., Mingari, M. C., and Moretta, A. (1979). *Immunol. Rev.* **45**, 163–191.

Moss, D. J., Rickinson, A. B., and Pope, J. H. (1978). *Int. J. Cancer* **22**, 662–668.

Moss, D. J., Rickinson, A. B., and Pope, J. H. (1979). *Int. J. Cancer* **23**, 618–625.

Nagington, J., and Gray, J. (1980). *Lancet* **1**, 536–537.

Neel, B. G., Hayward, W. S., Robinson, H. L., Fang, J., and Astrin, S. M. (1981). *Cell* **23**, 323–334.

Neubauer, R. H., and Rabin, H. (1979). *In* "Naturally Occurring Biological Immunosuppressive Factors and Their Relationship to Disease" (R. H. Neubauer, ed.), pp. 203–231. CRC, Boca Raton, Florida.

Neubauer, R. H., Wallen, W. C., and Rabin, H. (1975). *Infect. Immun.* **12**, 1021–1028.

Neubauer, R. H., Rabin, H., Strnad, B. C., Lapin, B. A., Yakovleva, L., and Indzie, E. (1979). *Int. J. Cancer* **23**, 186–192.

Neubauer, R. H., Rabin, H., Strnad, B. C., Nonoyama, M., Lapin, B. A., Dyachenko, A. G., Agrba, V. Z., Indzhiia, L., and Yakovleva, L. (1980a). *In* "Advances in Comparative Leukemia Research 1979" (D. S. Yohn, B. A. Lapin, and J. R. Blakeslee, eds.), pp. 411–412. Elsevier, Amsterdam.

Neubauer, R. H., Rabin, H., and Dunn, F. E. (1980b). *Infect. Immun.* **27**, 549–555.

Neubauer, R. H., Dunn, F. E., and Rabin, H. (1981). *Infect. Immun.* **32**, 698–706.

Neubauer, R. H., Marchalonis, J. J., Strnad, B. C., and Rabin, H. (1982). *J. Immunogenet.*, in press.

Nilsson, K., Giovanella, B. C., Stehlin, J. S., and Klein, G. (1977). *Int. J. Cancer* **19**, 337–344.

Nowinski, R. C., and Hays, E. F. (1978). *J. Virol.* **27**, 13–18.

Nowinski, R. C., Hays, E. F., Doyle, T., Linkhart, S., Medeiros, E., and Pickering, R. (1977). *Virology* **81**, 363–370.

O'Conor, G. T. (1970). *Am. J. Med. Sci.* **48**, 279–285.

Oliff, A., Linemeyer, D. L., Ruscetti, S. K., Lowy, R., Lowy, D. R., and Scolnick, E. M. (1980). *J. Virol.* **35**, 924–936.

Oppermann, H., Levinson, A. D., Varmus, H. E., Levintow, L., and Bishop, J. M. (1979). *Proc. Natl. Acad. Sci. U.S.A.* **76**, 1806–1808.

Orgel, L. C., and Crick, F. H. C. (1980). *Nature (London)* **284**, 601–603.

Oroszlan, S., and Gilden, R. V. (1980). *In* "Molecular Biology of RNA Tumor Viruses" (J. R. Stephenson, ed.), pp. 299–334. Academic Press, New York.

Oroszlan, S., Barbacid, M., Copeland, T. D., Aaronson, S., and Gilden, R. V. (1981a). *J. Virol.* **39**, 845–854.

Oroszlan, S., Sarngadharan, M. G., Copeland, T. D., Kalyanaraman, V. S., Gilden, R. V., and Gallo, R. C. (1982). *Proc. Natl. Acad. Sci. U.S.A.* **79**, 1291–1294.

Oskarsson, M., McClements, W. L., Blair, D. G., Maizel, J. V., and Vande Woude, G. F. (1980). *Science* **207**, 1222–1224.

Pawson, T., Guyden, J., Kung, T.-H., Radke, K., Gilmore, T., and Martin, G. S. (1980). *Cell* **22**, 767–775.

Payne, G. S., Courtneidge, S. A., Crittenden, L. B., Fadley, A. M., Bishop, J. M., and Varmus, H. E. (1981). *Cell* **23**, 311–322.

Pearson, G. R. (1980). *In* "Viral Oncology" (G. Klein, ed.), pp. 739–767. Raven, New York.

Pearson, G. R., Johansson, B., and Klein, G. (1978). *Int. J. Cancer* **22**, 120–125.

Pearson, G. R., Qualtiere, L. F., Klein, G., Norin, T., and Bal, I. S. (1979). *Int. J. Cancer* **24**, 402–406.

Penman, H. G. (1970). *J. Clin. Pathol.* **23**, 765–771.

Poiesz, B. J., Ruscetti, F. W., Gazdar, A. F., Bunn, P. A., Minna, J. D., and Gallo, R. C. (1980a). *Proc. Natl. Acad. Sci. U.S.A.* **77**, 7415–7419.

Poiesz, B. J., Ruscetti, F. W., Mier, J. W., Woods, A. M., and Gallo, R. C. (1980b). *Proc. Natl. Acad. Sci. U.S.A.* **77**, 6815–6819.

Poiesz, B. J., Ruscetti, F. W., Reitz, M. S., Kalyanaraman, V. S., and Gallo, R. C. (1981). *Nature (London)* **29**, 268–271.

Pope, J. (1979). *In* "The Epstein-Barr Virus" (M. A. Epstein and B. G. Achong, eds.), pp. 205–223. Springer-Verlag, Berlin and New York.

Posner, L. E., Robert-Guroff, M., Kalyanaraman, V. S., Poiesz, B. J., Ruscetti, F. W., Bunn, P. A., Minna, J. D., and Gallo, R. C. (1981). *J. Exp. Med.* **154**, 333–340.

Provisor, A. J., Iacuone, J. J., Chilcote, R. R., Neilburger, R. G., Crussi, F. G., and Baehner, R. L. (1975). *N. Engl. J. Med.* **239**, 62–65.

Purtilo, D. T. (1976). *Lancet* **2**, 882–885.

Purtilo, D. T. (1981). *Adv. Cancer Res.* **34**, 279–312.

Purtilo, D. T., Cassel, C. K., Yang, J. P. S., Harper, R., Stephenson, S. R., Landing, B. H., and Vawter, G. F. (1975). *Lancet* **1**, 935–941.

Purtilo, D. T., DeFlorio, D., Jr., Hutt, L. M., Bhawan, J. Yang, J. P. S., Otto, R., and Edwards, W. (1977). *N. Engl. J. Med.* **20**, 1077–1081.

Purtilo, D., Sakamoto, K., Barnabei, V., and Seeley, J. (1981). *Proc. Int. Symp. Comp. Res. Leukemia Related Dis., 10th, Los Angeles*, p. 163.

Quintrell, N., Hughes, S. H., Varmus, H. E., and Bishop, J. M. (1980). *J. Mol. Biol.* **142**, 363–393.

Rabin, H., Pearson, G., Chopra, H. C., Orr, T., Ablashi, D. V., and Armstrong, G. R. (1973). *In Vitro* **9**, 65–72.

Rabin, H., Wallen, W. C., Neubauer, R. H., and Epstein, M. A. (1975a). *In* "Comparative Leukemia Research 1973, Leukemogenesis" (Y. Ito and R. M. Dutcher, eds.), pp. 367–374. Karger, Basel.

Rabin, H., Neubauer, R. H., Pearson, G. R., Cicmanec, J. L., Wallen, W. C., Loeb, W. F., and Valerio, M. G. (1975). *J. Natl. Cancer Inst.* **54**, 499–502.

Rabin, H., Pearson, G. R., Wallen, W. C., Neubauer, R. H., Cicmanec, J. L., and Levy, B. (1976). *In* "Comparative Leukemia Research 1975, Bibl. Haemat." (J. Clemmesen and D. S. Yohn, eds.), No. 43, pp. 326–330. Karger, Basel.

Rabin, H., Neubauer, R. H., Hopkins, R. F., III, and Levy, B. M. (1977). *Int. J. Cancer* **20**, 44–50.

Rabin, H., Neubauer, R. H., Hopkins, R. F., III, and Nonoyama, M. (1978). *Int. J. Cancer* **21**, 762–767.

Rangan, S. R. S., and Gallagher, R. E. (1979). *Adv. Virus Res.* **24**, 1–123.

Rasheed, S., Rongey, R. W., Nelson-Rees, W. A., Rabin, H., Neubauer, R. H., Bruszweski, J., Esra, G., and Gardner, M. B. (1977). *Science* **198**, 407–409.

Rho, H. M., Poiesz, B. J., Ruscetti, F. W., and Gallo, R. C. (1981). *Virology* **112**, 355–360.

Rice, N. R., Bonner, T. I., and Gilden, R. V. (1981). *J. Virol.* **114**, 286–290.

Rickinson, A. B., Crawford, D., and Epstein, M. A. (1977). *Clin. Exp. Immunol.* **28**, 72–79.

Rickinson, A. B., Moss, D. J., and Pope, J. H. (1979). *Int. J. Cancer* **23**, 610–617.

Rickinson, A. B., Moss, D. J., Pope, J. H., and Ahlberg, N. (1980). *Int. J. Cancer* **25**, 59–65.

Robinson, J. E., Brown, N., Andiman, W., Halliday, K., Francke, U., Robert, M. F., Andersson-Anvret, M., Horstmann, D., and Miller, G. (1980). *N. Engl. J. Med.* **302**, 1293–1297.

Rommerlaere, J., Faller, D. V., and Hopkins, N. (1978). *Proc. Natl. Acad. Sci. U.S.A.* **75**, 495–499.

Ruscetti, S. K., Linemeyer, D., Field, J., Troxler, D., and Scolnick, E. M. (1979). *J. Virol.* **30**, 787–798.

Sakamoto, K., Freed, H. J., and Purtilo, D. T. (1980). *J. Immunol.* **125**, 921–925.

Saxon, A., Stevens, R. H., and Golde, D. W. (1979). *N. Engl. J. Med.* **300**, 700–704.

Schaadt, M., Kirchner, H., Fonatsch, C., and Diehl, V. (1979). *Int. J. Cancer* **23**, 751–761.

Schultz, A. M., Ruscetti, S. K., Scolnick, E. M., and Oroszlan, S. (1980). *Virology* **107**, 537–542.

Schwarz, H., Fischinger, P. J., Ihle, J. N., Thiel, H. J., Weiland, F., Bolognesi, D. P., and Schafer, W. (1979). *Virology* **93**, 159–174.

Scolnick, E. M., Papageorge, A. G., and Shih, T. Y. (1979). *Proc. Natl. Acad. Sci. U.S.A.* **76**, 5355–5359.

Sheiness, D. K., Hughes, S. H., Stubblefield, E., and Bishop, J. M. (1980). *Virology* **105**, 415–424.

Shibuya, M., Hanafusa, T., Hanafusa, H., and Stephenson, J. R. (1980). *Proc. Natl. Acad. Sci. U.S.A.* **77**, 6536–6540.

Shih, T. Y., Williams, D. R., Weeks, M. O., Maryak, J. M., Vass, W. C., and Scolnick, E. M. (1978). *J. Virol.* **27**, 45–55.

Shih, T. Y., Weeks, M. O., Young, H. A., and Scolnick, E. M. (1979a). *Virology* **96**, 64–79.

Shih, T. Y., Weeks, M. O., Young, H. A., and Scolnick, E. M. (1979b). *J. Virol.* **31**, 546–556.

Shih, C., Shilo, B. Z., Goldfarb, M. P., Dannenberg, A., and Weinberg, R. A. (1979c). *Proc. Natl. Acad. Sci. U.S.A.* **76**, 5714–5718.

Shih, C., Padsy, L. C., Murray, M., and Weinberg, R. A. (1981). *Nature (London)* **290**, 261–264.

Shilo, B. Z., and Weinberg, R. A. (1981). *Nature (London)* **289**, 607–609.

Shimotohno, K., Mizutani, S., and Temin, H. M. (1980). *Nature (London)* **285**, 550–554.

Shope, T. C. (1975). *In* "Oncogenesis and Herpesviruses II" (G. de-The, M. A. Epstein, and H. zur Hausen, eds.), Part 2, pp. 153–159. International Agency for Research on Cancer, Lyons, France.

Shope, T., Dechairo, D., and Miller, G. (1973). *Proc. Natl. Acad. Sci. U.S.A.* **70**, 2487–2491.

Simek, S., and Rice, N. R. (1980). *J. Virol.* **33**, 320–329.

Skinner, G. R. B. (1976). *Br. J. Exp. Pathol.* **57**, 361–376.

Spector, B. D., Perry, G. S., III, and Kersey, J. H. (1978a). *Clin. Immunol. Immunopathol.* **11**, 12–29.

Spector, D. H., Varmus, H. E., and Bishop, J. M. (1978b). *Proc. Natl. Acad. Sci. U.S.A.* **75**, 4102–4106.

Staal, S. F., Hartley, J. W., and Rowe, W. P. (1977). *Proc. Natl. Acad. Sci. U.S.A.* **74**, 3065–3067.

Steeves, R. A., Eckner, R. J., Bennett, M., Mirand, E. A., and Trudel, P. J. (1971). *J. Natl. Cancer Inst.* **46**, 1209–1217.

Steffen, D., and Weinberg, R. A. (1978). *Cell* **15**, 1003–1010.

Stehelin, D., Varmus, H. E., Bishop, J. M., and Vogt, P. K. (1976). *Nature (London)* **260**, 170–173.

Stephenson, J. R. (1980). *In* "Molecular Biology of RNA Tumor Viruses" (J. R. Stephenson, ed.), pp. 245–297. Academic Press, New York.

Strnad, B. C., Neubauer, R. H., and Rabin, H. (1979). *Int. J. Cancer* **23**, 76–81.

Sullivan, J. L., Byron, K. S., Brewster, F. E., and Purtilo, D. T. (1980). *Science* **210**, 543–545.

Sutcliffe, J. G., Shinnick, T. M., Verma, I. M., and Lerner, R. A. (1980). *Proc. Natl. Acad. Sci. U.S.A.* **77**, 3302–3306.

Svedmyr, E., and Jondal, M. (1975). *Proc. Natl. Acad. Sci. U.S.A.* **72**, 1622–1626.

Taylor, D. W., and Siddiqui, W. A. (1978). *Infect. Immun.* **21**, 147–150.

Theilen, G. H., Gould, D., Fowler, M., and Dungworth, D. L. (1971). *J. Natl. Cancer Inst.* **47**, 881–889.

Thorley-Lawson, D. A. (1979). *Nature (London)* **281**, 486–488.

Thorley-Lawson, D. A. (1980). *J. Immunol.* **124**, 745–751.

Thorley-Lawson, D. A., Chess, L., and Strominger, J. L. (1977). *J. Exp. Med.* **146**, 495–508.

Todaro, G. J. (1980). *In* "Molecular Biology of RNA Tumor Viruses" (J. R. Stephenson, ed.), pp. 47–76. Academic Press, New York.

Todaro, G. J., and Heubner, R. J. (1972). *Proc. Natl. Acad. Sci. U.S.A.* **69**, 1009–1015.

Todaro, G. J., Lieber, M. M., Benveniste, R. E., Sherr, C. J., Gibbs, C. J., Jr., and Gajdusek, D. C. (1975). *Virology* **67**, 335–343.

Troxler, D. H., and Scolnick, E. M. (1978). *Virology* **85**, 17–27.

Troxler, D. H., Boyars, J. K., Parks, W. P., and Scolnick, E. M. (1977). *J. Virol.* **22**, 361–372.

Troxler, D. H., Ruscetti, S. K., and Scolnick, E. M. (1980). *Biochim. Biophys. Acta* **605**, 305–324.

Tsichlis, P. M., and Coffin, V. M. (1980). *J. Virol.* **33**, 238–249.

Van Beveren, C. V., Galleshaw, J. A., Jonas, V., Berns, A. J. M., Doolittle, R. F., Donoghue, D. J., and Verma, I. M. (1981). *Nature (London)* **289**, 258–262.

Virelizier, J.-L., Lenoir, G., and Griscelli, C. (1978). *Lancet* **2**, 231–234.

Wallen, W. C., Neubauer, R. H., Rabin, H.,and Cicmanec, J. L. (1973). *J. Natl. Cancer Inst.* **51**, 967–975.

Wallen, W. C., Neubauer, R. H., and Rabin, H. (1974). *J. Med. Primatol.* **3**, 41–53.

Wedderburn, N. (1970). *Lancet* **2**, 1114–1116.

Weinberg, R. A. (1980). *Cell* **22**, 643–644.

Werner, J., Wolf, H., Apodaca, J., and zur Hausen, H. (1975). *Int. J. Cancer* **15**, 1000–1008.

Westin, E., Wong-Staal, F., Gelmann, E., Baluda, M., Papas, T., Eva, A., Reddy, E. P., Tronick, S., Aaronson, S., and Gallo, R. C. (1981). *Int. Symp. Comp. Res. Leukemia Related Dis., 10th, Los Angeles*.

Willingham, M. C., Jay, G., and Pastan, I. (1979). *Cell* **18**, 125–134.

Willingham, M. C., Pastan, I., Shih, T. Y., and Scolnick, E. M. (1980). *Cell* **19**, 1005–1019.

Witte, O., Rosenberg, N., and Baltimore, D. (1979). *J. Virol.* **31**, 776–784.

Witte, O. N., Dasgupta, A., and Baltimore, D. (1980a). *Nature (London)* **283**, 826–831.

Witte, O. N., Graff, S., Rosenberg, N., and Baltimore, D. (1980b). *Proc. Natl. Acad. Sci. U.S.A.* **77**, 4993–4997.

Wong-Staal, F., Favera, R. D., Gelmann, E., Westin, E., and Gallo, R. C. (1981). *Int. Symp. Comp. Res. Leukemia Related Dis., 10th, Los Angeles*.

Wright, J., Falk, L. A., and Deinhardt, F. (1974). *J. Natl. Cancer Inst.* **53**, 271–275.

Wright, J., Falk, L. A., Collins, D., and Deinhardt, F. (1976). *J. Natl. Cancer Inst.* **57**, 959–962.

Yakovleva, L. A., Bukaeva, I. A., and Markova, T. P. (1980). *In* "Advances in Comparative Leukemia Research 1979" (D. S. Yohn, B. A. Lapin, and J. R. Blakeslee, eds.), pp. 419–420. Elsevier, Amsterdam.

Young, H. A., Shih, T. Y., Scolnick, E. M., Rasheed, S., and Gardner, M. B. (1979). *Proc. Natl. Acad. Sci. U.S.A.* **76**, 3523–3527.

Young, H. A., Gonda, M. A., DeFeo, D., Ellis, R. W., Nagashima, K., and Scolnick, E. M. (1980). *Virology* **107**, 88–99.

Young, H. A., Rasheed, S. R., Sowder, R., Benton, C. V., and Henderson, L. E. (1981). *J. Virol.* **38**, 286–293.

Zech, L., Haglund, U., Nilsson, K., and Klein, G. (1976). *Int. J. Cancer* **17**, 47–56.

Zeng, Y., Lui, Y., Lui, C., Chen, S., Wei, J., Zhu, J., and Zai, H. (1980). *Intervirology* **13**, 162–168.

zur Hausen, H. (1976). *Cancer Res.* **36**, 678–680.

zur Hausen, H. (1980). *In* "DNA Tumor Viruses" (J. Tooze, ed.), pp. 747–795. Cold Spring Harbor Laboratory, Cold Spring Harbor, New York.

AUTHOR INDEX

Numbers in italics refer to the pages on which the complete references are listed.

Z

SUBJECT INDEX

A

Adenoviruses
  attachment proteins of, 191
  cell penetration by, 152, 191–192
  cellular receptors for, 191
Alfalfa mosaic virus, *Plantago* hosts and
    occurrence of, 116, 130
Alphaviruses, cellular receptors for, 147–
    148
Amantadine, as virus uncoating inhibi-
    tor, 163, 165
Anemone mosaic virus, *Plantago* host
    and occurrence of, 113
Antibiotics, bacteriophage resistance to,
    236
Arabis mosaic virus, *Plantago* hosts and
    occurrence of, 114, 126–127
Arildone, as virus uncoating inhibitor,
    165–166
Aster ringspot virus, *Plantago* host and
    occurrence of, 114
Aster yellows, *Plantago* hosts and occur-
    rence of, 116
Attachment proteins, of viruses, 148–
    150, 167–168, 175–176, 178–179,
    180–181, 185, 187, 189–190, 191,
    192, 196

B

BABIM as virus uncoating inhibitor,
    163, 166
Bacilliform viruses, *Plantago* hosts and
    occurrence of, 116, 130–131
Bacteriocins (particulate)
  evolution of, 243
  properties of, 244
Bacteriophages, 205–280
  antibiotic resistance genes in, 236
  basic properties of, 210
  coat proteins of, 230–233

comparative biology, of, 205–280
defective, evolutionary position of,
    243–247
distribution of, 211–219
in eukaryotes, 259–260
evolution of, 233–242
  DNA/DNA interactions, 234–239
  modular, 239–242
  by point mutations, 233–234
fundamental groups of, 209
genetic maps of, 221
genome economy in morphogenesis of,
    247–248
morphological evolution of, 247–254
morphological variation of, 248–249
nucleic acids of
  comparative anatomy, 219–233
  heteroduplex comparisons, 226–229
  hybridization, 229
  restriction endonuclease patterns,
    229
  sequence comparison, 224–226
  origins for, 260–261
patterns of macromolecular synthesis
    in, 230
plasmids and, 263–265
self-assembly of, 247–248
systematics of, 208–211
virus relationship to, 254–260
Bases, unusual, in phage DNA, 222–223
Beet curly top virus, *Plantago* hosts and
    occurrence of, 114, 125
Beet mild yellows virus, *Plantago* host
    and occurrence of, 114, 126
Beet ringspot virus, *Plantago* hosts and
    occurrence of, 115
Beet yellow virus, *Plantago* hosts and
    occurrence of, 112, 119–120
Bird seed, *Plantago* seed use in, 111
Brain tissue, virus detection in, 33–38
Breast milk, virus detection in, 32
Broad bean wilt virus, *Plantago* host and
    occurrence of, 116, 129–130

CONTENTS OF RECENT VOLUMES